Acoustic Waves Handbook

Acoustic Waves Handbook

Edited by **Sonny Lin**

CLANRYE INTERNATIONAL

New Jersey

Published by Clanrye International,
55 Van Reypen Street,
Jersey City, NJ 07306, USA
www.clanryeinternational.com

Acoustic Waves Handbook
Edited by Sonny Lin

International Standard Book Number: 978-1-63240-012-3 (Hardback)

Printed in the United States of America.

Contents

Preface

In this book, acoustic waves have been described comprehensively. The concept of acoustic wave is an extensive one; it materializes in a variety of mediums, from solids to plasmas, and at varying length and time scales which range from sub-micrometric layers in microdevices to seismic waves in the Sun's interior. This book discusses various approaches of the active research going on in this field. Theoretical attempts lead to a comprehensive understanding of this phenomenon. Such waves are useful in exploring the characteristics of a variety of structures, be it inorganic layers or other bio-structures. These waves can also function as a tool to influence matter, from the gentle evaporation of bio-molecules to be evaluated, to the phase conversions inferred by intense shock waves. Furthermore, a whole class of widespread micro tools, inclusive of filters and sensors, depends upon the behavior of these waves propagating in fine layers.

The researches compiled throughout the book are authentic and of high quality, combining several disciplines and from very diverse regions from around the world. Drawing on the contributions of many researchers from diverse countries, the book's objective is to provide the readers with the latest achievements in the area of research. This book will surely be a source of knowledge to all interested and researching the field.

In the end, I would like to express my deep sense of gratitude to all the authors for meeting the set deadlines in completing and submitting their research chapters. I would also like to thank the publisher for the support offered to us throughout the course of the book. Finally, I extend my sincere thanks to my family for being a constant source of inspiration and encouragement.

Editor

Part 1

Theoretical and Numerical Investigations of Acoustic Waves

1

Analysis of Acoustic Wave in Homogeneous and Inhomogeneous Media Using Finite Element Method

Zi-Gui Huang
Department of Mechanical Design Engineering, National Formosa University
Taiwan

1. Introduction

Even though the propagation of elastic/acoustic waves in inhomogeneous and layered media has been an active research topic for many decades already, new problems and challenges continue to be posed even up to now. In fact, during the last few years, renewed interests have been witnessed by researchers in the various fields of acoustics, such as acoustic mirrors, filters, resonators, waveguides, and other kinds of acoustic devices, in relation to wave propagation in periodic elastic media. In acoustics and applied mechanics, these developments have been triggered by the need for new acoustic devices in order to obtain quality control of elastic/acoustic waves.

What sort of material can allow us to have complete control over the elastic/acoustic wave's propagation? We would like to discuss and answer this question in this chapter. It is well known that the successful applications of photonic band-gap materials have hastened the related researches on phononic band-gap materials. *Analysis of Acoustic Wave in Homogeneous and Inhomogeneous Media Using Finite Element Method* explores the theoretical road leading to the possible applications of phononic band gaps. It should quickly bring the elastic/acoustic professionals and engineers up to speed in this field of study where elastic/acoustic waves and solid-state physics meet. It will also provide an excellent overview to any course in elastic/acoustic media.

Previous research on photonic crystals (Johnson & Joannopoulos, 2001, 2003; Joannopoulos et al., 1995; Leung & Liu, 1990) has sparked rapidly growing interest in the analogous acoustic effects of phononic crystals and periodic elastic structures. The various techniques for band structure calculations were introduced (Hussein, 2009). There are many well-known methods of calculating the band structures of photonic and phononic crystals in addition to the reduced Bloch mode expansion method: the plane-wave expansion (PWE) method (Huang & Wu, 2005; Kushwaha et al., 1993; Laude et al., 2005; Tanaka & Tamura, 1998; Wu et al., 2004 ; Wu & Huang, 2004), the multiple-scattering theory (MST) (Leung & Qiu, 1993; Kafesaki & Economou, 1999; Psarobas & Stefanou, 2000; Wang et al., 1993), the finite-difference (FD) method (Garica-Pabloset et al., 2000; Sun & Wu, 2005; Yang, 1996), the transfer matrix method (Pendry & MacKinnon, 1992), the meshless method (Jun et al., 2003), the multiple multipole method (Moreno et al., 2002), the wavelet method (Checoury & Lourtioz, 2006; Yan & Wang, 2006), the pseudospectral method (Chiang et al., 2007), the finite element method (FEM) (Axmann & Kuchment, 1999; Dobson, 1999; Huang & Chen,

2011; Wu et al., 2008), the mass-in-mass lattice model (Huang & Sun, 2010), and the micropolar continuous modeling (Salehian & Inman, 2010).
Many studies on phononic band structures from the past decade use the PWE, MST, and FD methods to analyze the frequency band gaps of bulk acoustic waves (BAW) in composite materials or phononic band structures. Studies adopting the PWE method investigate the dispersion relations and the frequency band-gap feathers of the BAW and surface acoustic wave (SAW) modes. Other studies use the layered MST to study the frequency band gaps of bulk acoustic waves in three-dimensional periodic acoustic composites and the band structures of phononic crystals consisting of complex and frequency-dependent Lame' coefficients. Other researchers applied the finite-difference time-domain method to predict the precise transmission properties of slabs of phononic crystals and analyze the mode coupling in joined parallel phononic crystal waveguides.
The techniques for tuning frequency band gaps of elastic/acoustic waves in phononic crystals are very important, and remain exciting research topics in the physics community. The filling fraction, rotation of noncircular rods, different cuts of anisotropic materials, and the temperature effect all produce large frequency band gaps in the BAW and SAW modes of periodic structures. A previous review paper (Burger et al., 2004) discusses the technique used to optimize the unit cell material distribution, achieving the largest possible band gap in photonic crystals for a given cell symmetry. Studies over the past decade focus on the theoretical and numerical analysis of phononic structures based on circular or square cylinders embedded in background materials. In this case, the PWE method can easily calculate the dispersion relations by constructing the structural functions with Bessel or Sinc functions. However, research on the more complicated problem of waves in the reticular and other special periodic band structures has not started until recently.
This chapter uses the 2D and 3D finite element methods to discuss the wave velocities of isotropic and anisotropic materials in homogeneous media. It also considers the tunable band gaps of acoustic waves in two-dimensional phononic crystals with reticular geometric structures (Huang & Chen, 2011). The concept of adopting a reticular geometric structure comes from the variations of similar geometry in bio-structural reticular formation and fibers. The PWE method used to calculate the structural functions of densities and elastic constants cannot numerically analyze the Gibbs phenomenon. Therefore, this chapter adopts the FEM to discuss this special periodic band structure. Changing the filling fraction, scale parameters, and rotating angles of reticular geometric structures can tune the frequency band gaps of mixed polarization modes. This technique is suitable for analyzing the phenomenon of frequency band gaps in special band structures.

2. Theory

In this chapter, based on the theorems of solid-state physics and the finite element method with Bloch calculations, equation of motion of the acoustic modes in two-dimensional inhomogeneous media, phononic band structures, are derived and discussed in detail. In the beginning, the concepts of the real space and **k** space are introduced while the Brillouin zone is also addressed in the text. Generalized techniques of Bloch calculations in finite element method are used to analyze the acoustic modes in two-dimensional homogeneous and inhomogeneous media, phononic band structures, consisting of materials with general anisotropy. The mixed and transverse polarization modes and quasi-polarization modes are investigated in the text.

2.1 Real space and k space

It is well-known that the analysis of wave motion in infinite periodic structures is difficult in real space. For dealing with the periodic structures, the Fourier series and Bloch's theorem are used to expand the periodic parameters such as the density, material constants, displacement fields, or potential. Regarding to the transformation of the real space and **k** space, the reciprocal lattice vectors (RLVs) are adopted from the solid-state physics. In general, we consider a three-dimensional phononic crystal with primitive lattice vectors $\mathbf{a}_1$, $\mathbf{a}_2$, and $\mathbf{a}_3$. The complete set of lattice vectors is written as $\{\mathbf{R} \mid \mathbf{R} = l_1\mathbf{a}_1 + l_2\mathbf{a}_2 + l_3\mathbf{a}_3\}$, where l_1, l_2, and l_3 are integers. The associated primitive reciprocal lattice vectors $\mathbf{b}_1$, $\mathbf{b}_2$, and $\mathbf{b}_3$ are determined by (Kittel, 1996)

$$\mathbf{b}_i = 2\pi \frac{\varepsilon_{ijk}\mathbf{a}_j \times \mathbf{a}_k}{\mathbf{a}_1 \cdot (\mathbf{a}_2 \times \mathbf{a}_3)}, \tag{1}$$

where ε_{ijk} is the three-dimensional Levi-Civita completely antisymmetric symbol. The complete set of reciprocal lattice vectors is written as $\{\mathbf{G} \mid \mathbf{G} = N_1\mathbf{b}_1 + N_2\mathbf{b}_2 + N_3\mathbf{b}_3\}$, where N_1, N_2, and N_3 are integers. Figure 1 shows the primitive unit cell in two-dimensional real space while the Fig. 2 shows the relationship between the real space and **k** space. A property between the primitive lattice vectors and associated primitive reciprocal lattice vectors is $\mathbf{b}_i \cdot \mathbf{a}_j = 2\pi\delta_{ij}$, where δ_{ij} is the kronecker symbol. Note that the associated primitive reciprocal lattice vectors are constructed as **k** space from the concept of crystal diffraction.

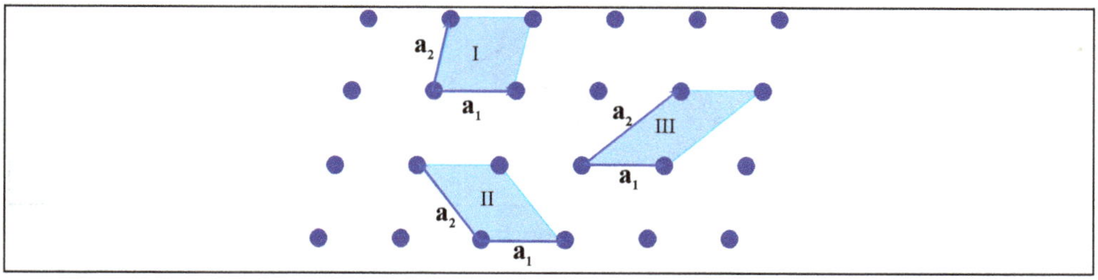

Fig. 1. Primitive unit cell in real space

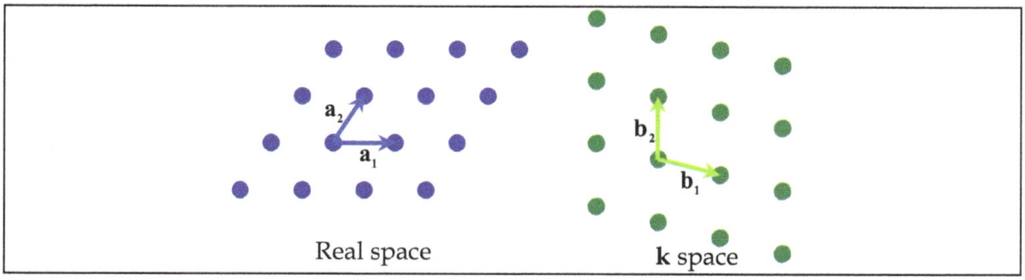

Fig. 2. Relationship between the real space and **k** space

We will find that, in following sections, the discrete translational symmetry of a phononic crystal allows us to classify the elastic/acoustic waves with a wave vector **k**. The propagating modes can be written in "Bloch form," consisting of a plane wave modulated by a function that shares the periodicity of the lattice (Joannopoulos et al., 1995):

$$\mathbf{P}_{\mathbf{k}}(\mathbf{r}) = e^{i\mathbf{k}\cdot\mathbf{r}}\mathbf{u}_{\mathbf{k}}(\mathbf{r}) = e^{i\mathbf{k}\cdot\mathbf{r}}\mathbf{u}_{\mathbf{k}}(\mathbf{r}+\mathbf{R}). \tag{2}$$

The important feature of the Bloch states is that different values of **k** do not necessarily lead to different modes. It is clear that a mode with wave vector **k** and a mode with wave vector **k+G** are the same mode, where **G** is a reciprocal lattice vector. The wave vector **k** serves to specify the phase relationship between the various cells that are described by **u**. If **k** is increased by **G**, then the phase between cells is increased by **G**·**R**, which we know is $2\pi n$ (n= $l_1N_1+l_2N_2+ l_3N_3$ is an integer) and not really a phase difference at all. So incrementing **k** by **G** results in the same physical mode. This means that we can restrict our attention to a finite zone in reciprocal space in which we *cannot* get from one part of the volume to another by adding any **G**. All values of **k** that lie outside of this zone, by definition, can be reached from within the zone by adding **G**, and are therefore redundant labels shown in Fig. 3. This zone is the so-called *Brillouin zone*.

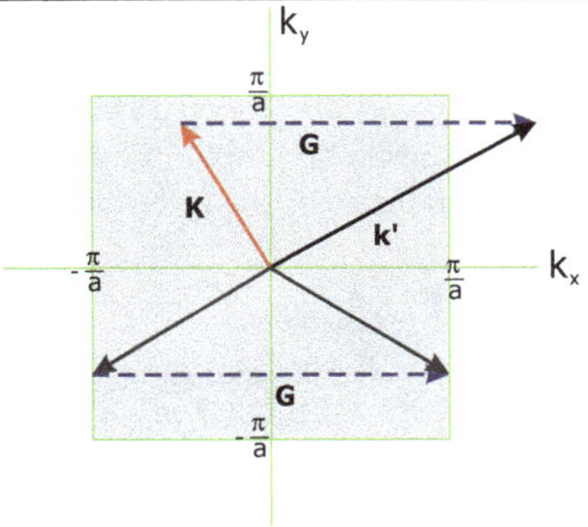

Fig. 3. All values of **k** that lie outside of this zone, by definition, can be reached from within the zone by adding **G**

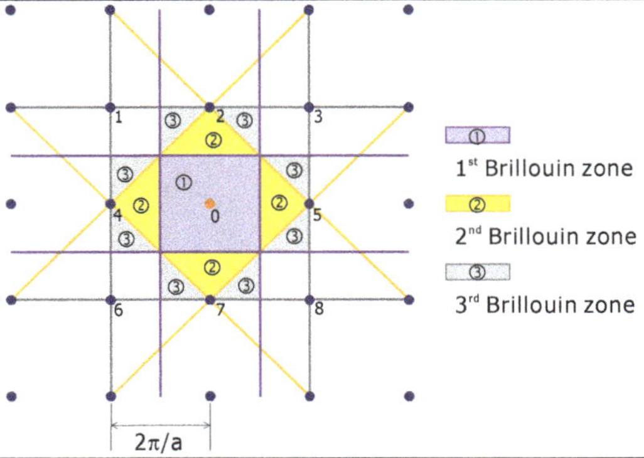

Fig. 4. Brillouin zones in a square lattice

By the periodicity of the reciprocal lattice, any reciprocal lattice point which represents a wave vector **k** outside the first Brillouin zone can be found a corresponding point in the first Brillouin zone. Therefore, the wave vectors **k** can always be confined in the first Brillouin zone. In the square lattice, only the wave vectors **k** in the region of the first Brillouin zone between $-\pi/a$ to π/a (the lattice constant is a) need to be considered. The Fig. 4 shows the *first, second,* and *third Brillouin zones*. For more details, it is best to consult the first few chapters of a solid-state physics text, such as Kittel, 1996, or consult the appendix of popular photonic text like Joannopoulos et al. 1995 and Johnson & Joannopoulos, 2001, 2003.

2.2 Equation of motion

This section provides a brief introduction of the theory of analyzing acoustic wave propagation in inhomogeneous media like as phononic band structures. The theory in this chapter can also be used to discuss acoustic wave propagation in homogeneous media because a homogeneous medium is symmetric with respect to any periodicity.

In an inhomogeneous linear elastic medium with no body force, the equation of motion of the displacement vector $\mathbf{u}(\mathbf{r},t)$ can be written as

$$\rho(\mathbf{r})\ddot{u}_i(\mathbf{r},t) = \partial_j[C_{ijmn}(\mathbf{r})\partial_n u_m(\mathbf{r},t)], \tag{3}$$

where $\mathbf{r}=(\mathbf{x},z)=(x,y,z)$ is the position vector, t is the time variable, and $\rho(\mathbf{r})$ and $C_{ijmn}(\mathbf{r})$ are the position-dependent mass density and elastic stiffness tensor, respectively. The following discussion considers a periodic structure consisting of a two-dimensional periodic array (x-y plane) of material A embedded in a background material B shown in Fig. 5. It is noted that when the properties of materials A and B tend to coincide, the homogeneous case is recovered.

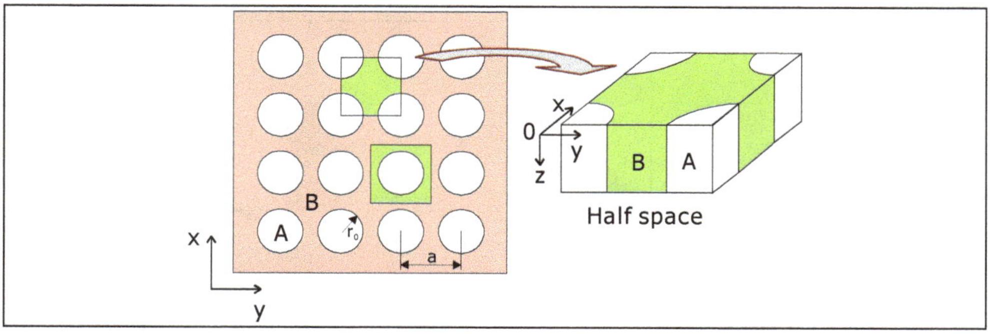

Fig. 5. Periodic structures with square lattice. When the properties of materials A and B tend to coincide, the homogeneous case is recovered

To calculate the dispersion diagrams of periodic structures, this study uses COMSOL Multiphysics software to apply the Bloch boundary condition to the unit cell domain in the FEM method. Based on the periodicity of phononic crystals, the displacement and stress components in the periodic structure are expressed as follows:

$$u_i(\mathbf{x},t) = e^{i\mathbf{k}\cdot\mathbf{x}}U_i(\mathbf{x},t), \tag{4}$$

$$\sigma_{ij}(\mathbf{x},t) = e^{i\mathbf{k}\cdot\mathbf{x}} T_{ij}(\mathbf{x},t), \tag{5}$$

where $\mathbf{k} = (k_1, k_2)$ is the Bloch wave vector, and $i = \sqrt{-1}$; $U_i(\mathbf{x},t)$ and $T_{ij}(\mathbf{x},t)$ are periodic functions that satisfy the following relation (Tanaka et al., 2000):

$$U_i(\mathbf{x}+\mathbf{R},t) = U_i(\mathbf{x},t), \tag{6}$$

$$T_{ij}(\mathbf{x}+\mathbf{R},t) = T_{ij}(\mathbf{x},t), \tag{7}$$

where $\mathbf{R}$ is a lattice translation vector with components of R_1 and R_2 in the x and y directions. The relationships between the original variables $u_i(\mathbf{x},t)$, $\sigma_{ij}(\mathbf{x},t)$, $u_i(\mathbf{x}+\mathbf{R},t)$, and $\sigma_{ij}(\mathbf{x}+\mathbf{R},t)$ about the Bloch boundary conditions are characterized as:

$$u_i(\mathbf{x}+\mathbf{R},t) = e^{i\mathbf{k}\cdot(\mathbf{x}+\mathbf{R})} U_i(\mathbf{x}+\mathbf{R},t) = e^{i\mathbf{k}\cdot\mathbf{R}} e^{i\mathbf{k}\cdot\mathbf{x}} U_i(\mathbf{x},t) = e^{i\mathbf{k}\cdot\mathbf{R}} u_i(\mathbf{x},t), \tag{8}$$

$$\sigma_{ij}(\mathbf{x}+\mathbf{R},t) = e^{i\mathbf{k}\cdot(\mathbf{x}+\mathbf{R})} T_{ij}(\mathbf{x}+\mathbf{R},t) = e^{i\mathbf{k}\cdot\mathbf{R}} e^{i\mathbf{k}\cdot\mathbf{x}} T_{ij}(\mathbf{x},t) = e^{i\mathbf{k}\cdot\mathbf{R}} \sigma_{ij}(\mathbf{x},t). \tag{9}$$

The Bloch calculations in this study record the variation of the displacements, stress fields, and eigen-frequencies as the wave vector increases. By using the FEM, the unit cell is meshed and divided into finite elements which connect by nodes, and is used to obtain the eigen-solutions and mechanical displacements. The types of finite elements used in this chapter are the default element types, Lagrange-quadratic, in COMSOL Multiphysics. In order to simulate the dispersion diagrams, the wave vectors are condensed inside the first Brillouin zone in the square lattice. According to the above theories, the results of dispersion relations in a band structure along the $\Gamma - \mathrm{X} - \mathrm{M} - \Gamma$ are characterized and presented in the following sections.

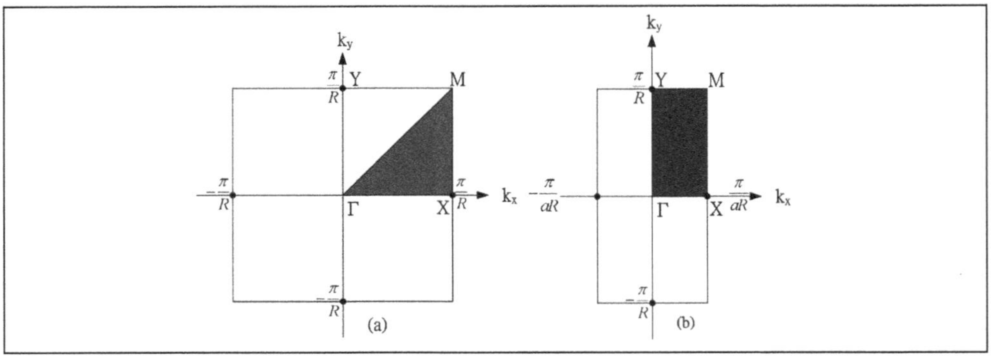

Fig. 6. Brillouin regions of the square and rectangular lattices

This chapter considers a periodic homogeneous medium with square lattice and phononic structures with square and rectangular lattices. These lattices consist of periodic structures that form two-dimensional lattices with lattice spacing R (square lattice) and lattice spacing aR (rectangular lattice). The term a is a scale from 0.1 to 2.0 in this chapter. The periodic structures are parallel to the z-axis. Figures 6(a) and 6(b) illustrate the Brillouin regions of the square lattice and rectangular lattice, respectively. In the square lattice, Fig. 6(a) shows

the irreducible part of the Brillouin zone, which is a triangle with vertexes Γ, X, and M. Similarly, Fig. 6(b) shows the irreducible part of the Brillouin zone of a rectangular lattice due to the geometric anisotropy, which is a rectangle with vertexes Γ, X, M, and Y, and the same as discussing the material anisotropy (Wu et al., 2004).
The finite element method divides a unit cell with a three-dimensional model into finite elements connected by nodes. The FEM obtains the eigen-solutions and contours of a mode shape. To simulate the dispersion diagrams, the wave vectors are condensed inside the first Brillouin zone in the square and rectangular lattices. Using the theories above, the following section presents the results of dispersion relations in a band structure for the $\Gamma-X-M-\Gamma$ square lattice or isotropic materials, and $\Gamma-X-M-Y-\Gamma$ rectangular lattice or anisotropic materials. Note that the 2D FEM model calculates the dispersion relations of mixed polarization modes, while the 3D FEM model describes the dispersion relations of mixed and transverse polarization modes.

3. Acoustic wave in homogeneous media

It can be noted that a homogeneous medium is symmetric with respect to any periodicity, and it can be shown that the results for an infinite homogeneous medium can be cast in the form appropriate for a periodic medium. In this section, we introduce the mixed polarization modes and transverse polarization modes in a homogeneous medium. Displacement fields (polarizations) are also investigated and used to distinguish the different modes in the dispersion relations. The aluminum and quartz are adopted for examples and discussed in the section. The wave velocities of different propogating modes are also observed and discussed.

3.1 Isotropic medium

In Fig. 5, when the properties of materials A and B tend to coincide, the homogeneous case is recovered. Consider a periodic structure consisting of aluminum (Al) circular cylinders embedded in a background material of Al forming a two-dimensional square lattice with lattice spacing R. It means this is a homogeneous medium in a 3D FEM model. Figure 7 shows the dispersion relations along the boundaries of the irreducible part of the Brillouin zone $\Gamma-X-M-\Gamma$. The vertical axis is the frequency (Hz) and the horizontal axis is the reduced wave vector $k^* = kR/\pi$. Here, k is the wave vector along the Brillouin zone. The Young's modulus E, Poisson's ratio ν, and density ρ of the material Al utilized in this example are E=70 GPa, ν =0.33, and ρ =2700 kg/m^3.
As the elastic waves propagate along the x axis, the nonvanishing displacement fields of the shear horizontal mode (SH), shear vertical mode (SV), and longitudinal mode (L) are u_y, u_z, and u_x respectively. It is noted that wave velocity $c_{S,L} = d\omega/dk = 2R * m_{S,L}$, so the slopes of dispersion curves in the $\Gamma-X$ section of Fig. 7 are exactly the straight lines and can be explained as the wave velocities of shear (S) and longitudinal (L) modes. Here, $m_{S,L}$ are the slopes of shear and longitudinal modes in Fig. 7. It is noted that the wave velocities of shear horizontal mode and shear vertical mode are the same in an isotropic material. From the results in Fig. 7, the wave velocities of shear and longitudinal modes are 3119 and 6174 m/s. As we know, the wave velocities of shear and longitudinal modes in an isotropic material can be obtain from

$$c_S = \sqrt{\frac{E}{\rho}\frac{1}{2(1+\nu)}} = 3122 \quad m/s, \tag{10}$$

$$c_L = \sqrt{\frac{E}{\rho}\frac{(1-\nu)}{(1+\nu)(1-2\nu)}} = 6031 \quad m/s. \tag{11}$$

Note that the FEM method can easily describe the mode characteristics. Figure 8 shows the vibration mode shapes of unit cell for shear and longuitudinal modes in X point. In this example, Fig. 8(a) is a shear horizontal mode with mode vibrating displacement along the y direction when the wave propagates along the x direction ($\Gamma - X$ direction). Also, Fig. 8(b) is a shear vertical mode with mode vibrating displacement along z direction, and Fig. 8(c) is a longitudinal mode with mode vibrating displacement along x direction. The arrows shown in Fig. 8 are the polarizations.

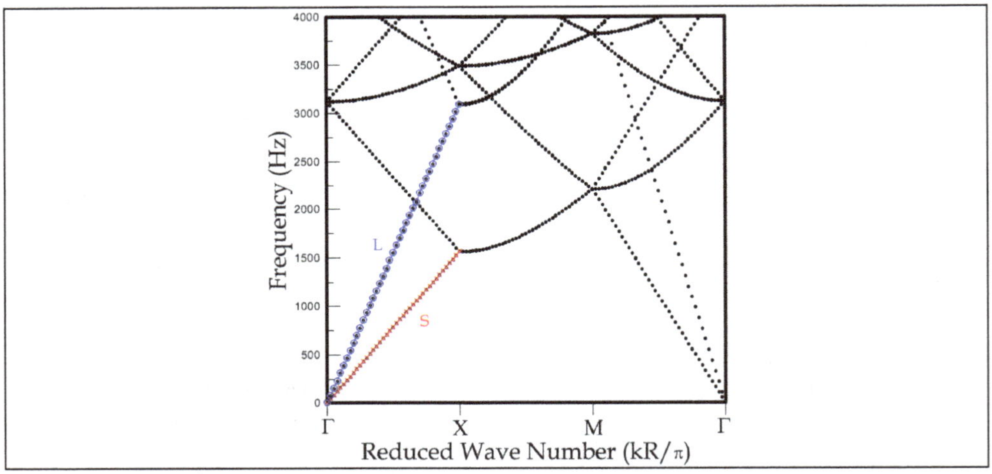

Fig. 7. The dispersion relations of homogeneous and isotropic material Al along the boundaries of the irreducible part of the Brillouin zone $\Gamma - X - M - \Gamma$

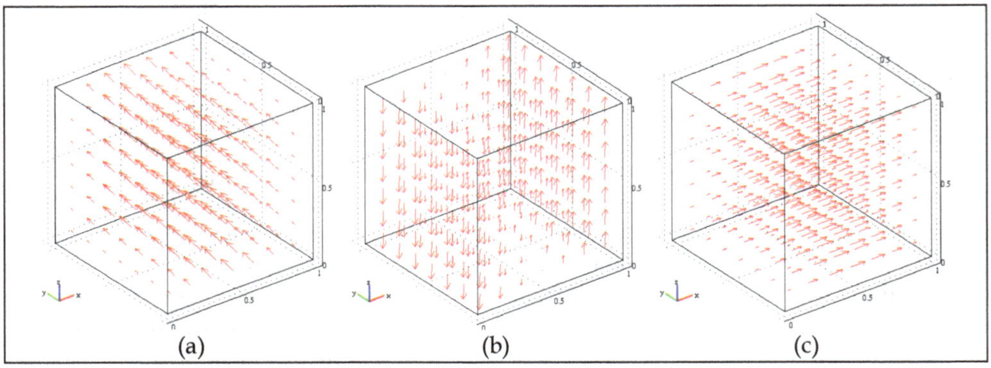

Fig. 8. (a) shear horizontal mode (b) shear vertical mode, (c) longitudinal mode in the Al

3.2 Anisotropic medium

Similarly, the method in this chapter is used to discuss the wave velocities of acoustic modes in an anisotropic material. Consider a periodic structure consisting of quartz circular cylinders embedded in a background material of quartz forming a two-dimensional square lattice with lattice spacing R. This is also a homogeneous medium. The quartz is a piezoelectric and anisotropic material. The density ρ =2651 kg/m^3. The elastic constants, piezoelectric constants, and relative permittivity of quartz utilized in this example are shown in Tables 1-3. The piezoelectric material, quartz, is a complete structural-electrical material, and thus all piezoelectric material properties were defined and entered into the FEM model. Figure 9 shows the dispersion relations along the boundaries of the irreducible part of the Brillouin zone $\Gamma - X - M - Y - \Gamma$ due to the material anisotropy. In the calculations, the x-y plane is parallel to the (001) plane and the x axis is along the [100] direction of quartz. The vertical axis is the frequency in Hz unit and the horizontal axis is the reduced wave vector.

86.7362	6.98527	11.9104	17.9081	0	0
6.98527	86.7362	11.9104	-17.9081	0	0
11.9104	11.9104	107.194	0	0	0
17.9081	-17.9081	0	57.9428	0	0
0	0	0	0	57.9492	17.9224
0	0	0	0	17.9224	39.9073

Table 1. The elastic constants of quartz in GPa unit

-0.19543	0.19543	0	-0.1212	0	0
0	0	0	0	0.12127	0.19558
0	0	0	0	0	0

Table 2. The piezoelectric constants of quartz in C/m^2 unit

4.4093	0	0
0	4.4092	0
0	0	4.68

Table 3. The relative permittivity of quartz

Shown in $\Gamma - X$ section of Fig. 9, the cross symbols represent the quasi shear horizontal (quasi-SH) mode. The square symbols represent the quasi shear vertical (quasi-SV) mode and the open circle symbols represent the quasi longitudinal (quasi-L) mode. The wave velocities of quasi-SH, quasi-SV, and quasi-L modes along x axis are 3306, 5116, and 5741 m/s. Similarly, The wave velocities of quasi-SH, quasi-SV, and quasi-L modes along y axis ($\Gamma - Y$ section) are 3922, 4311, and 6009 m/s respectively.

Figure 10 also shows the vibration mode shapes of unit cell for quasi-SH, quasi-SV, and quasi-L modes in X point. The arrows shown in Fig. 10 are the polarizations. In this example, the quasi-longitudinal and quasi-transverse waves are almost indistinguishable from the truly longitudinal and truly transverse waves of Fig. 8.

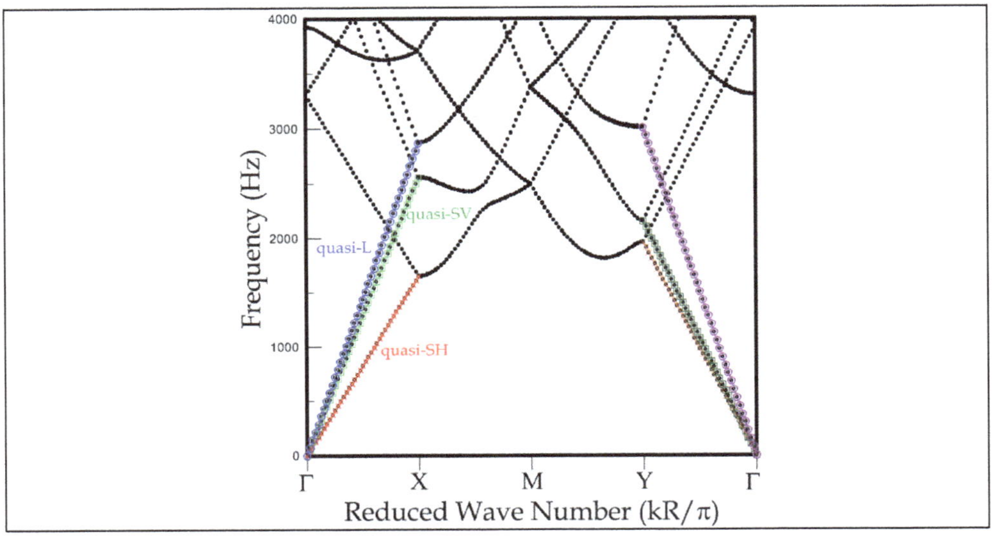

Fig. 9. The dispersion relations of homogeneous material quartz along the boundaries of the irreducible part of the Brillouin zone $\Gamma - X - M - Y - \Gamma$

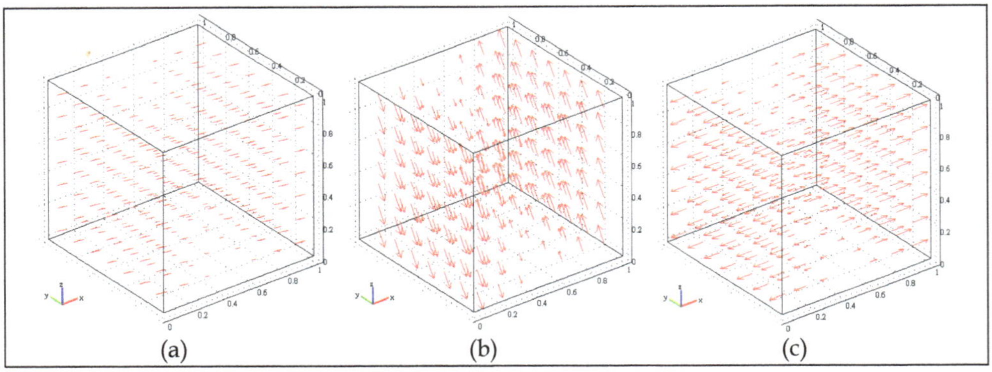

Fig. 10. (a) quasi shear horizontal mode (b) quasi shear vertical mode, (c) quasi longitudinal mode in the quartz

From the discussion, it shows that the method adopted in this chapter can be used to discuss the wave propagations in isotropic and anisotropic media.

4. Acoustic wave in inhomogeneous media

Previous studies on photonic crystals raise the exciting topic of phononic crystals. This section presents the results of acoustic waves in inhomogeneous media, Al/Ni periodic structures and phononic crystals with reticular geometric structures. It also discusses the tunable band gaps in the acoustic waves of two-dimensional phononic crystals with reticular geometric structures using the 2D and 3D finite element methods. This section calculates and discusses the band gap variations of the bulk modes due to different sizes of reticular geometric structures. Results show that adjusting the orientation of the reticular geometric structures can increase or decrease the total elastic band gaps for mixed polarization modes.

4.1 Periodic structure with two media

It is necessary and worthy to provide evidence supporting the FEM method's (COMSOL Multiphysics) ability to perform Bloch calculations with two media. This chapter compares the dispersion relations of Al/Ni band structure using the PWE method with the results of using the FEM method. Consider a phononic structure consisting of Al circular cylinders embedded in a background material of Ni to form a two-dimensional square lattice with lattice spacing R. Figure 11 shows the dispersion relations along the boundaries of the irreducible part of the Brillouin zone in Fig. 6(a) with filling ratio 0.6. The vertical axis represents the normalized frequency $\omega^* = \omega R / C_t$ and the horizontal axis represents the reduced wave number $k^* = kR / \pi$. Here, C_t and k are the shear velocity of Ni and the wave vector along the Brillouin zone, respectively. The Young's modulus E, Poisson's ratio ν, and density ρ of the material Ni utilized in this example are E=214 GPa, ν =0.336, and ρ =8905 kg/m^3.

The diamond symbols represent the dispersion relations of the transverse polarization modes (shear vertical modes), and the cross symbols represent the mixed polarization modes (shear horizontal mode coupled with longitudinal mode) in the PWE method. The open circles represent the dispersion relations of all modes in the FEM method with a 3D model. The results of the FEM method match well with those of the PWE method. In the similar cases, when the differences of mass densities and elastic constants between the two periodic materials are larger, the convergence of the PWE method is slower and costs more CPU time.

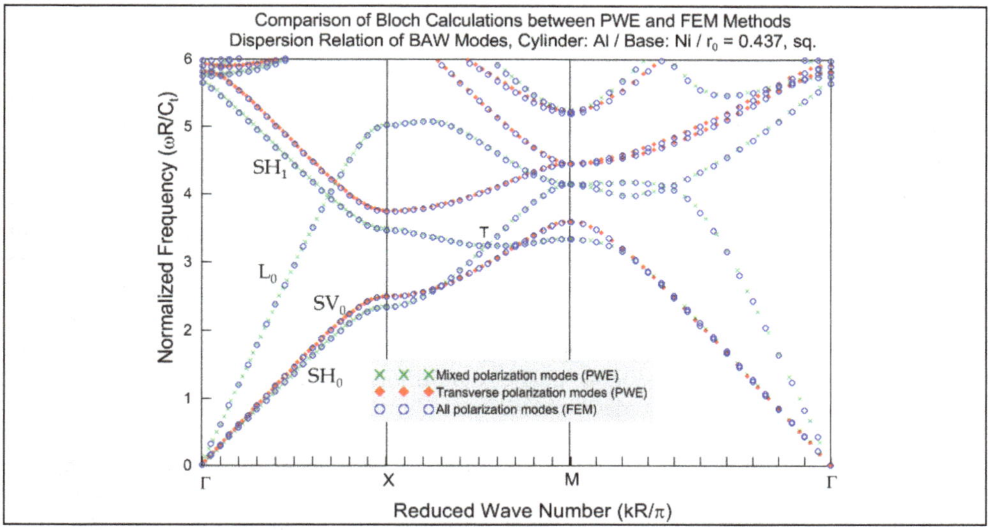

Fig. 11. Comparison of Bloch calculations between the PWE and FEM methods

As the elastic waves propagate along the x axis, the nonvanishing displacement fields of the shear horizontal mode, shear vertical mode, and longitudinal mode are u_y, u_z, and u_x respectively. For the sequence modes appear, the modes are always the same. When representing the whole wave vector space by the first Brillouin zone alone, they appear as further branches from higher Brillouin zones. In this example, the phase velocities of the SV_0 mode (diamond symbols) are larger than those of the SH_0 mode. The boundary of the Brillouin zone X-M of Fig. 11 represents the dispersion of the bulk waves with propagating direction varied 0 deg~ 45 deg counterclockwise away from the x direction.

4.2 Periodic structure with single medium

Figure 12(a) depicts a two-dimensional phononic crystal with the reticular geometric structures of square lattice. These reticular structures are parallel to the z-axis. In a perfect two-dimensional phononic crystal, the periodic structure is constant in the z direction and the size of the structure is infinite in the x and y directions. To analyze the dispersion relations of all bulk acoustic modes in this band structure, the FEM should consider the 3D model in Fig. 12(c). The dimensions of the unit cell in Fig. 12(a) are c=d=0.8R and R=h=1 in the calculations.

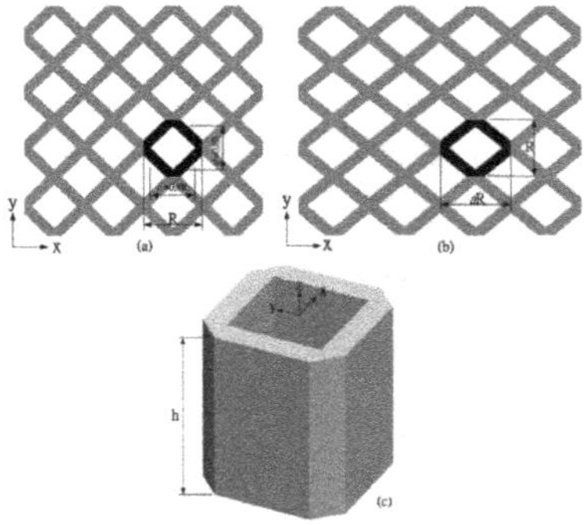

Fig. 12. (a) square lattice with lattice spacing R and (b) rectangular lattice with lattice spacing aR along x-axis and R along y-axis, (c) a unit cell with reticular structures in a 3D FEM model

The material of the reticular structures in the unit cell in this chapter is aluminum. Figure 12(c) shows a diagram of the unit square lattice in a 3D FEM model. The periodicity of phononic crystals along the z direction is used to calculate the dispersion relations of the mixed and transverse polarization modes. The types of finite elements used for the 2D and 3D cases are the default element types, Lagrange-Quadratic, in COMSOL Multiphysics. Figure 13 shows the dispersion relations of the mixed and transverse polarization modes along the boundaries of the irreducible part of the Brillouin zone in Fig. 6(b) with the scales R=h=1, c=0.8, and a=1.2. The horizontal axis represents the reduced wave number along $\Gamma-X-M-Y-\Gamma$ and the vertical axis represents the frequency (Hz). Note that this band structure shows no full band gap of the mixed and transverse polarization modes. Adopting the 2D FEM model to discuss the mixed polarization modes in this kind of band structure shows that there is only one full frequency band gap in Fig. 13, located at 3311 ~ 3400 Hz. Figure 13 compares the 3D and 2D FEM models. Open circles represent the dispersion relations of mixed polarization modes in the 2D FEM model, while solid circles represent the results of all bulk modes in the 3D FEM model. Figure 14 shows the eigenmode shapes with 4×4 supercell of total displacements for M_1 and M_2 modes indicated in Fig. 13. These figures clearly show the phenomena of wave localizations in this reticular geometric structure. Note that the FEM method can easily describe the mode characteristics. In this chapter, M_1 is a shear horizontal mode with mode vibrating displacement along the y direction when the wave propagates along the x direction ($\Gamma - X$ direction). Also, M_2 is a shear vertical mode with mode vibrating displacement along z direction, and it does not couple with the mixed polarization modes.

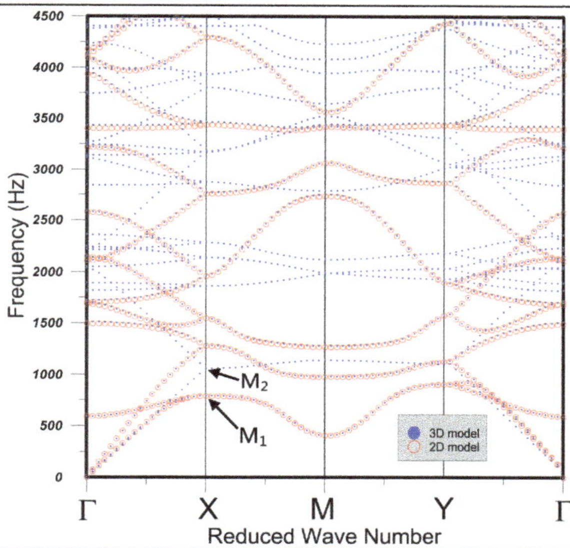

Fig. 13. The dispersion relations of the mixed and transverse polarization modes along the boundaries of the irreducible part of the Brillouin zone with the scales R=h=1, c=0.8, and a=1.2

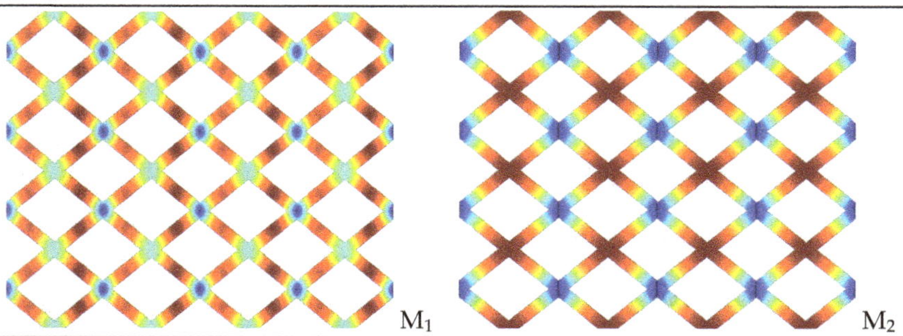

Fig. 14. The eigenmode shapes with 4×4 supercell of total displacements for M_1 and M_2 modes indicated in Fig. 13

The following discussion addresses several parameters of the reticular geometric in this chapter. First, the effect of filling fraction is discussed when the parameters c=d varied from 0.1 to 0.9 in Fig. 12(a). Figure 15 shows the distribution of the total band gaps of mixed polarization modes, in which only one total band gap appears at approximately 3560 ~ 3736 Hz in c=d=0.8. The horizontal axis represents the parameter c, and the vertical axis represents frequency (Hz). Figure 15 also shows the 2D diagrams of the reticular geometric structures with c=d=0.1, 0.5, and 0.8.

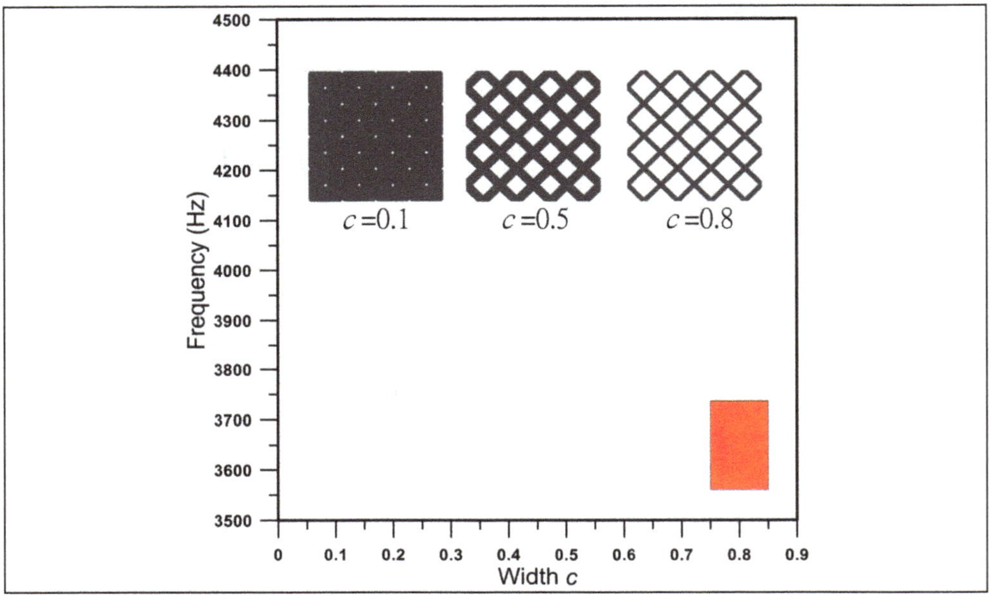

Fig. 15. The band gap width with parameters c=d varying from 0.1 to 0.9 when the vertical range is selected from 3500 to 4500 Hz

On the other hand, the scale a in Fig. 12(b) varies from 0.1 to 2.0 along the x direction and the width of the unit cell along y direction remains 1.0 in the Bloch calculations. Changing the scale a from 0.1 to 2.0 can tune the full frequency band gaps of mixed polarization modes. Using detailed calculations of dispersion relations of reticular geometric structures with scale a=0.1 to 2.0, Fig. 16 shows the band gap widths with the scale a from 0.1 to 2.0 when the vertical range ranges from 2400 to 5200 Hz. The horizontal axis ranges from 0 to 2.0, and the vertical axis represents frequency (Hz). No full frequency band gap exists when the scale a are 0.1, 0.2, 0.3, 0.4, 0.6, 0.7, 1.5, 1.6, and 1.7. These results clearly show that changing the scale a can increase or decrease the full frequency band gap.

It is noted that the unit cells with a=0.5 and 2.0 are the same in the Bloch calculations. However, the dispersion phenomena is similar except for the scalar of the eigenmode frequencies in the vertical axis of dispersion relations. In both cases, there is only one total band gap of the mixed polarization modes. The location of the band gap ranges from approximately 5009 to 5017.4 Hz with a=0.5, while that for a=2.0 ranges from approximately 2504.5 to 2508.7 Hz.

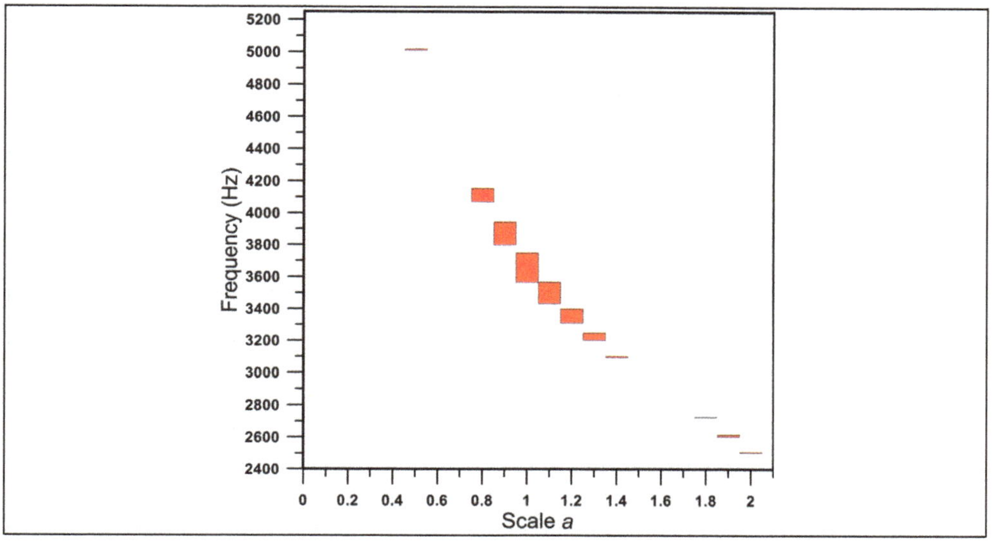

Fig. 16. The band gap widths with the scale a from 0.1 to 2.0

Finally, the rotating angles of reticular geometric structures were changed to analyze the distribution of total band gaps. Figure 17 shows the 2D diagrams of unit rectangular lattices in different rotating angles D=30 deg, 45 deg, 75 deg, and 90 deg. In these cases, the widths of aluminum remain constant, 0.14R, in the reticular geometric structures with different rotating angles in the calculations. Figure 18 shows the band gap widths of rectangular lattices with different rotating angles of reticular geometric structures. Based on the symmetry of the geometry, the different angles in the Bloch calculations were adopted from 15 deg ~ 90 deg. In the calculated results, no band gap is detected from D=5 deg to 65 deg.

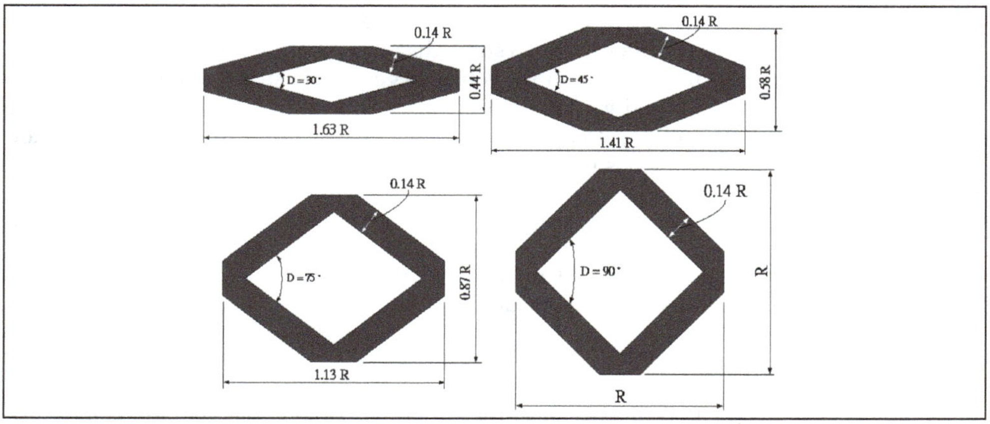

Fig. 17. 2D diagrams of unit rectangular lattices in different rotating angles D=30 deg, 45 deg, 75 deg, and 90 deg

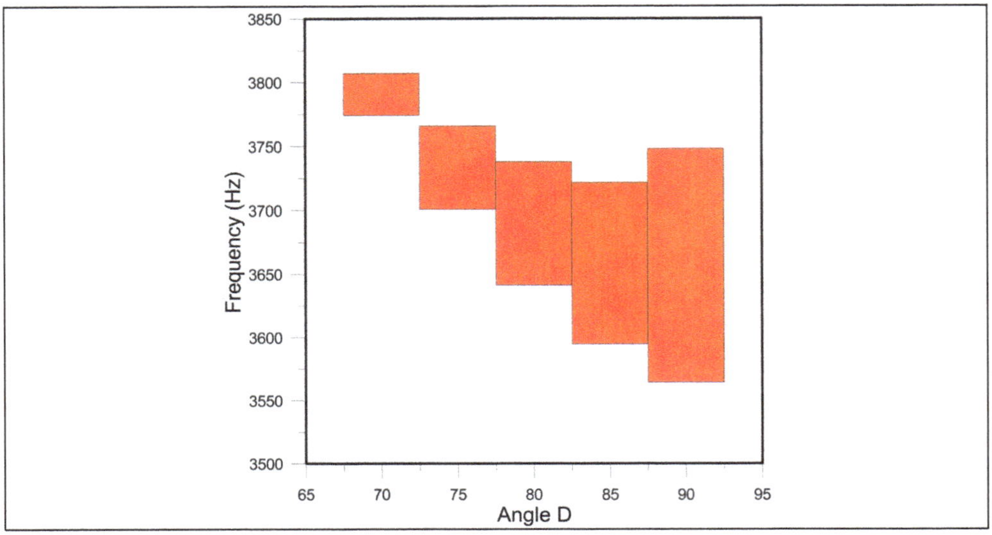

Fig. 18. The band gap widths of the rectangular lattices with different rotating angles of reticular geometric structures

5. Conclusion

This chapter examines and discusses the acoustic waves in homogeneous medium and inhomogeneous medium, periodic structures with two media and one medium with geometrical periodicity. The wave velocities of shear and longitudinal modes in an isotropic material and those of quasi-SV, quasi-SH, and quasi-L modes in an anisotropic material are obtained using the finite element method. This method also discusses the tunable frequency band gaps of bulk acoustic waves in two-dimensional phononic crystals with reticular geometric structures using the 2D and 3D finite element methods. This study adopts the finite element method to calculate dispersion relations, avoiding the numerical errors, Gibbs phenomenon, from the PWE method. Results show that changing the filling fraction, scale a, and the rotating angles of unit lattices in the reticular geometric structures can increase or decrease the elastic/acoustic band gaps. The effect discussed in this chapter can be utilized to enlarge the phononic band gap frequency and may enable the study of the frequency band gaps of elastic/acoustic modes in special phononic band structures.

6. Acknowledgment

The authors thank the National Science Council (NSC 97-2218-E-150-006, 98-2221-E-150-026, and 99-2628-E-150-001) of Taiwan for financial support.

7. References

Axmann, W. & Kuchment, P. (1999). An efficient finite element method for computing spectra of photonic and acoustic band-gap materials. J. Comput. Phys. Vol. 150, pp. 468, ISSN 0021-9991

Burger, M. S., Osher, J., & Yablonovitch, E. (2004). Inverse Problem Techniques for the Design of Photonic Crystals. IEICE Trans. Electron, E87C, 258-265.

Checoury, X. & Lourtioz, J. M. (2006). Wavelet method for computing band diagrams of 2D photonic crystals. Optics Communications Vol. 59, pp. 360, ISSN 0030-4018

Chiang, P. J., Yu, C. P., & Chang, H. C. (2007). Analysis of two-dimensional photonic crystals using a multidomain pseudospectral method. Phys. Rev. E Vol. 75, pp. 026703, ISSN 1539-3755

Dobson, D. C. (1999). An efficient method for band structure calculations in 2D photonic crystals. J. Comput. Phys. Vol. 149, pp. 363, ISSN 0021-9991

Garica-Pablos, D., Sigalas, M., Montero de Espinosa, F. R., Torres, M., Kafesaki, M., and Garcia, N. (2000). Theory and Experiments on Elastic Band gaps. Phys. Rev. Lett. Vol. 84, pp. 4349, ISSN 0031-9007

Huang, G. L. & Sun, C. T. (2010). Band Gaps in a Multiresonator Acoustic Metamaterial. ASME J. Vib. Acoust. Vol. 132, pp. 031003. ISSN 1048-9002

Huang, Z. G. & Chen, Z. Y. (2011). Acoustic Waves in Two-dimensional Phononic Crystals with Reticular Geometric Structures. ASME J. Vib. Acoust. Vol. 133(3), pp.031011, ISSN 1048-9002

Huang, Z. G. & Wu, T.-T. (2005). Temperature effects on bandgaps of surface and bulk acoustic waves in two-dimensional phononic crystals. IEEE Trans. Ultrason. Ferroelectr. Freq. Control Vol. 52, pp. 365, ISSN 0885-3010

Hussein, M. I. (2009). Reduced Bloch mode expansion for periodic media band structure calculations. Proceedings of the Royal Society A Vol. 465, pp. 2825-2848, ISSN 1364-5021

Joannopoulos, J. D., Meade, R. D. & Winn, J. N. (1995) . Photonic Crystals: Molding the flow of light, ISBN: 978-0691124568, Princeton University Press, Princeton, NJ.

Johnson, S. G. & Joannopoulos, J. D. (2001). Block-iterative frequency-domain methods for Maxwell's equations in a planewave basis. Optics Express Vol. 8, pp. 173, ISSN 1094-4087

Johnson, S. G. & Joannopoulos, J. D. (2003). PHOTONIC CRYSTALS: The road from theory to practice, ISBN 978-0792376095, Kluwer academic publishers, Boston.

Jun, S., Cho, Y. S., & Im, S. (2003). Moving least-square method for the band-structure calculation of 2D photonic crystals. Optics Express Vol. 11, pp. 541, ISSN 1094-4087

Kafesaki, M. & Economou, E. N. (1999). Multiple-scattering theory for three-dimensional periodic acoustic composites. Phys. Rev. B Vol. 60, pp. 11993, ISSN 1098-0121

Kittel, C. (1996). Introduction to Solid State Physics, ISBN 978-0-471-41526-8, 7th ed., John Wiley & Sons. Inc.

Kushwaha, M. S., Halevi, P., Dobrzynski, L. & Djafari-Rouhani, B. (1993). Acoustic Band Structure of Periodic Elastic Composites. Phys. Rev. Lett. Vol. 71, pp. 2022, ISSN 0031-9007

Laude, V., Wilm, M., Benchabane, S., Khelif, A. (2005). Full band gap for surface acoustic waves in a piezoelectric phononic crystal. Phys. Rev. E Vol. 71, pp. 036607, ISSN 1539-3755

Leung, K. M. & Liu, Y. F. (1990). Full vector wave calculation of photonic band structures in face-centered-cubic dielectric media. Phys. Rev. Lett. Vol. 65, pp. 2646, ISSN 0031-9007

Leung, K. M. & Qiu, Y. (1993). Multiple-scattering calculation of the two-dimensional photonic band structure. Phys. Rev. B Vol. 48, pp. 7767, ISSN 1098-0121

Moreno, E., Erni, D., & Hafner, C. (2002). Band structure computations of metallic photonic crystals with the multiple multipole method. Phys. Rev. B Vol. 65, pp. 155120, ISSN 1098-0121

Pendry, J. B. & MacKinnon, A. (1992). Calculation of photon dispersion relations. Phys. Rev. Lett. Vol. 69, pp. 2772, ISSN 0031-9007

Psarobas, I. E. & Stefanou, N. (2000). Scattering of elastic waves by periodic arrays of spherical bodies. Phys. Rev. B Vol. 62, pp. 278, ISSN 1098-0121

Salehian, A. & Inman, D. J. (2010). Micropolar Continuous Modeling and Frequency Response Validation of a Lattice Structure. ASME J. Vib. Acoust. Vol. 132, pp.011010, ISSN 1048-9002

Sun, J. H. & Wu, T.-T. (2005). Analyses of mode coupling in joined parallel phononic crystal waveguides. Phys. Rev. B Vol. 71, pp. 174303, ISSN 1098-0121

Tanaka, Y. & Tamura, S. (1998). Surface acoustic waves in two-dimensional periodic elastic structures. Phys. Rev. B Vol. 58, pp. 7958, ISSN 1098-0121

Tanaka, Y., Tomoyasu, Y., & Tamura, S. I. (2000). Band structure of acoustic waves in phononic lattices: Two-dimensional composites with large acoustic mismatch. Phys. Rev. B Vol.62, no. 11, 7387–7392.

Wang, X., Zhang, X. G., Yu, Q. & Harmon, B. N. (1993). Multiple-scattering theory for electromagnetic waves. Phys. Rev. B Vol. 47, pp. 4161, ISSN 1098-0121

Wu, T.-T. & Huang, Z. G. (2004). Level repulsion of bulk acoustic waves in composite materials. Phys. Rev. B Vol. 70, pp. 214304, ISSN 1098-0121

Wu, T.-T., Huang, Z. G., & Lin, S. (2004). Surface and bulk acoustic waves in two-dimensional phononic crystals consisting of materials with general anisotropy. Phys. Rev. B Vol. 69, pp. 094301, ISSN 1098-0121

Wu, T.-T., Huang, Z. G., Tsai, T. C. & Wu, T. C. (2008). Evidence of complete band gap and resonances in a plate with periodic stubbed surface. Applied Physics Letters Vol. 93, pp. 111902, ISSN 0003-6951

Yan, Z. Z. & Wang, Y. S. (2006). Wavelet-based method for calculating elastic band gaps of two-dimensional phononic crystals. Phys. Rev. B Vol. 74, pp. 224303, ISSN 1098-0121

Yang, H. Y. D. (1996). Finite difference analysis of 2-D photonic crystals. IEEE Trans. Microwave Theory Tech. Vol. 44, pp. 2688, ISSN 0018-9480

2

Topological Singularities in Acoustic Fields due to Absorption of a Crystal

V. I. Alshits[1,2], V. N. Lyubimov[1] and A. Radowicz[3]
[1]*A.V. Shubnikov Institute of Crystallography, Russian Academy of Sciences, Moscow,*
[2]*Polish-Japanese Institute of Information Technology, Warsaw,*
[3]*Kielce University of Technology, Kielce*
[1]*Russia*
[2,3]*Poland*

1. Introduction

The influence of energy dissipation on the properties of bulk elastic waves in crystals is not at all reduced to trivial decrease in their amplitudes along propagation. In anisotropic media the situation is much more complicated than it looks like at first glance, at least for such specific directions of propagation as acoustic axes. The latter are defined as directions $\mathbf{m}_0$ along which a degeneracy of the phase speeds of two isonormal waves occurs (Fedorov, 1968; Khatkevich, 1962a, 1964). The corresponding points of the contact of the degenerate sheets of the phase velocity surface P may be tangent or conical (Alshits & Lothe, 1979; Alshits, Sarychev & Shuvalov, 1985) (Fig.1).

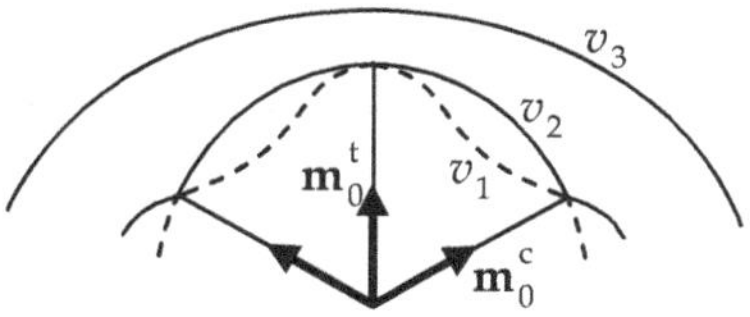

Fig. 1. Schematic plot of the section fragment of the three sheets of the phase velocity surface $v_\alpha(\mathbf{m})$ ($\alpha = 1, 2, 3$) containing one tangent and two conical points of degeneracy

Taking into account that formally the wave attenuation may be described as an imaginary perturbation of the phase speed, one could expect due to the damping either a shift or a split of the acoustic axis, of course if it is not created by a symmetry. As we shall see below, for an acoustic axis of general position it is just splitting what is realized, and with quite a radical transformation of the local geometry of the phase velocity surface. The other possible reason for sensitivity of the wave properties to a small attenuation is related to a polarization aspect. Indeed, it is known (Alshits & Lothe, 1979; Alshits, Sarychev & Shuvalov, 1985) that the acoustic axes indicate on the unit sphere of propagation directions $\mathbf{m}^2 = 1$ the singular points in the vector fields of polarizations which are characterized by the definite vector rotation around these points on $\pm 2\pi$ or $\pm \pi$, i.e. by the Poincarè indices $n= \pm 1$ or $\pm 1/2$ (Fig.2). It is clear that a split of such singular points must be quite catastrophic for the corresponding polarization distribution. And that really occurs.

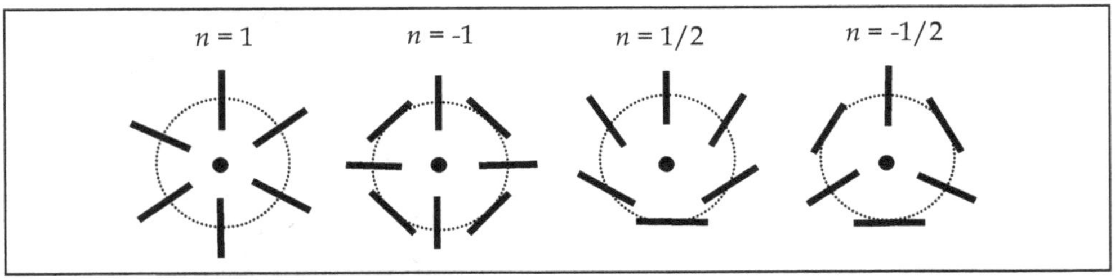

Fig. 2. Singular polarization distributions around the two types of tangent degeneracy points $n=\pm 1$ and the two types of conical degeneracies $n=\pm 1/2$

The above peculiarities are associated with space distribution of wave characteristic rather than with individual properties of bulk elastic waves. Meanwhile, as we shall see, in absorptive crystals the individual wave properties close to degeneracy directions also manifest quite unusual features, such as an almost circular polarization, in contrast to a quasi-linear one in the non-degenerate regions.

The theory of acoustic axes in non-absorptive anisotropic media is quite complete. For a review we address readers to the paper by Shuvalov (1998). The theory gives the general criteria of the degeneracy occurrence and describes all possible types of acoustic axes classifying them with respect to a local geometry of the degenerate velocity sheets and to specific features of the vector polarization fields around the degeneracy directions. This classification (Alshits & Lothe, 1979; Alshits, Sarychev & Shuvalov, 1985) includes more types than we presented in Figs. 1 and 2. However, apart from a line degeneracy known in hexagonal crystals, the rest additional types relate to the model media with accidentally coinciding or vanishing material constants (or some their combinations). Such media are beyond our interest in this paper. Note in addition that conical acoustic axes may exist in real crystals even in quite non-symmetric directions and always exist along the symmetry axis *3*. In contrast, tangent degeneracies are realized in practice only due to a high symmetry of the crystals and are known only along symmetry axes ∞ and *4*. As was shown in (Alshits & Lothe, 1979; Alshits, Sarychev & Shuvalov, 1985) , all the "model" acoustic axes together with any tangent or line degeneracies are unstable and must disappear, split or be transformed into other types under any small triclinic perturbation of the elastic moduli tensor $\hat{c}$. The only stable type of acoustic axes is the conical type. Under any real perturbation $\delta\hat{c}$ a conical degeneracy never split or disappear, but can only shift.

The wave attenuation can be interpreted as a perturbation of the tensor $\hat{c}$, however not real but imaginary. As was mentioned in (Alshits, Sarychev & Shuvalov, 1985), under such a non-hermitian perturbation even a conical degeneracy may lose its stability. Later Shuvalov & Chadwick (1997) rigorously investigated the stability of different acoustic axes with respect to a weak thermoelastic coupling. Their conclusion was: all types of degeneracies are unstable including a conical acoustic axis which splits into a pair. The same problem for viscoelastic and thermo-viscoelastic media has been studied by Shuvalov & Scott (1999, 2000) with similar conclusions.

It is evident that the considered physical mechanisms of the damping definitely do not disturb symmetry of the crystal and therefore cannot shift or split degeneracies along symmetry axes ∞, *4* and *3*. It means that any really existing tangent degeneracies and conical degeneracies along symmetry axes *3* must be stable under the damping perturbation. This statement was proved by Alshits & Lyubimov (1998) for viscoelastic media.

In this chapter we shall consider the attenuation in terms of viscoelasticity following to the approach of the papers (Alshits & Lyubimov, 1998, 2011). We shall analyse in detail the mentioned above geometrical peculiarities and polarization singularities related to a pair of the so-called singular acoustic axes representing a new type of stable degeneracy and arising as a result of the considered split of a conical acoustic axis. On this basis we shall develop an extension of the classical theory of internal conical refraction (Barry & Musgrave, 1979; De Klerk & Musgrave, 1955; Fedorov, 1968; Khatkevich, 1962b; Musgrave, 1957) for an absorptive crystal. As will be shown, the damping provides very radical and non-trivial modifications of fundamental features of the phenomenon.

2. Statement of the problem and general relations

Let us consider the viscoelastic medium characterized by the density ρ and the tensors of elastic moduli $\hat{c}$ and viscosity $\hat{\eta}$. The dynamic displacement field $\mathbf{u}(\mathbf{r},t)$ in such medium is described by the known equation (Landau & Lifshitz, 1986)

$$\rho\ddot{u}_i = c_{ijkl}u_{l,kj} - \eta_{ijkl}\dot{u}_{l,kj}, \tag{1}$$

where the vectors $\dot{\mathbf{u}}$ and $\ddot{\mathbf{u}}$ are the velocity and acceleration fields and the usual notation $..._{,k} \equiv \partial/\partial x_k ...$ is accepted. For the bulk wave

$$\mathbf{u}(\mathbf{r},t) = C\mathbf{A}\exp[ik(\mathbf{m}\cdot\mathbf{r} - vt)] \tag{2}$$

propagating along the wave vector $\mathbf{k} = k\mathbf{m}$ with the amplitude *C*, the frequency $\omega = kv$, the phase speed v and the polarization **A**, eqn. (1) is transformed into Christoffel`s equation

$$(\hat{Q}' - i\hat{Q}'')\mathbf{A} = \rho v^2\mathbf{A} \tag{3}$$

where $\hat{Q}'$ and $\hat{Q}''$ are the real symmetric matrices

$$Q'_{jk} = m_i c_{ijkl} m_l, \qquad Q''_{jk} = \omega m_i \eta_{ijkl} m_l. \tag{4}$$

Note, that the imaginary addition $-i\hat{Q}''$ to the usual acoustic tensor $\hat{Q}'$, in contrast to the latter, is dependent on the frequency. Eqn. (3) determines the three complex eigenvectors $\mathbf{A}_\alpha$ and the three corresponding complex eigenvalues ρv_α^2 (α = 1, 2, 3), i.e. the three phase speeds v_α as functions of the direction **m**:

$$\mathbf{A}_\alpha = \mathbf{A}'_\alpha + i\mathbf{A}''_\alpha, \qquad v_\alpha = v'_\alpha - iv''_\alpha. \tag{5}$$

Below the frequency will be supposed to be real. Hence, by (5), the value k_α should contain an imaginary addition determining the decay of the wave along its propagation:

$$k_a \equiv k'_a + ik''_a = \frac{\omega}{v'_a - iv''_a} \approx \frac{\omega}{v'_a}\left(1 + i\frac{v''_a}{v'_a}\right). \tag{6}$$

The complex phase speeds of eigenwaves are found from the equation

$$\rho v_a^2 = \frac{\mathbf{A}_a(\hat{Q}' - i\hat{Q}'')\mathbf{A}_a}{\mathbf{A}_a \cdot \mathbf{A}_a}. \tag{7}$$

The polarization vectors $\mathbf{A}_a$ as eigenvectors of the symmetric matrix $\hat{Q}' - i\hat{Q}''$ for non-degenerate directions **m** of propagation must be mutually orthogonal

$$\mathbf{A}_a \cdot \mathbf{A}_\beta = \delta_{a\beta}, \qquad a \neq \beta. \tag{8}$$

As regards to their normalization, we cannot use the customary condition $\mathbf{A}_a^2 = 1$, bearing in mind the possibility of a circular polarization for which $\mathbf{A}_a^2 = 0$. Instead, the normalizing factor will be chosen so that

$$|\mathbf{A}_a|^2 = \mathbf{A}_a'^2 + \mathbf{A}_a''^2 = 1. \tag{9}$$

For a further development let us divide the basic eqn. (3) on the real and imaginary parts

$$\begin{aligned} \hat{Q}'\mathbf{A}'_a + \hat{Q}''\mathbf{A}''_a &= \rho(v_a'^2 - v_a''^2)\mathbf{A}'_a + 2\rho v'_a v''_a \mathbf{A}''_a, \\ \hat{Q}'\mathbf{A}''_a - \hat{Q}''\mathbf{A}'_a &= \rho(v_a'^2 - v_a''^2)\mathbf{A}''_a - 2\rho v'_a v''_a \mathbf{A}'_a. \end{aligned} \tag{10}$$

Multiplying these equations by $\mathbf{A}'_{a,\beta}$ or $\mathbf{A}''_{a,\beta}$ ($\alpha \neq \beta$) and combining the results one obtains

$$2\rho v'_a v''_a = \mathbf{A}'_a \cdot \hat{Q}''\mathbf{A}'_a + \mathbf{A}''_a \cdot \hat{Q}''\mathbf{A}''_a, \tag{11}$$

$$\rho(v_a'^2 - v_a''^2) = \mathbf{A}'_a \cdot \hat{Q}'\mathbf{A}'_a + \mathbf{A}''_a \cdot \hat{Q}'\mathbf{A}''_a, \tag{12}$$

$$\rho(v_\beta'^2 - v_a'^2 + v_a''^2 - v_\beta''^2)\mathbf{A}''_a \cdot \mathbf{A}'_\beta = \mathbf{A}'_a \cdot \hat{Q}''\mathbf{A}'_\beta + \mathbf{A}''_a \cdot \hat{Q}''\mathbf{A}''_\beta - 2\rho(v'_a v''_a + v'_\beta v''_\beta)\mathbf{A}''_a \cdot \mathbf{A}''_\beta. \tag{13}$$

Eqns. (11)-(13) are exact. The first two of them show that the imaginary part of the phase speed v''_a being linear in small viscosity $\hat{\eta}$ is small compared to v'_a independently of the direction **m**. In accordance with Eqn. (13), one can also conclude that $A''_a \ll A'_a$ however not for any **m**, but only far enough from acoustic axes, when the difference $v'_\beta - v'_a$ is not small. In this case the value $A''_a = |\mathbf{A}''_a|$ is also linear in $\hat{\eta}$ and therefore small. Let us decompose the vector $\mathbf{A}''_a$ on the two components: $\mathbf{A}''_a = \mathbf{A}''^{\perp}_a + \mathbf{A}''^{\parallel}_a$, where $\mathbf{A}''^{\perp}_a \perp \mathbf{A}'_a$ and $\mathbf{A}''^{\parallel}_a \parallel \mathbf{A}'_a$. Thus, the ellipticity $\varepsilon = A''^{\perp}_a / A'_a$ of the wave polarization due to the damping is also small almost everywhere beyond small domains around acoustic axes. Let us estimate this ellipticity to the first order in $\hat{\eta}$.

Being perpendicular to $\mathbf{A}'_a$, the vector $\mathbf{A}''^{\perp}_a$ may be expressed in leading approximation as a superposition of two isonormal vectors $\mathbf{A}'_\beta$ and $\mathbf{A}'_\gamma$ which are almost orthogonal to $\mathbf{A}'_a$,

$$\mathbf{A}''^{\perp}_a \approx (\mathbf{A}''^{\perp}_a \cdot \mathbf{A}'_\beta)\mathbf{A}'_\beta + (\mathbf{A}''^{\perp}_a \cdot \mathbf{A}'_\beta)\mathbf{A}'_\beta \approx (\mathbf{A}''_a \cdot \mathbf{A}'_\beta)\mathbf{A}'_\beta + (\mathbf{A}''_a \cdot \mathbf{A}'_\beta)\mathbf{A}'_\beta. \tag{14}$$

In view of eqn. (13) this gives far from degeneracies

$$\mathbf{A}_a''^{\perp} = \frac{\mathbf{A}_a' \cdot \hat{Q}'' \mathbf{A}_\beta'}{\rho(v_\beta'^2 - v_a'^2)} \mathbf{A}_\beta' + \frac{\mathbf{A}_a' \cdot \hat{Q}'' \mathbf{A}_\gamma'}{\rho(v_\gamma'^2 - v_a'^2)} \mathbf{A}_\gamma' . \tag{15}$$

In fact, for considered non-singular directions the component $\mathbf{A}_a''^{\parallel}$ is physically unimportant. Indeed, the vector amplitude of the wave (2) in the accepted linear approximation is equal

$$C\mathbf{A}_a = C(\mathbf{A}_a' + i\mathbf{A}_a''^{\parallel} + i\mathbf{A}_a''^{\perp}) = C(1 + i\varepsilon^{\parallel})\left(\mathbf{A}_a' + \frac{i\mathbf{A}_a''^{\perp}}{1 + i\varepsilon^{\parallel}} \right) \approx C'(\mathbf{A}_a' + i\mathbf{A}_a''^{\perp}) , \tag{16}$$

where the notations $\varepsilon^{\parallel} = A_a''^{\parallel} / A_a'$ and $C' = C(1 + i\varepsilon^{\parallel})$ are introduced.
With (15) and (9), the wave ellipticity is readily estimated as $\varepsilon \approx A_a''^{\perp} = |\mathbf{A}_a''^{\perp}|$. For similar non-singular directions the speeds v_a'' and v_a' determined by eqns. (11) and (12) may be expressed in the same leading approximation as

$$v_a'' = \frac{\mathbf{A}_a' \cdot \hat{Q}'' \mathbf{A}_a'}{2\rho v_a'} , \qquad \rho v_a'^2 = \mathbf{A}_a' \cdot \hat{Q}' \mathbf{A}_a' . \tag{17}$$

On the other hand, eqn. (15) demonstrates the tendency to increasing ellipticity ε when the wave normal $\mathbf{m}$ approaches the degeneracy direction ($v_a' = v_\beta'$ or $v_a' = v_\gamma'$) and one of the denominators in (15) decreases becoming singular. Of course, in the vicinity of the degeneracy it is necessary to replace eqns. (15) and (17) by some other relations.

3. General formalism for the neighbourhood of an acoustic axis

In fact, eqns. (15) and (17) quite hold for the description of the non-degenerate wave branch even along the direction where two other branches are degenerate. In the further development we shall choose for the non-degenerate wave characteristics the number $a = 3$. In this notation, by eqn. (15), the vector $\mathbf{A}_3''$ must be small addition to $\mathbf{A}_3'$. In view of the orthogonality condition (8) this allows us in the leading approximation to replace the complex polarization vectors $\mathbf{A}_{1,2}$ by their projections on the plane orthogonal to the vector $\mathbf{A}_3'$. This must work even close to acoustic axes where the imaginary components of $\mathbf{A}_{1,2}$ might be comparable in the length with their real counterparts. We are following here the ideology developed in the theory of acoustic axes for the case of zero damping (Alshits & Lothe, 1979; Alshits, Sarychev & Shuvalov, 1985).
Thus, let $\mathbf{m}_0$ is the direction of the acoustic axis in the crystal with the "switched off" attenuation. By definition, along $\mathbf{m}_0$ there must be $v_1 = v_2 \equiv v_0$ and, apart from the non-degenerate wave with the speed v_3 and the polarization $\mathbf{A}_{03}$, any polarization in the degeneracy plane $D \perp \mathbf{A}_{03}$ is permissible (Fig.3).
Let us choose in the D plane an arbitrary pair of unit orthogonal vectors $\mathbf{A}_{01}$ and $\mathbf{A}_{02}$ forming with $\mathbf{A}_{03}$ the orthonormal right-handed basis $\{\mathbf{A}_{01}, \mathbf{A}_{02}, \mathbf{A}_{03}\}$ (Fig. 3).
Now "switch on" the damping and consider eqn. (3) close to $\mathbf{m}_0$ at $\mathbf{m} = \mathbf{m}_0 + \delta\mathbf{m}$:

$$(\mathbf{m}_0 + \delta\mathbf{m})(\hat{c} - i\omega\hat{\eta})(\mathbf{m}_0 + \delta\mathbf{m})\mathbf{A}_a = \rho(v_{0a} + \delta v_a)^2 \mathbf{A}_a . \tag{18}$$

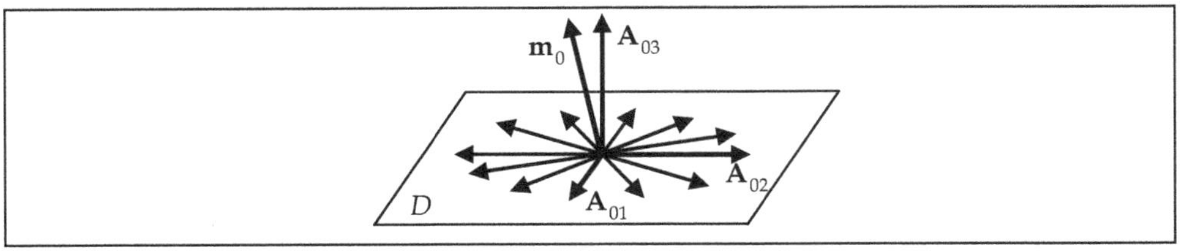

Fig. 3. Allowed polarizations along the acoustic axis $\mathbf{m}_0$ at "switched off" attenuation

In the linear approximation eqn. (18) is transformed to

$$(\hat{Q}_0 + \delta\hat{Q})\mathbf{A}_a = \rho(v_{0a}^2 + 2v_{0a}\delta v_a)\mathbf{A}_a , \tag{19}$$

where

$$\hat{Q}_0 = \hat{Q}'(\mathbf{m}_0) , \qquad \delta\hat{Q} = \mathbf{m}_0\hat{c}\delta\mathbf{m} + \delta\mathbf{m}\hat{c}\mathbf{m}_0 - i\hat{Q}_0'' , \qquad \hat{Q}_0'' = \hat{Q}''(\mathbf{m}_0) . \tag{20}$$

The complex polarization vectors $\mathbf{A}_a$ may be decomposed in the basis $\{\mathbf{A}_{01}, \mathbf{A}_{02}, \mathbf{A}_{03}\}$ as

$$\mathbf{A}_a = a_{a\beta}\mathbf{A}_{0\beta} , \tag{21}$$

where a, β = 1, 2, 3 and the summation over β is assumed. Substituting the linear superpositions (21) at a = 1 and a = 2 into eqn. (19) one obtains

$$\begin{aligned} \delta\hat{Q}\mathbf{A}_1 &= 2\rho v_0 \delta v_1 \mathbf{A}_1 + \rho(v_0^2 - v_{03}^2)a_{13}\mathbf{A}_{03}, \\ \delta\hat{Q}\mathbf{A}_2 &= 2\rho v_0 \delta v_2 \mathbf{A}_2 + \rho(v_0^2 - v_{03}^2)a_{23}\mathbf{A}_{03}. \end{aligned} \tag{22}$$

Eqns. (22) show that the coefficients a_{13} and a_{23} must be linearly small. So, indeed in the leading approximation one can replace the polarization vectors $\mathbf{A}_1$ and $\mathbf{A}_2$ by their projections on the D-plane

$$\mathbf{A}_1 \approx a_{11}\mathbf{A}_{01} + a_{12}\mathbf{A}_{02} , \qquad \mathbf{A}_2 \approx a_{21}\mathbf{A}_{01} + a_{22}\mathbf{A}_{02} . \tag{23}$$

Multiplying eqns. (22) by $\mathbf{A}_1$ or $\mathbf{A}_2$ we obtain the two linear systems determining the coefficients $a_{a\beta}$ in (23):

$$\begin{cases} (\delta Q_{11} - 2v_0\rho\delta v_1)a_{11} + \delta Q_{12}a_{12} = 0, \\ \delta Q_{12}a_{11} + (\delta Q_{22} - 2v_0\rho\delta v_1)a_{12} = 0; \end{cases} \qquad \begin{cases} (\delta Q_{11} - 2v_0\rho\delta v_2)a_{21} + \delta Q_{12}a_{22} = 0, \\ \delta Q_{12}a_{21} + (\delta Q_{22} - 2v_0\rho\delta v_2)a_{22} = 0; \end{cases} \tag{24}$$

where

$$\delta Q_{ij} = \mathbf{A}_{0i} \cdot \delta\hat{Q}\mathbf{A}_{0j} , \qquad i, j = 1, 2 . \tag{25}$$

The conditions for the existence of nontrivial solutions of the systems (24) give the common quadratic equation determining both δv_1 and δv_2

$$(\delta Q_{11} - 2v_0\rho\delta v)(\delta Q_{22} - 2v_0\rho\delta v) - \delta Q_{12}^2 = 0 \tag{26}$$

with the roots determining the unknown additions $\delta v_{1,2}$ to the degenerate speed v_0 :

$$\delta v_{1,2} = \mathbf{s}_0 \cdot \delta\mathbf{m} - is'' \mp R\ . \tag{27}$$

Here the notations are introduced

$$R = \sqrt{(\mathbf{p}\cdot\delta\mathbf{m} - ip'')^2 + (\mathbf{q}\cdot\delta\mathbf{m} - iq'')^2}\ ; \tag{28}$$

$$\left.\begin{matrix}\mathbf{s}_0\\ \mathbf{p}\end{matrix}\right\} = \frac{1}{2\rho v_0}(\mathbf{A}_{01}\hat{c}\mathbf{A}_{01} \pm \mathbf{A}_{02}\hat{c}\mathbf{A}_{02})\mathbf{m}_0, \qquad \mathbf{q} = \frac{1}{2\rho v_0}(\mathbf{A}_{01}\hat{c}\mathbf{A}_{02} + \mathbf{A}_{02}\hat{c}\mathbf{A}_{01})\mathbf{m}_0\ ; \tag{29}$$

$$\left.\begin{matrix}s''\\ p''\end{matrix}\right\} = \frac{Q''_{11} \pm Q''_{22}}{4\rho v_0}, \qquad q'' = \frac{Q''_{12}}{2\rho v_0}; \tag{30}$$

$$Q''_{ij} = \mathbf{A}_{0i} \cdot \hat{Q}''_0 \mathbf{A}_{0j}\ . \tag{31}$$

The introduced vectors $\mathbf{s}_0$, $\mathbf{p}$ and $\mathbf{q}$ have the following projections on $\mathbf{m}_0$:

$$\mathbf{s}_0 \cdot \mathbf{m}_0 = v_0, \qquad \mathbf{p}\cdot\mathbf{m}_0 = \mathbf{q}\cdot\mathbf{m}_0 = 0\ . \tag{32}$$

Note, that the vectors $\mathbf{s}_0$, $\mathbf{p}$ and $\mathbf{q}$ were first introduced by Fedorov (1968) in his theory of internal conical refraction. Then the same vectors were used in the theory of acoustic axes (Alshits, Sarychev & Shuvalov, 1985). With (27), systems (24) are easily solved which allows us to find the polarization vectors $\mathbf{A}_{1,2}$ (23) (not normalized at this stage):

$$\mathbf{A}_{1,2} = -(\mathbf{q}\cdot\delta\mathbf{m} - iq'')\mathbf{A}_{01} + (\mathbf{p}\cdot\delta\mathbf{m} - ip'' \pm R)\mathbf{A}_{02}\ . \tag{33}$$

It is easily checked that $\mathbf{A}_1 \cdot \mathbf{A}_2 = 0$, i.e. the orthogonality property, eqn. (8), is fulfilled. Actually, eqns. (27) and (33) contain all necessary information for our further analysis. However, in the next section we shall have to make preliminary "step aside".

4. On the acoustic axes along directions of high symmetry

Note, that the above formalism linear in small parameters does not work for the case of tangent acoustic axes along which $\mathbf{p} = \mathbf{q} = 0$ (Alshits, Sarychev & Shuvalov, 1985) and one should keep the higher order terms in all expansions. The above criterion for a tangent degeneracy can be satisfied either because of an accidental vanishing of some combinations of material parameters (i.e. in model crystals) or due to a high symmetry of the direction $\mathbf{m}_0$. That is why tangent degeneracies are known in real crystals only along 4- and ∞-fold symmetry axes. In the first case the both Poincarè indices $n = \pm 1$ (Fig. 2) are possible, in the latter case only the index n =+1 can occur (Alshits, Sarychev & Shuvalov, 1985). We already mentioned that model media are beyond our interest in this paper. As to "symmetrical" tangent acoustic axes, their reaction to "switching on" the damping is predictable without any calculations. The answer is rather natural: existing due to a symmetry which is not disturbed by the attenuation, they keep their directions and linear polarizations of the elastic waves propagating along them also retain, though the phase speeds v_a of these waves certainly take small imaginary components.

Indeed, the tensors $\hat{c}$ and $\hat{\eta}$ have completely the same symmetrical structure. It is well known, that along the direction $\mathbf{m}_0$ of the symmetry axis ∞ or 4 the tensor $\hat{Q}_0 = \mathbf{m}_0\hat{c}\mathbf{m}_0$ has eigenvectors $\mathbf{A}_{01}, \mathbf{A}_{02}$ and $\mathbf{A}_{03}$ coinciding with the basis vectors of the crystallographic coordinate system. Clearly, the tensor $\hat{Q}_0'' = \omega\mathbf{m}_0\hat{\eta}\mathbf{m}_0$ (20) must have the same eigenvectors. Hence, the combined complex Christoffel tensor $\hat{Q}' - i\hat{Q}'' = \hat{Q}_0 - i\hat{Q}_0''$ along the direction $\mathbf{m}_0$ admits purely real polarizations of three isonormal eigenwaves: one longitudinal ($\mathbf{A}_{03}$ parallel to z) and two transverse ($\mathbf{A}_{01}$ parallel to x and $\mathbf{A}_{02}$ parallel to y). It is easy to check that the degeneracy along $\mathbf{m}_0$ also retains. In accordance with eqn. (7)

$$\begin{aligned} \rho v_1^2 &= c_{1331} - i\omega\eta_{1331} = c_{55} - i\omega\eta_{55}, \\ \rho v_2^2 &= c_{2332} - i\omega\eta_{2332} = c_{44} - i\omega\eta_{44}, \\ \rho v_3^2 &= c_{3333} - i\omega\eta_{3333} = c_{33} - i\omega\eta_{33}. \end{aligned} \tag{34}$$

But the symmetry axes ∞ or 4 present only in tetragonal, cubic and hexagonal crystals where

$$c_{44} = c_{55}, \qquad \eta_{44} = \eta_{55} \tag{35}$$

and therefore $v_1 = v_2$. Accordingly, the degenerate tensor $\hat{Q}_0 - i\hat{Q}_0''$ has the spectral representation

$$\hat{Q}_0 - i\hat{Q}_0'' = [c_{44} - c_{33} - i\omega(\eta_{44} - \eta_{33})]\hat{I} + (c_{33} - i\,\omega\eta_{33})\mathbf{A}_{03} \otimes \mathbf{A}_{03} \tag{36}$$

where $\hat{I}$ denotes the unit tensor and the symbol $\otimes$ means a dyadic product. So, one can see, that any linear combination $a\mathbf{A}_{01} + \beta\mathbf{A}_{02}$ is also an eigenvector of $\hat{Q}_0 - i\hat{Q}_0''$ and any transverse wave may propagate along $\mathbf{m}_0$ (Fig.3). Note in addition, that in the considered situation the nominator $\mathbf{A}_{01} \cdot \hat{Q}''\mathbf{A}_{02}$ of the singular term in eqn. (15) vanishes together with η_{45}. This explains why our qualitative expectations of increasing imaginary components $\mathbf{A}_{1,2}''$ close to $\mathbf{m}_0$ have not been realized.

The same arguments are equally applicable for the directions of 3-fold symmetry axes along which the conical acoustic axes necessarily occur (Alshits, Sarychev & Shuvalov, 1985) being characterized by the polarization singularity with the Poincarè index $n = -1/2$ (Fig. 2). However in this case $\mathbf{p} \neq \mathbf{q} \neq 0$ and the formalism developed in the previous section does work and may be used for the demonstration of the validity of the above considerations. For example, in the case of the 3-fold symmetry axis in trigonal crystal one has: $p'' \propto \eta_{1331} - \eta_{2332} = \eta_{55} - \eta_{44} = 0$ and $q'' \propto \eta_{1332} = \eta_{54} = 0$ but $s'' \propto \eta_{1331} + \eta_{2332} = 2\eta_{44} \neq 0$. As a result, the phase speed perturbations $\delta v_{1,2}$, eqn. (27), have the equal imaginary components and the difference between δv_1 and δv_2 remains purely real and vanishes at $\delta\mathbf{m} \to 0$. Thus, again there is neither a split nor a shift of the degeneracy. And in accordance with eqn. (33) the degenerate polarization fields $\mathbf{A}_{1,2}$ in the neighbourhood of the acoustic axis in the leading approximation remain linear, i.e. their imaginary components are small and vanish at $\delta\mathbf{m} \to 0$.

But such a trivial situation occurs only along the symmetry axes ∞, 4 and 3. As we shall see, any other point of degeneracy, even in a symmetry plane, manifests instability with respect to attenuation and singular behaviour of basic wave parameters close to new acoustic axes.

Let us consider the other "symmetric" case known in real crystals: the line of degeneracy which occurs in some transversely isotropic media. According to (Alshits, Sarychev & Shuvalov, 1985), along such line: $\mathbf{p}\times\mathbf{q}=0$ i.e. the vectors **p** and **q** must be parallel or one of them should vanish (let $\mathbf{q}=\gamma\mathbf{p}$ say). Note that at $\mathbf{p}\times\mathbf{q}=0$ the point degeneracy is also possible (in model crystals, of course). In this case for its description one should keep in expansions the terms of the higher order. But for a line degeneracy the leading approximation used above is completely sufficient and eqns. (27) and (33) may be applied for an analysis. The condition of the degeneracy $\delta v_1 = \delta v_2$ is equivalent to the requirement of the vanishing square root in (27) which brings us to the following system

$$(\mathbf{p}\cdot\delta\mathbf{m})^2 + (\mathbf{q}\cdot\delta\mathbf{m})^2 - p''^2 - q''^2 = 0\,, \tag{37}$$

$$p''(\mathbf{p}\cdot\delta\mathbf{m}) + q''(q\cdot\delta\mathbf{m}) = 0\,. \tag{38}$$

At $\mathbf{q}=\gamma\mathbf{p}$ this system becomes clearly contradictive,

$$\mathbf{p}\cdot\delta\mathbf{m} = \sqrt{(p''^2+q''^2)/(1+\gamma^2)}\,,$$

$$\mathbf{p}\cdot\delta\mathbf{m} = 0\,,$$

and has no solutions unless the both parameters p'' and q'' simultaneously vanish which does not occur in hexagonal crystals. Thus the line degeneracy $\mathbf{p}\cdot\delta\mathbf{m}=0$ under the damping perturbation must completely disappear which coincide with the corresponding conclusion in (Shuvalov & Chadwick, 1997). But looking at eqns. (27) and (33) one can say more. It follows from (27) that at $\mathbf{p}\cdot\delta\mathbf{m}=0$ the perturbations $\delta v_{1,2}$ are purely imaginary. This means that real components of the phase speeds v_{01} and v_{02} coincide as before on the same line $\mathbf{p}\cdot\delta\mathbf{m}=0$. However, the imaginary components $\delta v''_{1,2}$ are different on this line which eliminates the degeneracy. As regards to polarizations, the only peculiarity of the polarization field on the line $\mathbf{p}\cdot\delta\mathbf{m}=0$ is the lack of even a symbolic ellipticity: by (33) it is purely linear.

5. Split of acoustic axes of general positions

At the switched off attenuation eqn. (27) transforms to the known equation (Alshits, Sarychev & Shuvalov, 1985) describing local geometry of sheets of the phase velocity surface P: $v_{1,2}(\mathbf{m}) = v_0 + \delta v_{1,2}(\mathbf{m}_0 + \delta\mathbf{m})$ in the vicinity of the degeneracy point $v_1(\mathbf{m}_0) = v_2(\mathbf{m}_0) = v_0$:

$$\delta v_{1,2} = \mathbf{s}_0\cdot\delta\mathbf{m} \mp \sqrt{(\mathbf{p}\cdot\delta\mathbf{m})^2 + (\mathbf{q}\cdot\delta\mathbf{m})^2}\,. \tag{39}$$

If the vectors $\mathbf{p}$ and $\mathbf{q}$ are neither vanishing nor parallel to each other, $\mathbf{p}\times\mathbf{q}\neq 0$, then eqn.(39) describes a conical contact of the sheets $v_{1,2}(\mathbf{m})$ and simultaneously of the sheets $1/v_{1,2}(\mathbf{m})$ of the slowness surface S. This is a conical degeneracy of a general type not related to symmetry of a crystal.
As we know, a "switching on" the attenuation causes a small imaginary addition to a phase speed of the wave: $v = v' - iv''$. As a result, apart from the wave surfaces P and S the new surface of attenuation, $v''(\mathbf{m})$, arises. And the real components of the phase speeds $v'_{1,2}(\mathbf{m})$ also manifest important changes providing a topological transformation of the wave surfaces P and S.
Let us return to eqns. (37), (38). At $\mathbf{p}\times\mathbf{q}\neq 0$ this system describes the split of conical acoustic axis due to a damping. The new positions of the degeneracies $\delta\mathbf{m}_1$ and $\delta\mathbf{m}_2$ are given by the intersection of the ellipse

$$\left(\frac{\delta\mathbf{m}\cdot\mathbf{p}}{r}\right)^2 + \left(\frac{\delta\mathbf{m}\cdot\mathbf{q}}{r}\right)^2 = 1, \qquad r = \sqrt{p''^2 + q''^2}, \tag{40}$$

with the straight line $\delta\mathbf{m}\cdot\mathbf{M} = 0$ passing through the end of the vector $\mathbf{m}_0$ perpendicularly to the vector (Fig. 4)

$$\mathbf{M} = p''\mathbf{p} + q''\mathbf{q} = r(\mathbf{p}\sin a + \mathbf{q}\cos a), \tag{41}$$

where the angle a is introduced by the expressions

$$\sin a = p''/r, \qquad \cos a = q''/r. \tag{41a}$$

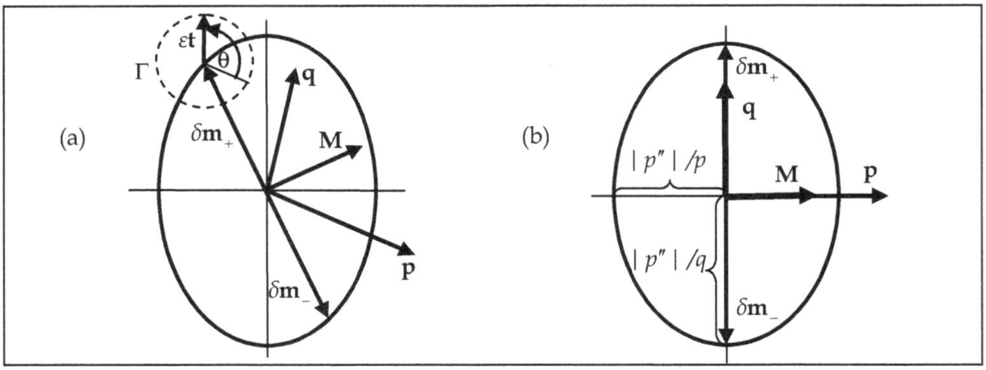

Fig. 4. Schematic plot of the ellipse, eqn. (40), in the general case (a), and for the case of the conical acoustic axis $\mathbf{m}_0$ splitting from the symmetry plane of a crystal (b)

In accordance with eqn. (32) the both vectors $\mathbf{p}$ and $\mathbf{q}$ are orthogonal to $\mathbf{m}_0$. Therefore the ellipse (40) (Fig.4) belongs to the plane tangent to the unit sphere $\mathbf{m}\cdot\mathbf{m} = 1$ at the point $\mathbf{m} = \mathbf{m}_0$ which indicates the center of the ellipse. Thus, "switching on" the damping causes the split of the conical axis $\mathbf{m}_0$ into the two singular axes directed along the wave normals $\mathbf{m}_\pm = \mathbf{m}_0 + \delta\mathbf{m}_\pm$ where

$$\delta \mathbf{m}_{\pm} = \pm \frac{\mathbf{m}_0 \times (p''\mathbf{p} + q''\mathbf{q})}{\mathbf{m}_0 \cdot (\mathbf{p} \times \mathbf{q})}. \tag{42}$$

Note that the projections of $\delta \mathbf{m}_{\pm}$ (42) on $\mathbf{p}$ and $\mathbf{q}$ vectors look rather simple

$$\delta \mathbf{m}_{\pm} \cdot \mathbf{p} = \mp q'', \qquad \delta \mathbf{m}_{\pm} \cdot \mathbf{q} = \pm p''. \tag{43}$$

Let us consider the example of splitting of a conical axis belonging to the symmetry plane S of the crystal. It is evident that in this case the polarization vector $\mathbf{A}_{03}$ also belongs to the plane S. The other vectors of our basis may be chosen so that, say, the vector $\mathbf{A}_{01}$ is directed along the normal to the plane S, and the vector $\mathbf{A}_{02}$ belongs to the same plane S together with the vectors $\mathbf{m}_0$ and $\mathbf{A}_{03}$ (Fig. 5a). It is easily checked that in the given case due to a crystal symmetry, which is not less than monoclinic, there must be

$$q'' = 0, \quad \mathbf{q} \parallel \mathbf{A}_{01}, \quad \mathbf{p} \parallel \mathbf{A}_{01} \times \mathbf{m}_0 \tag{44}$$

(Fig. 5b). By eqns. (43), (44), the split from the symmetry plane is determined by the vectors

$$\delta \mathbf{m}_{\pm} = \pm \frac{|p''|}{q} \mathbf{A}_{01}. \tag{45}$$

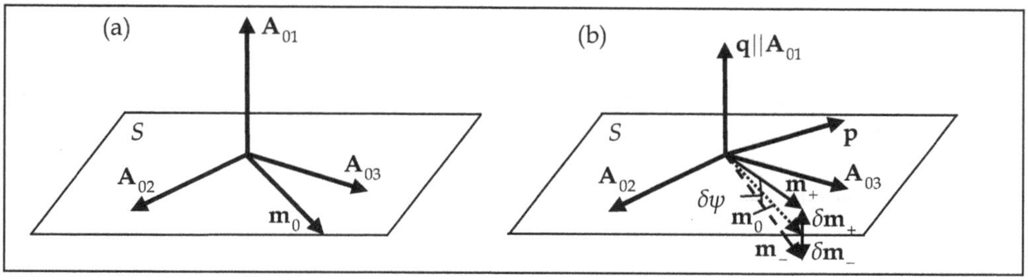

Fig. 5. Acoustic axis $\mathbf{m}_0$ in symmetry plane S (a), and its splitting due to the damping (b)

For the found mutual orthogonality of the vectors $\mathbf{p}$ and $\mathbf{q}$ ellipse (40) looks very symmetric (Fig. 4b). Thus, in the considered particular case the split of the acoustic axis occurs in the plane orthogonal to S and the angle $\delta\psi$ of splitting is proportional to the damping (Fig. 5b)

$$\delta\psi \approx 2|\delta m_{\pm}| = 2|p''|/q. \tag{46}$$

6. Local geometry of the velocity surfaces in the vicinity of split axes

Let us now return to eqns. (27), (28). We shall not divide eqn. (27) on the real and imaginary parts. It is more convenient to analyse this equation in its combined form. First of all, let us note that the expression under square root in eqn. (28) along the line $\delta \mathbf{m} \cdot \mathbf{M} = 0$ is purely real, being negative between the degeneracy points (i.e. inside the ellipse, eqn. (40) and Fig.4) and positive beyond them (i.e. outside the ellipse). But this means that on the part of this line which is inside of the ellipse, the square root is purely imaginary. Accordingly, on this part of the line the real components of phase speed $v_1'(\mathbf{m})$ and $v_2'(\mathbf{m})$ must coincide which creates the lines of self-intersection of the wave surfaces $v_{1,2}'(\mathbf{m})$ and $1/v_{1,2}'(\mathbf{m})$. Quite similarly, we come to the conclusion that the corresponding sheets of the attenuation surface $v_{1,2}''(\mathbf{m})$ must intersect each other over the line $\delta\mathbf{m}\cdot\mathbf{M}=0$ outside the ellipse (40). Fig. 6 gives a schematic illustration of such self-intersection of the slowness surface.

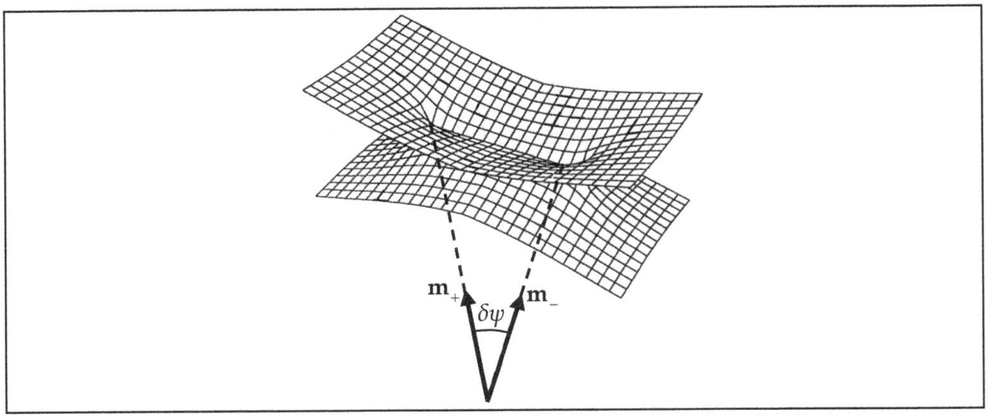

Fig. 6. Self-intersection of the slowness surface $1/v_{1,2}'(\mathbf{m})$ and split acoustic axes

Now let us study the above surfaces close to one of the new degeneracies, say, $\mathbf{m}_+=\mathbf{m}_0+\delta\mathbf{m}_+$. We are going to find the phase speeds of isonormal waves at the contour Γ (Fig. 4a): $\mathbf{m}=\mathbf{m}_+ +\delta\mathbf{m}_\varepsilon(\theta)$. The contour lies in the plane orthogonal to $\mathbf{m}_0$ and its radius is supposed to be small: $\varepsilon=|\delta\mathbf{m}_\varepsilon|<<|\delta\mathbf{m}_+|$. Denote

$$\delta\mathbf{m}_\varepsilon(\theta)=\varepsilon\mathbf{t}(\theta), \tag{47}$$

where **t** is the unit vector making the angle θ with the vector **p**:

$$\mathbf{t}(\theta)=[\mathbf{p}\cos\theta+(\mathbf{m}_0\times\mathbf{p})\sin\theta]/p. \tag{48}$$

Thus, by changing θ from 0 to 2π, the vector $\delta\mathbf{m}_\varepsilon$ (47) path-traces the contour Γ around the degeneracy point $\delta\mathbf{m}_+$ (Fig. 4a).
With (47), (48), eqn. (27) gives in the leading approximation over ε at the contour Γ:

$$\delta v_{1,2}'(\varepsilon,\theta)=v_{1,2}'-v_{0+}'=\mp\sqrt{\varepsilon}\mathrm{Re}[f(\theta)], \tag{49}$$

where

$$f(\theta)=\sqrt{2\mathbf{t}\cdot(\mathbf{N}-i\mathbf{M})}, \qquad \mathbf{N}=-q''\mathbf{p}+p''\mathbf{q}. \tag{50}$$

As is seen from (49), the dependence $\delta v_{1,2}'(\varepsilon)\propto\sqrt{\varepsilon}$ at $\varepsilon\to0$ is characterized by an infinite derivative over ε in any section $\theta\neq\theta_0$, where the angle θ_0 relates to a transition of the vector $\delta\mathbf{m}_\varepsilon$ through the self-intersection line, $\mathrm{Re}[f(\theta_0)]=0$. This singularity of the function

$\delta v'_{1,2}(\varepsilon)$ at the end of the wedge of self-intersection corresponds to a sharpening tip of the slowness surface $1/v'_{1,2}(\mathbf{m})$ and to a plane fan of the normals to this surface at the contour Γ when $\varepsilon \to 0$ (Fig. 7).

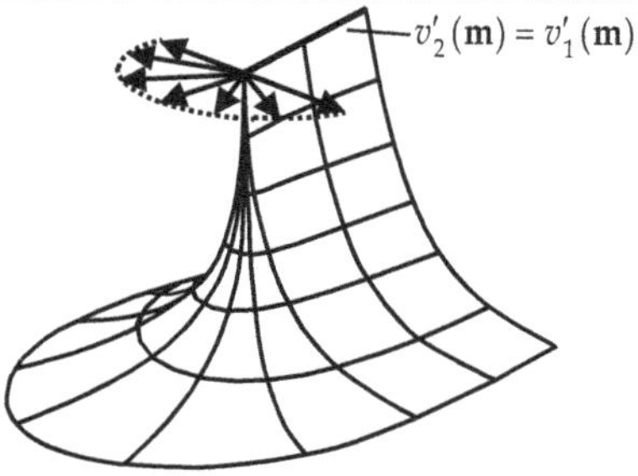

Fig. 7. The fragment of the internal degenerate sheet $1/v'_2(\mathbf{m})$ of the slowness surface close to the singular point at the end of the wedge of the self-intersection and the plane fan of normals to the surface at this point

7. Polarization field singularities around the split acoustic axes

At the same contour Γ (47) polarization vectors (33) up to normalizing factors are equal

$$\mathbf{A}_{1,2} = \mathbf{A}_{01} + i\left(1 \mp \frac{\sqrt{\varepsilon} f(\theta)}{q'' + ip''}\right)\mathbf{A}_{02}\,. \tag{51}$$

For the further analysis the function $f(\theta)$ here should be concretize to the form

$$f(\theta) = A\sqrt{\mathbf{p}\cdot\mathbf{q}\cos\theta + g\sin\theta - ip^2\cos\theta}\,, \tag{52}$$

where

$$A = \sqrt{2(p'' - iq'')/p}\,, \qquad g = \mathbf{m}_0 \cdot (\mathbf{p}\times\mathbf{q})\,. \tag{53}$$

It is easily checked, that at the rotation of unit vector $\mathbf{t}$, (48), over the whole circuit, i.e. at varying θ from 0 to 2π, the function $f(\theta)$ (52), changes its sign. Indeed, the phase of the complex function

$$f(\theta)/A \equiv R(\theta)\exp[i\Psi(\theta)] \tag{54}$$

must be twice less than the phase of its square

$$R^2\exp(2i\Psi) = \mathbf{p}\cdot\mathbf{q}\cos\theta + g\sin\theta - ip^2\cos\theta\,. \tag{55}$$

On the other hand, one can find from (55) the following relations

$$\Psi(\theta) = -\frac{1}{2}\left(\mathrm{Arctg}\frac{p^2}{\mathbf{p}\cdot\mathbf{q} + g\,\mathrm{tg}\theta}\right), \qquad \frac{\partial\Psi}{\partial\theta} = \frac{gp^2\cos^2 2\Psi}{(\mathbf{p}\cdot\mathbf{q}\cos\theta + g\sin\theta)^2}\,. \tag{56}$$

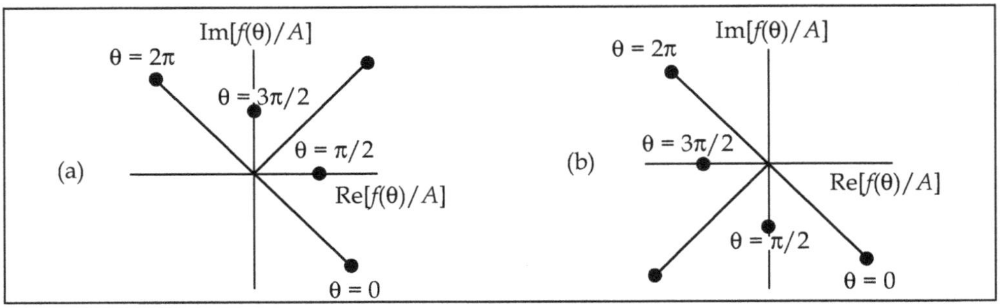

Fig. 8. The function $f(\theta)/A$ in the complex plane at $g > 0$ (a) and $g < 0$ (b)

This gives (see also Fig. 8)

$$\Psi(2\pi) - \Psi(0) = \pi \operatorname{sgn} g\,, \tag{57}$$

$$f(2\pi) = -f(0)\,. \tag{58}$$

Thus, after the whole turn over the contour Γ around the degeneracy point at $\delta\mathbf{m}_+$ (Fig. 4a) one has the identical transformation of the polarization field (51) in itself in the form

$$\mathbf{A}_1(2\pi) = \mathbf{A}_2(0)\,, \quad \mathbf{A}_2(2\pi) = \mathbf{A}_1(0)\,. \tag{59}$$

In other words, each of two orthogonal polarization ellipses rotates exactly on $\pi/2$ being transformed into the polarization of the isonormal wave (Fig. 9). And simultaneously the complex velocities $v_{1,2} = v'_{1,2} - iv''_{1,2}$ also are interchanging with their counterparts (Fig. 10).

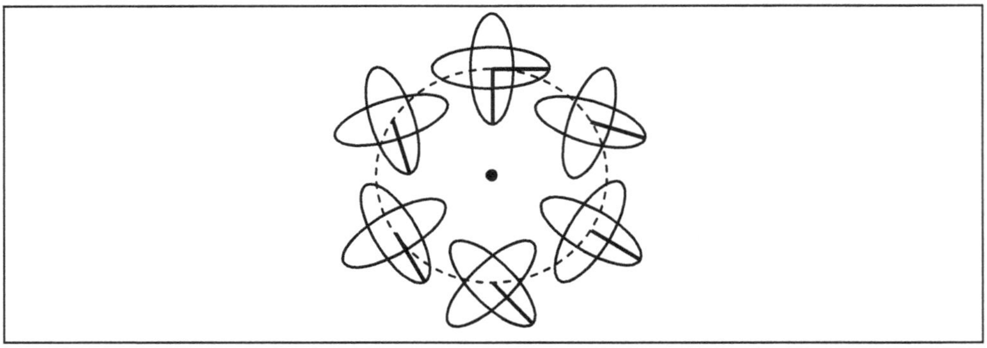

Fig. 9. The rotation of the polarization ellipses $\mathbf{A}_{1,2}$ in the degeneracy plane D when the wave normal $\mathbf{m}$ is scanning the contour Γ. The case $g > 0$ is shown when $n = 1/4$.

The found singularity of the polarization field at the degeneracy point $\mathbf{m}_+$ (Fig. 9) may be characterized by the Poincarѐ index defined as the value of the total polarization rotation (in the 2π units) at a complete path-tracing over the contour Γ around this point. The found turn of the polarization ellipses is equal $\pi/2$, and the direction of the rotation, by eqn. (57), is determined by the sign of the parameter g (53). Hence, one has (Alshits & Lyubimov, 1998)

$$n = \frac{1}{4}\text{sgn}[\mathbf{m}_0 \cdot (\mathbf{p} \times \mathbf{q})] \tag{60}$$

It is easily verified that the same relation is valid for the second degeneracy point $\mathbf{m}_-$.

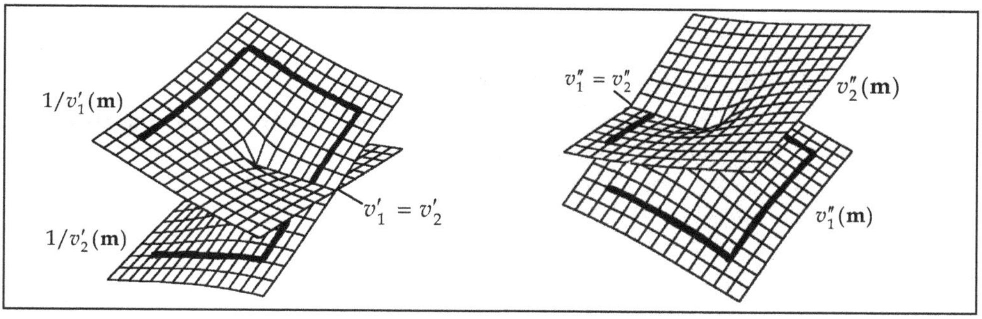

Fig. 10. The surfaces $1/v'_{1,2}(\mathbf{m})$ and $v''_{1,2}(\mathbf{m})$ in the vicinity of the acoustic axis $\mathbf{m}_+$. The transition between the sheets of the surfaces when $\mathbf{m}$ is scanning the contour around $\mathbf{m}_+$

Thus the physical equivalence of two pictures at $\theta = 0$ and $\theta = 2\pi$ is realized not by a coincidence of the wave characteristic inside each of the branches, as it occurs at zero damping, but by the identity of their superpositions. This becomes topologically possible due to such a new feature of the slowness surfaces as their self-intersections (Fig. 10). In the absence of damping, when the degenerate wave sheets locally have the only contact point, one of the branches along any direction is always "faster" than the other. And the related polarization cross, contained of isonormal linear non-directed vectors, has non-equivalent "differently coloured" crosspieces. Hence for a coincidence of such cross with itself it is required its turn on the minimum angle π, instead of $\pi/2$, as in the above case (Figs. 9, 10). The turn on $\pi/2$ is sufficient only when the change of "colours" of crosspieces occurs during the turn.

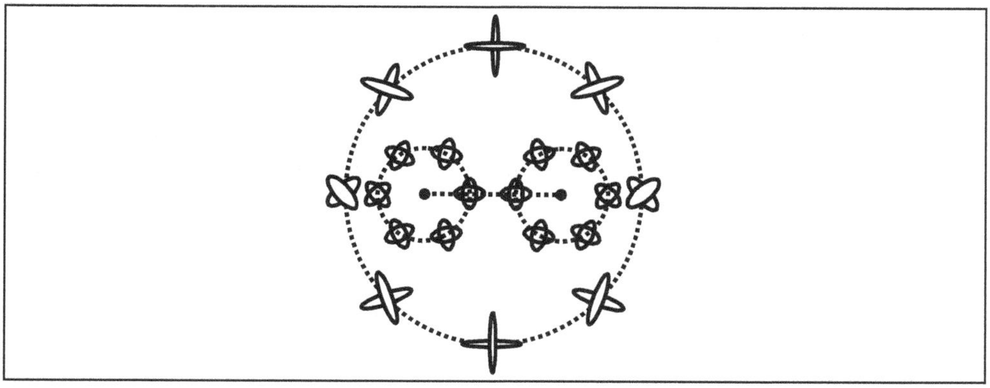

Fig. 11. The field of elliptic polarizations of degenerate branches in the vicinity of split axes of an absorptive crystal for $g < 0$. The Poincarè indices at small contours are $n = -1/4$, and the combined index at the external contour is $n = -1/2$

That is why (Alshits, Sarychev & Shuvalov, 1985) in the absence of the damping a conical axis along $\mathbf{m}_0$ is characterized by the Poincarè index $n = (1/2)\text{sgn}g$. This is the minimal index for a real polarization field. Its splitting into the two singularities (60) due to "switching on" attenuation satisfies the index conservation law. On the other hand, the same combined index $\pm 1/2$ arises at the path-tracing of the both points $\mathbf{m}_\pm$ (Fig. 11).

8. Conical refraction in absorptive crystals

Internal conical refraction of elastic waves in crystals is a good example of a non-trivial role of anisotropy, which may create new phenomena principally impossible in isotropic media. The energy flux **P** of the wave in crystal is, as a rule, non-parallel to its direction **m** of propagation. For any wave normal **m** the direction of the Poynting vector **P** is determined by the orientation of the normal **n** to the slowness surface. At the choice of the wave normal along a conical acoustic axis each polarization vector in the degeneracy plane *D* (Fig. 3) relates to the definite Poynting vector, i.e. to the definite normal to a cone. Rotation of the polarization in the plane *D* (e.g. in a circularly polarized wave) should create a precession of the energy flux **P**.

This phenomenon called the internal conical refraction was theoretically predicted and experimentally discovered by De Klerk & Musgrave (1955). They found a circular cone of refraction along the 3-fold symmetry axis in the cubic crystal Ni. Later on the more general cases of the refraction cones of elliptic section were theoretically studied (Barry & Musgrave, 1979; Khatkecich, 1962b; Musgrave, 1957) and experimentally found (Aleksandrov & Ryzhova, 1964). The complete theory of this phenomenon is presented in the monographs (Fedorov, 1968; Sirotin & Shaskolskaya, 1983). Below we shall develop an extension of this theory for absorptive crystals following to the recent paper (Alshits & Lyubimov, 2011).

8.1 Conical refraction in the absence of attenuation

As we have seen, in a crystal without damping along the acoustic axis $\mathbf{m}_0$, apart from the non-degenerate wave with the polarization vector $\mathbf{A}_{03}$, an infinite number of elastic waves may propagate with arbitrary polarization in the degeneracy plane *D* (Fig. 3). Thus in the basis $\{\mathbf{A}_{01}, \mathbf{A}_{02}\}$ belonging to the same plane, for any angle β the vector

$$\mathbf{A}(\beta) = \mathbf{A}_{01}\cos\beta + \mathbf{A}_{02}\sin\beta \tag{61}$$

determines polarization of the eigenwave propagating along $\mathbf{m}_0$ with the phase speed v_0. Certainly, the wave with a circular polarization $\mathbf{A} = \mathbf{A}_{01} + i\mathbf{A}_{02}$ can also propagate along the same direction.

Consider a monochromatic plane wave propagating along the acoustic axis $\mathbf{m}_0$ with the polarization **A** and the phase speed v_0:

$$\mathbf{u}(\mathbf{r},t) = C\mathbf{A}\exp(i\Phi_0), \qquad \Phi_0 = k(\mathbf{m}_0 \cdot \mathbf{r} - v_0 t). \tag{62}$$

The Poynting vector of such wave is equal (Fedorov, 1968)

$$\mathbf{P} = (\text{Re}\dot{\mathbf{u}})\hat{c}(\text{Re}\dot{\mathbf{u}})\mathbf{m}_0 / v_0. \tag{63}$$

For linear and circular polarizations one has, respectively,

$$\mathrm{Re}\mathbf{u}_{\mathrm{lin}} = C\mathbf{A}(\beta)\cos\Phi_0 , \tag{64}$$

$$\mathrm{Re}\mathbf{u}_{\mathrm{cir}} = C[\mathbf{A}_{01}\cos\Phi_0 - \mathbf{A}_{02}\sin\Phi_0] . \tag{65}$$

In these two cases the Poynting vectors are given by different expressions:

$$\mathbf{P}_{\mathrm{lin}} = C^2\rho\omega^2(\mathbf{s}_0 + \mathbf{p}\cos 2\beta + \mathbf{q}\sin 2\beta)\sin^2\Phi_0 , \tag{66}$$

$$\mathbf{P}_{\mathrm{cir}} = C^2\rho\omega^2(\mathbf{s}_0 - \mathbf{p}\cos 2\Phi_0 + \mathbf{q}\sin 2\Phi_0) . \tag{67}$$

Quite similarly one finds the elastic energy densities $W = \rho(\mathrm{Re}\dot{\mathbf{u}})^2$:

$$W_{\mathrm{lin}} = C^2\rho\omega^2\sin^2\Phi_0 , \qquad W_{\mathrm{cir}} = C^2\rho\omega^2 . \tag{68}$$

With eqns. (66)-(68), the ray velocities of the considered waves are equal

$$\mathbf{s}_{\mathrm{lin}} = \mathbf{P}_{\mathrm{lin}}/W_{\mathrm{lin}} = \mathbf{s}_0 + \mathbf{p}\cos 2\beta + \mathbf{q}\sin 2\beta , \tag{69}$$

$$\mathbf{s}_{\mathrm{cir}} = \mathbf{P}_{\mathrm{cir}}/W_{\mathrm{cir}} = \mathbf{s}_0 - \mathbf{p}\cos 2\Phi_0 + \mathbf{q}\sin 2\Phi_0 . \tag{70}$$

During the period of the circularly polarized wave at a complete turn of the polarization vector in the degeneracy plane D, the ray velocity vector $\mathbf{s}_{\mathrm{cir}}$ (70) twice circumscribes a cone (Fig. 12). At that the end of the vector $\mathbf{s}_{\mathrm{cir}}$ twice path-traces the ellipse

$$\Delta\mathbf{s} = \mathbf{s} - \mathbf{s}_0 = -\mathbf{p}\cos 2\Phi_0 + \mathbf{q}\sin 2\Phi_0 . \tag{71}$$

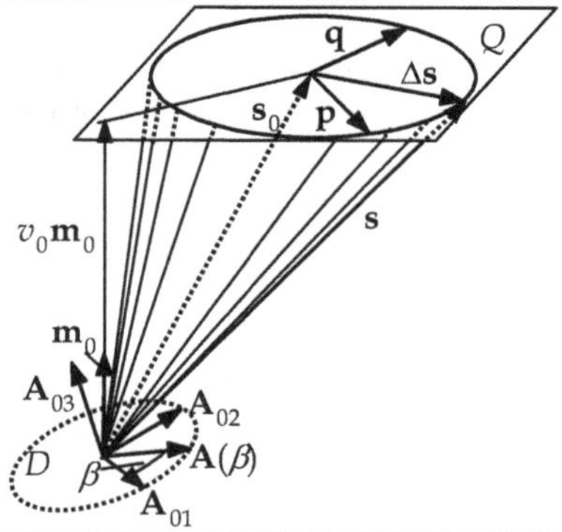

Fig. 12. The cone of the internal conical refraction

In view of (32), the plane Q of the ellipse is orthogonal to $\mathbf{m}_0$, and the directions of path-tracing of the vectors $\Delta\mathbf{s}$ and $\mathrm{Re}\mathbf{u}_{\mathrm{cir}}$ are the same when $g > 0$ and opposite when $g < 0$.

For a linearly polarized wave the same refraction cone is described by the vector $\mathbf{s}_{\mathrm{lin}}$ (69) when the angle β changes within the interval $0 \le \beta \le 2\pi$ (Fig. 12). This particular scheme was realized in the first experiments of De Klerk & Musgrave (1955).

8.2 The polarization ellipses at the ridge of the wedge of self-intersection

Consider now the wave characteristics of an absorptive crystal at the ridge of the wedge of self-intersection of the slowness surface. For a description of the set of wave normals related to the ridge between the two degeneracy points at the slowness surface let us introduce the vector $\delta\mathbf{m}_\xi = \delta\mathbf{m}_+ \sin\xi$. At changing ξ from $-\pi/2$ to $+\pi/2$ the vector $\delta\mathbf{m}_\xi$ moves through all the ridge from one degeneracy ($\delta\mathbf{m}_-$) to another one ($\delta\mathbf{m}_+$). Substitution $\delta\mathbf{m} = \delta\mathbf{m}_\xi$ into (33) gives the polarization vectors at any point of the line of self-intersection. Making use of relations (43) one obtains

$$\mathbf{A}_{1,2} = \frac{(\sin a \sin\xi - i\cos a)\mathbf{A}_{01} + [\cos a \sin\xi + i(\sin a \mp \cos\xi)]\mathbf{A}_{02}}{\sqrt{2(1 \mp \sin a \cos\xi)}}, \tag{72}$$

where the normalizing (9) is fulfilled and notations (41a) are used.

We remind that at the ridge of the wedge the real components of the phase speeds $v_{1,2}$ coincide: $v_1' = v_2' = v_\xi'$. The imaginary components $v_{1,2}''$ coincide only at the end points of the ridge, $\xi = \pm\pi/2$. In view of (6), the real components of the displacement vectors $\mathbf{u}_{1,2}$ take the form

$$\mathrm{Re}\mathbf{u}_{1,2}(\mathbf{r},\ t) = C_{1,2}\exp(-k_{1,2}''\mathbf{m}_\xi \cdot \mathbf{r})\mathrm{Re}[\mathbf{A}_{1,2}\exp(i\Phi_\xi)] \equiv C_{1,2}\exp(-k_{1,2}''\mathbf{m}_\xi \cdot \mathbf{r})\mathbf{U}_{1,2}\,. \tag{73}$$

We introduced here the wave normal $\mathbf{m}_\xi$ and the real phase Φ_ξ at the ridge,

$$\mathbf{m}_\xi = \mathbf{m}_0 + \delta\mathbf{m}_\xi\,, \qquad \Phi_\xi = k_\xi'\mathbf{m}_\xi \cdot \mathbf{r} - \omega t\,, \tag{74}$$

and the dimensionless displacement vectors

$$\mathbf{U}_{1,2} = \mathrm{Re}[\mathbf{A}_{1,2}\exp(i\Phi_\xi)]\,. \tag{75}$$

It is essential that in eqn. (73) a trivial damping of the wave $\propto \exp(-k_{1,2}''\mathbf{m}_\xi \cdot \mathbf{r})$ is separated from the vectors $\mathbf{U}_{1,2}$ describing much more important for us effects of attenuation.

In the considered stationary problem a choice of the time origin is certainly unessential and may be different for isonormal waves, independent from each other. Hence, the vectors $\mathbf{U}_{1,2}$ as well as the polarization vectors $\mathbf{A}_{1,2}$ are defined to the sign. Below this sign will be chosen so that our expression would be more compact.

Note, that at scanning the ridge by the wave normal $\mathbf{m}_\xi$, the elliptic polarization determined by eqns. (72), (75) is sharply changing. It is easily checked that this ellipticity provides rotations of the vectors $\mathbf{U}_{1,2}$ along the same directions corresponding to the right-hand screw along the propagation, until $\sin\xi > 0$, and to the left-hand screw, when $\sin\xi < 0$. At the ridge ends ($\xi = \pm\pi/2$) where the degeneracies occur, the isonormal waves, naturally, coincide: $\mathbf{U}_1 = \mathbf{U}_2 \equiv \mathbf{U}_0$. In both cases the polarization is circular however with different rotation "signs":

$$\begin{aligned} \mathbf{U}_0\big|_{\xi=\pi/2} &= \frac{1}{\sqrt{2}}(\mathbf{A}_{01}\cos\Phi_\xi - \mathbf{A}_{02}\sin\Phi_\xi), \\ \mathbf{U}_0\big|_{\xi=-\pi/2} &= \frac{1}{\sqrt{2}}(\mathbf{A}_{01}\cos\Phi_\xi + \mathbf{A}_{02}\sin\Phi_\xi). \end{aligned} \tag{76}$$

Here the angle α is excluded from the arguments by the choice of the time origin.
In any other points of the ridge the polarization ellipses of isonormal waves are different. In the middle point $\xi = 0$ the isonormal waves have linear polarizations orthogonal to each other:

$$\begin{aligned} \mathbf{U}_1\big|_{\xi=0} &= \left\{\mathbf{A}_{01}\cos\left(\frac{\pi}{4}-\frac{a}{2}\right)+\mathbf{A}_{02}\sin\left(\frac{\pi}{4}-\frac{a}{2}\right)\right\}\sin\Phi_0, \\ \mathbf{U}_2\big|_{\xi=0} &= \left\{\mathbf{A}_{01}\sin\left(\frac{\pi}{4}-\frac{a}{2}\right)-\mathbf{A}_{02}\cos\left(\frac{\pi}{4}-\frac{a}{2}\right)\right\}\sin\Phi_0. \end{aligned} \tag{77}$$

One can show that linear polarization retain on a whole line passing through the middle of the ridge ($\xi = 0$) perpendicular to it (at the unit sphere $\mathbf{m}^2 = 1$ this line passes through point $\mathbf{m}_0$ with local orientation along vector **M**).
Expressions for the polarization ellipses of isonormal waves at the ridge are remarkably simplified in the considered above particular case related to the unperturbed acoustic axis $\mathbf{m}_0$ situated in the symmetry plane of the crystal. In this case $q'' = 0$. Supposing for definiteness that $p'' > 0$, one can put $a = \pi/2$. Then, instead of (72), the polarization vectors of the isonormal waves are equal

$$\begin{aligned} \mathbf{A}_1 &= \mathbf{A}_{01}\cos(\xi/2) + i\mathbf{A}_{02}\sin(\xi/2), \\ \mathbf{A}_2 &= \mathbf{A}_{01}\sin(\xi/2) + i\mathbf{A}_{02}\cos(\xi/2). \end{aligned} \tag{78}$$

And the rotation of the displacement vectors $\mathbf{U}_{1,2}$ (75) over the ellipses is now described by

$$\begin{aligned} \mathbf{U}_1 &= \mathbf{A}_{01}\cos(\xi/2)\cos\Phi - \mathbf{A}_{02}\sin(\xi/2)\sin\Phi, \\ \mathbf{U}_2 &= \mathbf{A}_{01}\sin(\xi/2)\cos\Phi - \mathbf{A}_{02}\cos(\xi/2)\sin\Phi. \end{aligned} \tag{79}$$

These expressions represent ellipses in a parametric form. The lengths of the horizontal and vertical semi-axes of the first ellipse are equal $|\cos(\xi/2)|$ and $|\sin(\xi/2)|$, respectively. For the second ellipse the same length relate to the vertical and horizontal semi-axes. At the ridge ends $\xi = \pm\pi/2$ the above lengths of the semi-axes are equal to each other, and the polarization becomes circular. With a displacement of the "observation" point $\delta\mathbf{m}_\xi$ from the ridge ends to its middle the large semi-axes increase and the small semi-axes decrease to zero at $\xi = 0$.
Thus, both general expressions (76), (77) and the particular example (79) lead to the same picture of polarization distribution at the ridge of wedge of self-intersection. At passing along this line the isonormal waves, starting from a circular polarization of definite sign, monotonously decrease their ellipticity to zero in the middle of the ridge, where ellipses are transformed into non-directed vectors. At the second half of the ridge the ellipticity changes its sign and monotonously increases becoming circular at the other degeneracy point. Fig. 13 illustrates this behavior of polarization at the line of self-intersection of the slowness surface.
Consider now the kinematics of the motion of the displacement vectors of isonormal waves along the polarization ellipses. Express the radius-vectors $\mathbf{U}_a$ $(a = 1,2)$ at the ellipse in polar coordinates (U_a, φ_a):

$$\mathbf{U}_a = U_a(\mathbf{A}_{01}\cos\varphi_a + \mathbf{A}_{02}\sin\varphi_a). \tag{80}$$

Certainly, the lengths U_a of these radius-vectors at the ellipse depend on the azimuth φ_a. Comparing eqns. (72), (75) and (80), one has

$$U_{1,2}^2 = \frac{1}{2}\left(1 \pm \cos\xi \frac{(\sin a \mp \cos\xi)\cos 2\Phi_\xi + \sin\xi \cos a \sin 2\Phi_\xi}{1 \mp \sin a \cos\xi}\right), \tag{81}$$

$$\mathrm{tg}\,\varphi_{1,2} = \frac{\cos a \sin\xi - (\sin a \mp \cos\xi)\mathrm{tg}\Phi_\xi}{\sin a \sin\xi + \cos a\ \mathrm{tg}\Phi_\xi}. \tag{82}$$

Differentiating the latter expression with respect to time, it is easy to find the angular velocities $\dot{\varphi}_{1,2}$ of the radius-vectors $\mathbf{U}_{1,2}$ at their ellipses:

$$\dot{\varphi}_{1,2} = \frac{\omega \sin\xi}{2U_{1,2}^2}, \tag{83}$$

where we put $\dot{\Phi}_\xi = -\omega$. As is seen from (83), the angular velocities differently behave in time at different "observation" points at the ridge. Along the acoustic axes ($\xi = \pm\pi/2$), when the isonormal ellipses coincide into one circle (76), the denominator in (83) is equal to 1, and the circular motion has a constant angular velocity: $\dot{\varphi}_1 = \dot{\varphi}_2 = \pm\omega$. Here the upper and lower signs relate to different directions of the rotation at $\xi = \pm\pi/2$ (Fig. 13).

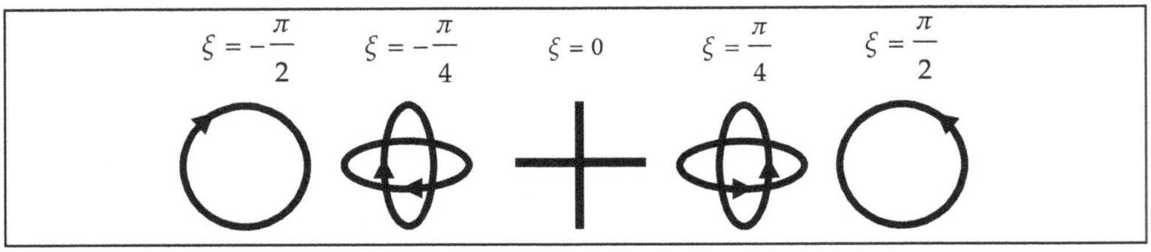

Fig. 13. Polarization distribution for isonormal waves at the line of self-intersection of the slowness surface

With decreasing $|\xi|$ a non-uniformity of the motion increases and at $|\xi| << 1$ acquires a singular character, when during the most part of the period the velocities $\dot{\varphi}_{1,2}$ are very small, and the azimuth angles $\varphi_{1,2}$ related to them are almost fixed. In this regime, the vectors $\mathbf{U}_{1,2}$ pass the most part of the ellipse in a short time with very high velocity. This is clearly seen from the analytical formulae related to the discussed above particular case of the acoustic axis splitting from the symmetry plane (Fig. 14):

$$\begin{aligned} \mathrm{tg}\,\varphi_1 &= -\mathrm{tg}(\xi/2)\mathrm{tg}\Phi_\xi, \\ \mathrm{tg}\,\varphi_2 &= -\mathrm{ctg}(\xi/2)\mathrm{tg}\Phi_\xi; \end{aligned} \tag{84}$$

$$\dot{\varphi}_{1,2} = \frac{\omega \sin\xi}{1 \pm \cos\xi \cos 2\Phi_\xi}. \tag{85}$$

Eqns. (81)-(85) and Fig. 14 show that the functions $\varphi_{1,2}(\Phi_\xi)$ and $\dot{\varphi}_{1,2}(\Phi_\xi)$ have the period twice less than the period of the wave. This means that the half-turn of the displacement vector over the polarization ellipse exhausts all its physically different orientations.

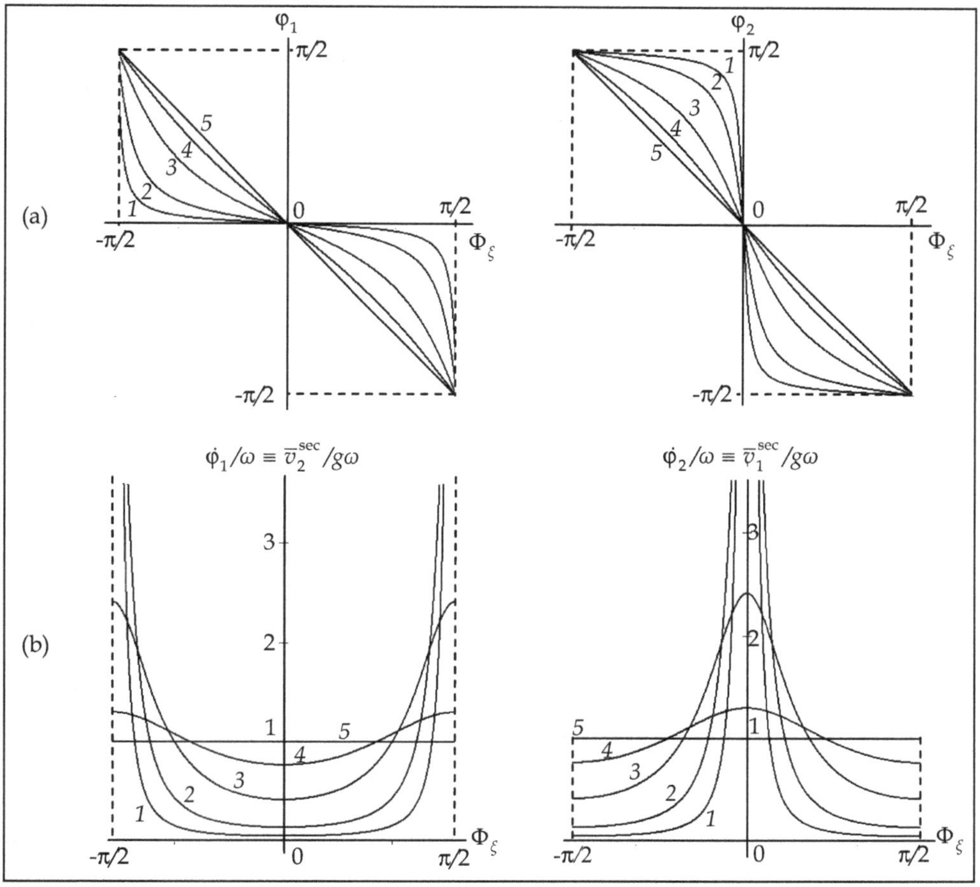

Fig. 14. Numerical plot of the azimuths $\varphi_{1,2}$ (84) (a) and the coinciding normalized speeds $\dot{\varphi}_{1,2}/\omega \equiv \overline{v}^{sec}_{2,1}/g\omega$ (85), (106) (b) versus the phase Φ_ξ for the series of "observation" points at the line of self-intersection of the slowness surface for a particular case of the acoustic axis splitting from the symmetry plane. *1* - $\xi = 5^\circ$, *2* - 15°, *3* - 45°, *4* - 75°, *5* - 90°

The other kinematic characteristics of the considered motion could be the so-called sectorial velocities defined as area circumscribed by a rotating vectors $\mathbf{U}_{1,2}$ per unit time:

$$v^{sec}_{1,2} = \frac{1}{2}U^2_{1,2}\dot{\varphi}_{1,2} = \frac{1}{4}\omega\sin\xi\,. \tag{86}$$

The found expression is valid for unrestricted anisotropy. It is identical for the both isonormal waves being independent of the time. However velocity (86) strongly depends on the position of the "observation" point at the line of self-intersection. In particular, it vanishes in the center of the ridge ($\xi = 0$), where the polarization becomes linear.

8.3 Universal refraction cone at the line of self-intersection and kinematics of ray velocity precession on this cone

Let us find the ray velocities of isonormal waves (73) at the ridge of self-intersection of the slowness surface $1/v'_{1,2}(\mathbf{m})$. Substituting (73) into (63) one obtains the energy flux

$$\mathbf{P}_{1,2} = |C_{1,2}\exp(-k''_{1,2}\mathbf{m}_{\xi}\mathbf{r})|^2 \rho\omega^2[\mathbf{s}_0(F^2 + G_{\mp}^2) - \mathbf{p}(G_{\mp}^2 - F^2) + 2\mathbf{q}FG_{\mp}], \quad (87)$$

where

$$F = \sin a \sin\xi \sin\Phi_{\xi} - \cos a \cos\Phi_{\xi} \quad (88)$$

$$G_{\mp} = \cos a \sin\xi \sin\Phi_{\xi} + (\sin a \mp \cos\xi)\cos\Phi_{\xi}. \quad (89)$$

The energy density in the isonormal waves in the same terms is equal

$$W_{1,2} = \rho(\mathrm{Re}\dot{\mathbf{u}}_{1,2})^2 = |C_{1,2}\exp(-k''_{1,2}\mathbf{m}_{\xi}\mathbf{r})|^2 \rho\omega^2(F^2 + G_{\mp}^2). \quad (90)$$

Accordingly, the ray velocities of these waves are given by

$$\mathbf{s}_{1,2} = \mathbf{P}_{1,2} / W_{1,2} = \mathbf{s}_0 - \mathbf{p}\cos 2\Theta_{1,2} + \mathbf{q}\sin 2\Theta_{1,2}, \quad (91)$$

where

$$\Theta_{1,2} = \Theta_{1,2}(a, \xi, \Phi_{\xi}), \qquad \mathrm{tg}\Theta_{1,2} = F / G_{\mp}. \quad (92)$$

Eqn. (91) is transformed from classic expression (70) for crystals without damping after the substitution in the latter $\Phi_0 \to \Theta_{1,2}$. This means that in an absorptive crystal at any point of the ridge of self-intersection the ends of the ray velocity vectors $\mathbf{s}_{1,2}$ move along the same trajectories, described by the universal ellipse

$$\Delta\mathbf{s}_{1,2} = -\mathbf{p}\cos 2\Theta_{1,2} + \mathbf{q}\sin 2\Theta_{1,2}. \quad (93)$$

The form of this ellipse is completely determined by the vectors **p** and **q**, and is independent of the parameters $\Theta_{1,2}$. In other words, it is insensitive neither to the phase Φ_{ξ} of the wave, nor to the angles α and ξ, related to parameters of damping and to a position of the "observation" point. The principal semi-axes of universal ellipse (93), coinciding with ellipse (71) for a non-attenuating medium, are equal

$$\lambda_{1,2} = \frac{1}{2}\left(\mathbf{p}^2 + \mathbf{q}^2 \pm \sqrt{(\mathbf{p}^2 + \mathbf{q}^2)^2 - 4(\mathbf{p}\times\mathbf{q})^2}\right). \quad (94)$$

Though the vectors **p** and **q** (29) do depend on a choice of the basis $\{\mathbf{A}_{01}, \mathbf{A}_{02}\}$, one can easily check that their combinations $\mathbf{p}^2 + \mathbf{q}^2$ and $\mathbf{p}\times\mathbf{q}$ determining semi-axes (94) are invariant with respect to orientation of this basis in the degeneracy plane D (Fig. 3).

With identical trajectories of the ray velocities precession at the whole ridge, the kinematics of their motion is very sensitive to the position ξ of the "observation" point. It may be shown that at the ridge ends $\xi = \pm\pi/2$ the values $\Theta_0(\pi/2)$ and $\Theta_0(-\pi/2)$ differ by only signs:

$$\Theta_0(\pm\pi/2) = \pm\Phi_\xi, \tag{95}$$

which gives, by (91),

$$\begin{aligned} \mathbf{s}(\pi/2) &= \mathbf{s}_0 - \mathbf{p}\cos 2\Phi_\xi + \mathbf{q}\sin 2\Phi_\xi, \\ \mathbf{s}(-\pi/2) &= \mathbf{s}_0 - \mathbf{p}\cos 2\Phi_\xi - \mathbf{q}\sin 2\Phi_\xi. \end{aligned} \tag{96}$$

This shows that the precession of the ray velocity vector along one of split acoustic axes is identical with the analogous process for a circularly polarized wave (70) propagating along the unsplit acoustic axis in the crystal without damping. Directions of rotation of the ray velocities $\mathbf{s}(\pm\pi/2)$ (96) have different signs. It is easily checked that at $g > 0$ they coincide with corresponding directions of circular polarization (76), and at $g < 0$ – are opposite to them.
In spite of the found identity of cones (70) and (96), there is an important difference between the related to them pictures of conical refraction. In the crystal without damping the ray velocities forming the refraction cone are directed along the appropriate normals to the slowness surface at the conical point of degeneracy. And the normals to the analogous surface in the vicinity of one of the split axes, as we have seen (Fig. 7), form a plane fan, which has nothing to do with a cone of ray velocities (96) (Fig. 12).
With passing of the "observation" point from the end of the ridge to its center, the motion of the ray velocity around universal cone (96) becomes less and less uniform. At the center point $\xi = 0$ the motion deceases at all: the isonormal vectors $\mathbf{s}_{1,2}$ are "frozen" at definite positions. Indeed, at $\xi = 0$ eqns. (88), (89) and (92) give the values $\Theta_{1,2}$ independent of time:

$$\Theta_{1,2} = \frac{a}{2} \pm \frac{\pi}{4}. \tag{97}$$

The corresponding fixed vectors of ray velocity are equal

$$\mathbf{s}_{1,2}(0) = \mathbf{s}_0 \pm (\mathbf{p}\sin a + \mathbf{q}\cos a) = \mathbf{s}_0 \pm \mathbf{M}/r. \tag{98}$$

As it would be expected, this result relates to expression (77) for a linear polarization of isonormal waves in the same way, as eqn. (69) to expression (64) from the refraction theory for crystals without damping. One can show that in this point ($\xi = 0$) of the ridge the two normals to the slowness surface are parallel not to vectors (98), but to their components belonging to the plane $\{\mathbf{m}_0, \mathbf{M}\}$ orthogonal to the ridge.

One should note that the fixed in time positions of the ray velocities (98) on ellipse (93) depend on the attenuation (the angle a, eqn. (41a)), whereas the universal ellipse does not "know" about a. This means that points (98), generally speaking, do not coincide with the ends of the principal semi-axes of the ellipse. Of course, in more symmetric situations the coincidence may occur, as it happens in the case of the splitting of the acoustic axis from a symmetry plane. What is more important that the vectors $\Delta\mathbf{s}_{1,2} = \mathbf{s}_{1,2}(0) - \mathbf{s}_0$, as it follows from eqns. (98), (41) and Fig. 4, remain universally orthogonal to the ridge of the wedge at any changes of the angle a.

It is evident that at any small deviation of ξ from zero the fixed vectors (98) acquire some increments dependent on the phase Φ_ξ. This will renew a motion of the ray velocities $\mathbf{s}_{1,2}$ over the cone. However, if not to pass far from the middle point $\xi = 0$, the most part of the period the vectors $\mathbf{s}_{1,2}$ will retain orientations close to directions (98). And the time-averaged vectors $\overline{\mathbf{s}}_{1,2}$ in these points will be close to directions (98). This means that in the middle domain of the self-intersection line of the slowness surface, the refraction will have rather a wedge than a cone character.

In the considered above particular case of the acoustic axis splitting from the symmetry plane, one can put $a = \pi / 2$ which remarkably simplifies expressions (88) and (89), and together with them also the formulae for angle parameters $\Theta_{1,2}$ (92):

$$\mathrm{tg}\Theta_1 = \mathrm{ctg}(\xi / 2)\mathrm{tg}\Phi_\xi, \qquad \mathrm{tg}\Theta_2 = \mathrm{tg}(\xi / 2)\mathrm{tg}\Phi_\xi . \tag{99}$$

The discussed problem of kinematics of the precession of ray velocities at the line of self-intersection of slowness surface may be quantitatively described. Introduce the polar coordinates ($S_{1,2}, \overline{\varphi}_{1,2}$) of the positions of the ends of the radius-vectors $\Delta\mathbf{s}_{1,2}$ at ellipses (93):

$$\Delta\mathbf{s}_{1,2} = S_{1,2}\left(\frac{\mathbf{p}}{p}\cos\overline{\varphi}_{1,2} + \frac{\mathbf{m}_0 \times \mathbf{p}}{p}\sin\overline{\varphi}_{1,2}\right). \tag{100}$$

Comparing (100) with (93) one obtains

$$S_{1,2}^2 = (\mathbf{q}\sin\Theta_{1,2} - \mathbf{p}\cos\Theta_{1,2})^2 = \frac{q^2F^2 - 2\mathbf{p}\cdot\mathbf{q}G_\mp + p^2G_\mp^2}{F^2 + G_\mp^2}, \tag{101}$$

$$\mathrm{ctg}\overline{\varphi}_{1,2} = (\mathbf{p}\cdot\mathbf{q} - p^2\mathrm{ctg}\Theta_{1,2}) / g . \tag{102}$$

Differentiating the latter equation gives the angular velocities

$$\dot{\overline{\varphi}}_{1,2} = -\frac{g}{S_{1,2}^2}\dot{\Theta}_{1,2} . \tag{103}$$

Here the derivatives $\dot{\Theta}_{1,2}$ are found from (92):

$$\dot{\Theta}_{1,2} = \frac{\dot{F}G_\mp - F\dot{G}_\mp}{F^2 + G_\mp^2} = -\frac{\omega\sin\xi}{U_{1,2}^2(\Phi_\xi + \pi / 2)} = -2\dot{\varphi}_{1,2}(\Phi_\xi + \pi / 2), \tag{104}$$

where $U_{1,2}^2$ and $\dot{\varphi}_{1,2}$ are given by functions (81) and (83) shifted in phase: $\Phi_\xi \to \Phi_\xi + \pi / 2$. The sectorial velocities $\overline{v}_{1,2}^{sec}$ of the motion of the vectors $\Delta\mathbf{s}_{1,2}$ over universal ellipse (93) are found in analogy with eqn. (86):

$$\overline{v}_{1,2}^{sec} = \frac{1}{2}S_{1,2}^2\dot{\overline{\varphi}}_{1,2} = -\frac{1}{2}g\dot{\Theta}_{1,2} = g\dot{\varphi}_{1,2}(\Phi_\xi + \pi / 2) . \tag{105}$$

Thus, this velocity differs from the angular velocity of the polarization, eqn. (83), only by the dimensional factor g and by the retardation $\pi/2$. Substituting into (105) the angular velocity $\dot{\varphi}_{1,2}$ (85) for the considered above symmetric example, one obtains the more compact form for the sectorial velocity:

$$\overline{v}_{1,2}^{sec}(\Phi_{\xi}) = \frac{g\omega \sin\xi}{1 \mp \cos\xi \cos 2\Phi_{\xi}} \equiv g\dot{\varphi}_{2,1}(\Phi_{\xi}) . \tag{106}$$

Here it is bearing in mind that the phase shift of the velocity $\dot{\varphi}_{1,2}$ in simplified variant (85) is equivalent to the transition at the counterpart branch: $\dot{\varphi}_{1,2}(\Phi_{\xi} + \pi/2) = \dot{\varphi}_{2,1}(\Phi_{\xi})$. The found relation (106) allowed us to use in Fig. 14 the same curves for a characterization of both angular velocities of polarization and sectorial velocities of ray speeds. The shown dependencies adequately reflect the discussed above properties of the ray velocity precession at the line of self-intersection of the slowness surface. Angle velocities (103) behave in a similar way, especially in the region of small ξ. With closing to acoustic axes $\xi = \pm\pi/2$, variations of angular velocities in time are smoothing, but retain finite until $p \neq q$, in contrast to the velocities $\overline{v}_{1,2}^{sec}$, which tend to constant at these limits.

9. Conclusions

Thus, we have found that specific features of the influence of attenuation on the basic wave properties are associated with two main qualities of the damping: i) it does not disturb the symmetry of a crystal, and ii) formally, it provides an imaginary, i.e. non-hermitian, perturbation of the acoustic tensor. Due to the first quality there is almost no influence of the damping on the acoustic axes which exist due to symmetry of the crystal (tangent degeneracies along ∞ and 4-fold symmetry axes and conical degeneracies along 3-fold axes). On the other hand, the conical acoustic axes of any other orientations manifest instability with respect to an imaginary perturbation of the acoustic tensor. They split into pairs of degeneracies of new type (the so-called singular acoustic axes), which never occurs without damping. In the neighbourhood of split acoustic axes the polarization of elastic waves proves to be strongly elliptical becoming almost circular close to the degeneracy points. A rotation of the polarization ellipses around those points is described by the Poincarè index $n = \pm 1/4$. The slowness surface acquires lines of self-intersection connecting the split singular acoustic axes. Only the end points of these lines correspond to true degeneracies where the imaginary components of phase speeds of isonormal waves also coincide. The latter coincidence also occurs on the whole equi-damping lines at the attenuation surface. These self-intersection lines at the two different surfaces (Fig. 10) after their projection on the unit sphere $\mathbf{m}^2 = 1$ of propagation directions continue each other at the degeneracy points.
Topological transformations of wave surfaces and polarization fields create new features of the phenomenon of internal conical refraction. Still an extension of the theory may be done in terms of the same classic refraction cone bounded by the universal ellipse. As we have seen, in crystals without damping the classic picture of conical refraction automatically arises for a circularly polarized wave propagating along conical acoustic axis. In an absorptive crystal the same cone and universal ellipse as a trajectory of precession of the ray velocity vectors retain at the whole self-intersection line of the slowness surface between split degeneracy points.

Along singular axes the refraction does not differ from the classical picture: the isonormal waves degenerate into one circularly polarized wave with the ray precession of constant sectorial velocity $\overline{v}^{\text{sec}} = g\omega$ at the ellipse. A screen "illumination" related to such precession would look as a completely drawn ellipse (Fig. 15a). Some increase of intensity in the vicinity of large semi-axes ($S_{\max}$) is explained by a slower motion of the vector $\mathbf{s}_0$ in this region (its linear speed at the ellipse is equal $2\overline{v}^{\text{sec}} / S$). When the "observation" point passes along the ridge of the wedge to its middle, both the precession of the vectors $\mathbf{s}_{1,2}$ and the "illumination" pattern become less and less uniform (Fig. 15b,c). And in the center (ξ = 0) only two points (Fig. 15d) will turn to be "illuminated". They relate to the isonormal waves with linear polarization: the refraction becomes purely wedge-like. Thus, with scanning the ridge by the wave normal **m** the refraction continuously transforms from purely conic to purely wedge type.

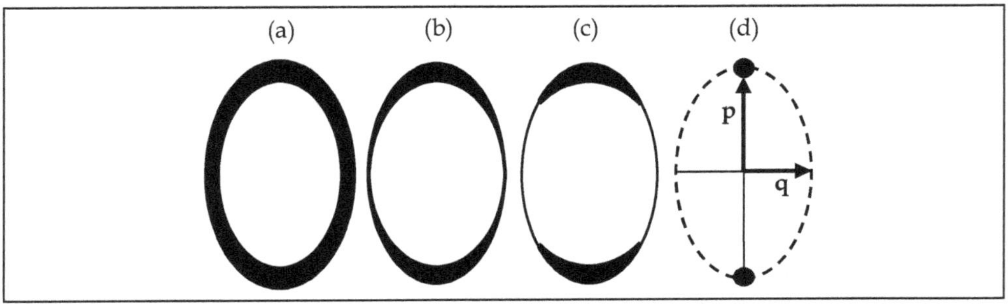

Fig. 15. Schematic plot of gradual transition from purely conical refraction (a) along a singular acoustic axis to purely wedge refraction in the middle of the ridge (d) for a particular case of the acoustic axis splitting from the symmetry plane

In conclusion, let us discuss the observability of the above beautiful and nontrivial physical effects. In principal, there is no threshold level of damping for the split of acoustic axes. Just the less damping, the less is the solid angle inside of which all the peculiarities manifest themselves. If this angle is less than the angle of the acoustic beam divergence, then we shall not observe neither splitting of acoustic axes, nor any accompanied effects. Thus, for the observability of our predictions the split angle $\delta\psi$ (46) must exceed the divergence of the beam. The best experimentally realizable collimation of sound beams is limited by the diffraction divergence, which is estimated as $\sim\lambda/d$, where λ is the wave length and d is the diameter of the beam. So, with increasing frequency ω the angle $\delta\psi$ increases and the beam divergence, on the contrary, decreases. Thus, we deal here with a frequency threshold from below. The order of the splitting angle is determined by the estimate $\delta\psi = 2\delta m_0 \sim \omega\eta / \mu$ (46), where μ is the shear modulus and η is the viscousity. Substituting this estimate to the inequality $\delta\psi > \lambda / d$, one obtains the following lower threshold for the frequency $\nu = \omega/2\pi$

$$\nu > \nu_{\text{th}} \sim \sqrt{\frac{c_s \mu}{2\pi\eta d}}, \tag{107}$$

where c_s is the sound speed. There are known a series of physical mechanisms of the sound attenuation η. Often phonons play in it an important role. The phonon viscosity at room (or high) temperature T may be estimated as

$$\eta_{\text{ph}} \sim \tau_{\text{ph}}(3k_B T / a^3)\,. \tag{108}$$

Here τ_{ph} is the phonon relaxation time, k_B is the Boltzmann constant, and a is the lattice parameter. Substituting into eqns. (107), (108) $c_s \sim 3 \cdot 10^5$ cm/s, $\mu \sim 10^{11}$ dyn/cm^2, $d \sim 0.5$ cm, $T \approx 300$ K, $a \sim 3 \cdot 10^{-8}$ cm, $\tau_{\text{ph}} \sim 10^{-9}$ s, we come to the estimate $\nu_{\text{th}} \sim 100$ MHz. Thus, at rather high-frequencies, which however belong to experimentally available ultrasound range, the properties and effects described in this chapter appear quite observable.

10. Acknowledgment

Authors are grateful to A.L. Shuvalov for helpful discussions and to W. Gerulski for the help in computations related to the illustrations. The support of the Polish Foundation MNiSW (grant No N N 501252334) is gratefully acknowledged. V.I.A. is also grateful to the Kielce University of Technology for a hospitality and support.

11. References

Aleksandrov, K.S. & Ryzhova, T.V. (1964). Internal conical refraction of elastic waves in ammonium dihydrogen phosphate. *Kristallografiya*, Vol. 9, No. 3 (June 1964) 373-376, ISSN 0023-4761 [*Sov. Phys. Crystallography*, Vol. 9, No. 3 (1964) 298-300, ISSN 1063-7745]

Alshits, V.I. & Lothe, J. (1979). Elastic waves in triclinic crystals. *Kristallografiya*, Vol. 24, No. 4, 6 (Aug., Dec. 1979) 972-993, 1122-1130, ISSN 0023-4761 [*Sov. Phys. Crystallography*, Vol. 24, No. 4, 6 (1979) 387-398, 644-648, ISSN 1063-7745]

Alshits, V.I. & Lyubimov, V.N. (1998). Elastic waves in absorptive media: peculiarities of wave surfaces and singularities in the polarization fields. In: *Dissipation in Physical Systems*, A. Radowicz (Ed.), pp. 15-43, Politechnika Swietokrzyska, ISSN 0239-4979, Kielce. *Proceedings of 2nd Workshop on Dissipation in Physical Systems*, Borkow, Poland, September 1-3, 1997

Alshits, V.I. & Lyubimov, V.N. (2011). Conical refraction of elastic waves in absorptive crystals. *Zh. Eksp. Teor. Fiz.*, Vol. 140, No. 2(8) (Aug. 2011) [JETP, Vol. 113, No. 2 (2011), ISSN 1063-7761]

Alshits, V.I.; Sarychev, A.V. & Shuvalov, A.L. (1985). Classification of degeneracies and analysis of their stability in the theory of elastic waves in crystals. *Zh. Eksp. Teor. Fiz.*, Vol. 89, No. 3(9) (Sept. 1985) 922-938, ISSN 0044-4510 [Sov. Phys. JETP, Vol. 62, No. 3 (1985) 531-539, ISSN 1063-7761]

Barry P.A. & Musgrave, M.J.P. (1979). On elliptical conical refraction of elastic waves in tetragonal crystals. *Q. J. Mech. & Appl. Math.*, Vol. 32, No. 3 (March 1979) 205-214, ISSN 0033-5614

De Klerk, J. & Musgrave, M.J.P. (1955). Internal conical refraction of transverse elastic waves in a cubic crystal. *Proc. Phys. Soc. Lond. B*, Vol. 68, No. 2 (Feb. 1955) 81-88, ISSN 1088-0370

Fedorov, F.I. (1968). *Theory of Elastic Waves in Crystals*, Plenum Press, ISBN, New York
Khatkievich, A.G. (1962a). The acoustic axes in crystals. *Kristallografiya*, Vol. 7, No. 5 (Oct. 1962) 742-747, ISSN 0023-4761 [*Sov. Phys. Crystallography*, Vol. 7, No. 5 (1963) 601-604, ISSN 1063-7745]
Khatkievich, A.G. (1962b). Internal conical refraction of elastic waves. *Kristallografiya*, Vol. 7, No. 6 (Dec. 1962) 916-920, ISSN 0023-4761 [*Sov. Phys. Crystallography*, Vol. 7, No. 6 (1963) 742-745, ISSN 1063-7745]
Khatkievich, A.G. (1964). Special directions for elastic waves in crystals. *Kristallografiya*, Vol. 9, No.5 (Oct. 1964) 690-695, ISSN 0023-4761 [*Sov. Phys. Crystallography*, Vol. 9, No. 5 (1964) 579-582, ISSN 1063-7745]
Landau L.D. & Lifshitz, E.M. (1986). *Theory of elasticity*. Pergamon Press, ISBN, London
Musgrave, M.J.P. (1957). On an elliptic cone of internal refraction for quasi-transverse waves in tetragonal crystals. *Acta Crystallogr.*, Vol. 10, No. 4 (Apr. 1957) 316-318, ISSN
Shuvalov, A.L. (1998). Topological features of polarization fields of plane acoustic waves in anisotropic media. *Proc. R. Soc. Lond. A*, Vol. 454, (Nov. 1998) 2911-2947, ISSN 1471-2946
Shuvalov, A.L. & Chadwick, P. (1997). Degeneracies in the theory of plane harmonic wave propagation in anisotropic heat-conducting elastic media. *Phil. Trans. R. Soc. Lond. A*, Vol. 355 (Jan. 1977) 156-188, ISSN 1471-2962
Shuvalov, A.L. & Scott, N.H. (1999). On the properties of homogeneous viscoelastic waves. *Q. J. Mech. Appl. Math.*, Vol. 52 (Sept. 1999) 405-417, ISSN 0033-5614
Shuvalov, A.L. & Scott, N.H. (2000). On singular features of acoustic wave propagation in weakly dissipative anisotropic thermoviscoelasticity. *Acta Mechanica*, Vol. 140, No 1-2 (March 2000) 1-15, ISSN
Sirotin Yu.I. & Shaskolskaya, M.P. (1979). *Fundamentals of Crystal Physics* (in Russian), Nauka, Moscow [(1982) translation into English, Mir, ISBN, Moscow]

3

Acoustic Wave

P. K. Karmakar
Department of Physics, Tezpur University, Napaam, Tezpur, Assam India

1. Introduction

An acoustic wave basically is a mechanical oscillation of pressure that travels through a medium like solid, liquid, gas, or plasma in a periodic wave pattern transmitting energy from one point to another in the medium [1-2]. It transmits sound by vibrating organs in the ear that produce the sensation of hearing and hence, it is also called acoustic signal. This is well-known that air is a fluid. Mechanical waves in air can only be longitudinal in nature; and therefore, all sound waves traveling through air must be longitudinal waves originating in the transmission form of compression and rarefaction from vibrating matter in the medium. The propagation of sound in absence of any material medium is always impossible. Therefore, sound does not travel through the vacuum of outer space, since there is nothing to carry the vibrations from a source to a receiver. The nature of the molecules making up a substance determines how well or how rapidly the substance will carry sound waves. The two characteristic variables affecting the propagation of acoustic waves are (1) the inertia of the constituent molecules and (2) the strength of molecular interaction. Thus, hydrogen gas, with the least massive molecules, will carry a sound wave at 1,284.00 ms^{-1} when the gas temperature is 0^0 C [1]. More massive helium gas molecules have more inertia and carry a sound wave at only 965.00 ms^{-1} at the same temperature. A solid, however, has molecules that are strongly attached, so acoustic vibrations are passed rapidly from molecule to molecule. Steel, for an instant example, is highly elastic, and sound will move rapidly through a steel rail at 5,940.00 ms^{-1} at the same temperature. The temperature of a medium influences the phase speed of sound through it. The gas molecules in warmer air thus have a greater kinetic energy than those of cooler air. The molecules of warmer air therefore transmit an acoustic impulse from molecule to molecule more rapidly. More precisely, the speed of a sound wave increases by 0.60 ms^{-1} for each Celcius degree rise in temperature above 0^0 C.

Acoustic waves, or sound waves, are defined generally and specified mainly by three characteristics: wavelength, frequency, and amplitude. The wavelength is the distance from the top of one wave's crest to the next (or, from the top of one trough to the next). The frequency of a sound wave is the number of waves that pass a point each second [1]. Sound waves with higher frequencies have higher pitches than sound waves with lower frequencies and vice versa. Amplitude is the measure of energy in a sound wave and affects volume. The greater the amplitude of an acoustic wave, the louder the sound and vice versa. An acoustic wave is what makes humans and other animals able to hear. A person's ear perceives the vibrations of an acoustic wave and interprets it as sound [1]. The outer ear, the visible part, is shaped like a funnel that collects sound waves and sends them into the ear canal where they hit the ear drum, which is a tightly stretched piece of skin that vibrates in time with the wave. The ear drum starts a chain reaction and sends the vibration through three little bones in the middle ear that amplify sound. Those bones are called the hammer, the anvil, and the stirrup.

Furthermore, acoustic waves from a purely hydrodynamic point of view are small-amplitude disturbances that propagate in a compressible medium (like a fluid) through the interplay between fluid inertia, and the restoring force of fluid pressure. The propagation of small-amplitude disturbances in homogeneous medium is observed as acoustic waves such as water waves, and in self-gravitationally stratified medium like stellar atmosphere [36-37, 41-44], acoustic-gravity waves such as p-modes, g-modes, f-modes, etc., as found by helio- and astero-seismological studies. Acoustic waves propagating through a dispersive medium may get dynamically converted into *solitons* or *shocks* depending on the physical mechanisms responsible for their saturation. When fluid nonlinearity (convective effect) is balanced by dispersion (geometrical effect), solitons usually result [4]. Conversely, shocks are formed if fluid nonlinearity is balanced by dissipation (damping effect). The nonlinear hydrodynamic equations of various forms (like KdV equation, Burger equation, NLS equation, BO equation, etc.) in the context of the generation, structure, propagation, self-organization and dissipation of solitons or shocks have long been developed applying the hydrodynamic views of the usual conservation laws of flux, momentum and energy [4]. Similar outlook is needed to understand the formation of other nonlinear localized structures of low frequency acoustic waves like double layers, vortices, etc. They are important in a wide variety of space, astrophysical and laboratory problems for the investigation of dynamical stability against perturbation [3-4]. In addition, these equations have wide applications to study a nonlinear, radial, energetic, and steady-flow problem that provides a first rough approximation to the physics of stellar winds and associated acoustic wave kinetics, which are responsible for stellar mass-loss phenomena via supersonic flow into interstellar space [2].

Acoustic mode in plasmas of all types [2-44], similarly, is actually a pressure driven longitudinal wave like the ordinary sound mode in neutral gas. In normal two-component plasmas, the electron thermal pressure drives the collective ion oscillations to propagate as the ion sound (acoustic) wave. Here the electron thermal pressure provides the restoring force to allow the collective ion dynamics in the form of ionic compression and rarefaction to propagate in the plasma background and ionic mass provides the corresponding inertial force. Thermal plasma species (like electrons) are free to carry out thermal screening of the electrostatic potential. In absence of any dissipative mechanism, the ion sound wave moves with constant amplitude. For mathematical description of the ion sound kinetics, the plasma electrons are normally treated as inertialess species and the plasma ions, with full inertial dynamics. However, recent finding of ion sound wave excitation in *transonic plasma* condition of hydrodynamic equilibrium offers a new physical scope of acoustic turbulence due to weak but finite electron inertial delay effect [5-12]. Qualitative and quantitative modifications are introduced into its nonlinear counterpart as well, under the same transonic plasma equilibrium configuration [12]. The *transonic transition* of the plasma flow motion quite naturally occurs in the neighborhood of boundary wall surface of laboratory plasmas, self-similar expansion of plasmas into vacuum, in solar wind plasmas and different astrophysical plasmas, etc. The self-similar plasma expansion model predicts supersonic motion of plasma flow into vacuum. This model is widely used to describe the motion of intense ion plasma jets produced by short time pulse laser interaction with solid target [17-

23]. Recently, the self-similar plasma expansion into vacuum is modeled by an appropriate consideration of space charge separation effect on the expanding front [13].

According to the recently proposed inertia-induced ion acoustic excitation theory [5-8], the large-scale plasma flow motion feeds the energy to the short scale fluctuations near the pre-sheath termination at sonic point. This is a kind of energy transfer process from large-scale flow energy to wave energy through short scale instability of cascading type. In order to maintain the turbulence type of hydrodynamic equilibrium, there must be some source to feed large-scale flow and sink to arrest the infinite growth of the excited short waves. The growing wave energy could be used to re-modify the global transonic equilibrium such that the transonic transition becomes a natural equilibrium with smooth change in flow motion from subsonic to supersonic regime. Of course, this is a quite involved problem to handle the self-consistent turbulence theory of transonic plasma in terms of anomalous transport [5]. Now one may ask how to produce such boundary layer with sufficient size of the transonic plasma layer for laboratory experimentations?

This, in fact, is an experimental challenge to design and set up such experiments to produce extended length of the transonic zone to sufficient extent to resolve the desired unstable wave spectral components. Creation of a thick boundary layer of transonic flow dynamics is, no doubt, an important task. This zone lies between subsonic and supersonic domains, and is naturally bounded by low supersonic and high subsonic speeds. It should be mentioned here that the sonic velocity corresponds to the phase velocity of the bulk plasma mode of the dispersionless ion acoustic wave. In case of sheath edge boundary, transonic layer could be probed by high-resolving diagnosis of the Debye length order. The desired experiments of spectral analysis of the unstable ion acoustic waves in *transonic plasma condition* may be quite useful to resolve the mystery of sheath edge singularity. Using *de-Lavel nozzle mechanism* of hydrodynamic flow motion, experiments could be designed to produce transonic transition layer of desired length and characteristics [6-8].

Study of the ambient acoustic spectrum associated with plasma flow motion can be termed as the *acoustic spectroscopy* of equilibrium homogeneous plasma flows [6, 26]. This may be useful for expanding background plasmas [13], solar wind plasmas and also in space plasmas through which the space vehicles' motion and aerodynamic motion occur [3, 25, 28]. Basic principles of the acoustic spectroscopy have concern to the linear and non-linear ion acoustic wave turbulence theory and properties of the transonic plasma equilibrium [5-12, 26]. These properties may be used to develop the required diagnostic tools to study and describe the hydrodynamic equilibrium states of plasma flows by suitable observations and analysis of the waves and instabilities they exhibit. In fact, the ambient turbulence-driven plasma flow is quite natural to occur in toroidal and poloidal directions of the magnetic confinement of tokamak device. Similar physical mechanism is supposed to be operative in the transonic transition behavior of equilibrium plasma flow motion [5-12]. Thorough investigations of acoustic wave turbulence theory in transonic plasma condition will be needed to explore transonic flow dynamics on a concrete footing.

Recently, there has been an outburst of interest in plasma states where the assumption of static equilibrium practically is violated [28-30]. Great deals of research activities are now going on in transonic and supersonic magnetohydrodynamic (MHD) flows in laboratory and astrophysical plasmas. Similar activities are also important for understanding the designing of supersonic aerodynamics having relevance in spacecraft-based laboratory experimentations of space plasma research as well [8, 30]. This is also argued that future tokamak reactors need the consideration of rotation of fusion plasma with high speeds that do not permit the assumption of static equilibrium to hold good. This may be brought about due to neutral beam heating and pumped divertor action for the extraction of heat and exhaust.

In astrophysics [3, 28-32, 35-44], the primary importance of plasma flows is revealed in such diverse situations as coronal flux tubes, stellar winds, rotating accretion disks, torsional modes, and jets emitted from radio galaxies. This is to argue that the basic understanding of the acoustic wave dynamics in transonic plasma system constitutes an important subject of future interdisciplinary research [5-12, 26-30]. This may be useful for development of the appropriate diagnostics for acoustic spectroscopy to measure and characterize the hydrodynamic equilibrium of flowing transonic plasmas [8-10]. Such concepts of acoustic wave dynamics in a wider horizon may also applied to understand some helio- and astero-seismic observations in astrophysical contexts.

Most of the plasma devices of industrial applications like dense plasma focus machine, plasma torches, etc. depend on the plasma flows that violate the static equilibrium [26-30]. In fusion plasmas of future generation too, the static approximation of the equilibrium plasma description may not be suitable to describe the acoustic wave behavior. In future course of fusion research, rotational motions of fusion plasmas in poloidal and toroidal directions may decide the equilibrium. This is important to state that in toroidal plasmas, the geodesic acoustic mode becomes of fundamental importance in comparison to the ordinary sound modes [30]. This may be more important when these rotational motions are in the defined range of the transonic limit. Simplicity is correlated to the local mode approximation of the acoustic wave description in transonic limit of uniform and unidirectional plasma flow motion without magnetic field.

The lowest order nonlinear wave theory of the ion acoustic wave dynamics predicts that the usual KdV equation is not suitable to describe the kinetics of the nonlinear traveling ion acoustic waves in transonic plasma condition [8-9, 12, 26]. A self-consistent linear source driven KdV equation, termed as *d*-KdV equation, is prescribed as a more suitable nonlinear differential equation to describe the nonlinear traveling ion acoustic wave dynamics in transonic plasma condition. By mathematical structure of the derived *d*-KdV equation, it looks analytically non-integrable and physically non-conservative dynamical system [8-9]. Due to linear source term, an additional class of nonlinear traveling wave solution of oscillatory shock-like nature is obtained. This is more prominent in the shorter scale domain of the unstable ion acoustic wave spectrum, but within the validity limit of weak nonlinearity and weak dispersion.

If there is multispecies ionic composition in a plasma system, varieties of plasma sound waves are likely to exist depending on, in principle, the number of inertial ionic species. In plasmas containing two varieties of dust or fine suspended particles, two distinct kinds of natural plasma sound modes are possible [15-16]. Such plasmas, termed as the *colloidal plasmas* [16], have become the subject of intensive study in various fields of physics and engineering such as in space, astrophysics, plasma physics, plasma-aided manufacturing technique, and lastly, fusion technology [14-23]. The dust grains or the solid fine particles suspended in low temperature gaseous plasmas are usually negatively charged. It is also observed that plasmas including micro scale-sized and nano scale-sized suspended particles exist in many natural conditions of technological values. Such plasmas have been generated in laboratories with a view to investigate the dust grain charging physics, plasma wave physics as well as some acoustic instability phenomena.

The two distinct sound modes, however, in bi-ion colloidal plasma are well-separated in space and time scales due to wide range variations of mass scaling of the normal ions and the charged dust grains and free electrons' populations. The charged dust grains are termed as the *Dust Grain Like Impurity Ions* (DGLIIs) [15] to distinguish from the normal impurity ions. The present contributory chapter, additionally, applies the inertia-induced acoustic excitation theory to nonlinear description of plasma sound modes in colloidal plasma [15-16] under different configurations. Two separate cases of ion flow motion and dust grain motions are considered. It is indeed found that the *modified Ion Acoustic Wave* (m-IAW) or *Dust Ion Acoustic* (DIA) *wave* and the *so-called (Ion) Acoustic Wave* (s-IAW) or *Dust Acoustic* (DA) *wave* both become nonlinearly unstable due to an active role of weak but finite inertial correction of the respective plasma thermal species [15, 26]. Proper mass domain scaling of the dust grains for acoustic instability to occur is estimated to be equal to that of the asymptotic mass ratio of plasma electron to ion as the lowest order inertial correction of background plasma thermal species. This contributory chapter is thus a review organized to aim at some illustrative examples of linear and nonlinear acoustic wave propagation dynamics through transonic plasma fluid, particularly, under the light of current scenario. Some important reported findings on nonlinear acoustic modes found in space and astrophysical situation [31-44], like in solar plasma system [10-11, 31-44], will also be presented in concise to understand space phenomena. Incipient future scopes of the presented contribution on transonic flow dynamics in different astrophysical situations will also be briefly pointed out.

2. Physical model description

A simple two-component non-isothermal, field-free and collisionless plasma system under fluid limit approximation is assumed. The plasma ions are supposed to be drifting with uniform velocity at around the sonic phase speed under field-free approximation. Global plasma equilibrium flow motion over transonic plasma scale length at hydrodynamic equilibrium is assumed to satisfy the global quasi-neutrality. Such situations are realizable in the transonic region of the plasma sheath system as well as in solar and other stellar wind plasmas [3, 10-11, 35-41]. Its importance has previously been discussed [1, 5-12], where the ion-beam driven wave phenomena are supposed to be involved in *Q*-machine or in unipolar/ bipolar ion rich sheath formed around an electrode wall or grid in Double Plasma Device (DPD) experiments of plasma sheath driven low frequency instabilities of relaxation type [5]. The unstable situation is equally likely to occur on both the sides of the sheath structures with plasma ion streamers [12].

3. Linear normal acoustic mode analyses

3.1 Basic governing equations

The basic set of governing dynamical evolution equations for the linear normal mode behavior of fluid acoustic wave consist of electron continuity equation, electron momentum equation, ion continuity equation, and ion momentum equation [5-6]. The set is closed by coupling the plasma thermal electron dynamics with that of plasma inertial ion dynamics through a single Poisson's equation for electrostatic potential distribution due to localized ambipolar effects. Applying Fourier's wave analysis for linear normal mode behavior of ion acoustic wave over the basic set of governing dynamical evolution equations [5-6], the linear dispersion relation is derived as follows

$$(\Omega + k.v_0)^2 = \frac{k^2 v^2_{te}(1 + k^2\lambda^2_{De})}{\Omega^2}(\Omega_a^2 - \Omega^2). \tag{1}$$

All the notations in the equation (1) are usual and conventional. Here Ω is the Doppler-shifted frequency of the ion acoustic wave, Ω_a is the ion acoustic wave frequency in laboratory frame of reference, k is the angular wave number of the ion acoustic wave such that $k\lambda_{De}$ is a measure of the acoustic wave dispersion scaling and v_{te} is the electron thermal velocity. Now the kinematics of any mode can be analyzed in two different ways: one in lab-frame and the other, in Doppler-shifted frame of reference. This is to note that the obtained dispersion relation differs from those of the other known normal modes of low frequency relaxation type of instability, ion plasma oscillations and waves. This is due to the weak but finite electron inertial delay effect in the dispersion relation of the wave fluctuations. This is mathematically incorporated by a weak inertial perturbation over electron inertial dynamics over the leading order solution obtained by virtue of electron fluid equations neglecting electron inertial term.

It is thus obvious from the mathematical construct of equation (1) that the LHS is a non-resonant term whereas RHS is a resonant term. The RHS gets artificially transformed into a resonant term if and only if $k.v_{i0} < 0$. Now, it can be inferred that equation (1) represents a resonantly unstable situation at Doppler shifted resonance frequency of $\Omega \approx |k.v_0| \geq \Omega_a$, if and only if $k.v_0$ <0. This means that only the mode counter moving with respect to the plasma beam mode gets resonantly unstable. The resonance growth rate for this resonant instability [5-6] is found to be of the following form

$$\gamma = \sqrt{\frac{m_i}{m_e} 2\Omega_a (1 + k^2\lambda^2_{De})}\left|(\Omega - |k.v_0|)^{1/2}\right|. \tag{2}$$

This is important to add that the resonance condition required by equation (1) dictates the propagation direction of the unstable ion acoustic wave (counter moving with respect to plasma ion streams) at reduced frequencies. It is clear from equation (2) that there is the physical appearance of two distinct classes of eigen mode frequencies of the resonantly coupled mode-mode system of linearly growing ion acoustic oscillations in lab-frame: near-zero frequency (standing mode pattern) and non-zero frequency (propagating mode pattern). These two distinct eigen modes are generated by the process of repeated Doppler-shifting of the ion acoustic wave frequency under the unique mathematical compulsion of the hydrodynamic tailoring of the electron fluid density perturbation over ion acoustic time scale. The unstable condition decides the resonant acoustic excitation threshold value for the onset of the instability in terms of normalized value of the eigen mode frequency of the acoustic fluctuations.

3.2 Graphical analysis

It is well-known that the graphical method is a more informative, simple and quick tool for analyzing the stability behavior of a plasma-beam system even without solving dispersion relation. To depict the clear-cut picture of the poles, relation (1) is rewritten as,

$$F(\Omega, k) = \frac{1}{(1 + k^2\lambda^2_{De})} = k^2 v^2_{te}\left[\frac{\Omega_a^2}{\Omega^2(\Omega + k.v_0)^2} - \frac{1}{(\Omega + k.v_0)^2}\right]. \tag{3}$$

It is clear from the equation (3) that two poles are possible to exist in Ω-space at $\Omega=0$ and $\Omega=|k.v_0|$ for $k.v_0<0$. According to graphical method, the beam-plasma system will exhibit instability only when the curve of $F(\Omega,k)$ versus Ω has multiple singular values in Ω-space having finite minima in between the two successive singularities, which do not intersect with the line $F(\Omega,k)=1/(1+k^2\lambda_{De}^2)$. The required condition for minimization of $F(\Omega,k)$ in Ω-space can be obtained by equating $dF/d\Omega=0$. Now this condition, when applied to equation (3), results into the following equality to derive the value of Ω where dispersion function is supposed to be minimum

$$\Omega_a^2(\Omega+k.v_0)+\Omega(\Omega_a^2-\Omega^2)=0. \tag{4}$$

In principle, equation (4) is to be solved to determine the value of Ω. This is obvious to note that this equality is satisfied at resonance value of $\Omega\sim|k.v_0|\sim\Omega_a$ for $k.v_0<0$. Now to indemnify the complex nature of Ω, the functional value of $F(\Omega,k)>1/(1+k^2\lambda_{De}^2)$. This can, however, be further simplified to yield the following inequality to determine the threshold value for the onset of the inertia-induced instability

$$k^2v_{te}^2(\Omega_a^2-\Omega^2)>(\Omega-|k.v_0|)^2/(1+k^2\lambda_{De}^2). \tag{5}$$

The threshold condition for the instability is satisfied for equality sign at resonance frequency $\Omega\sim|k.v_0|\sim\Omega_a$ that characterizes the case of a marginal instability. A few typical plots of the function $F(\Omega,k)$ in Ω-space for shorter and longer acoustic wavelengths (perturbation scale lengths) are represented in Fig. 1.

3.3 Numerical analysis

Numerical techniques for solving polynomials over years have developed to a vast extent for solving polynomials even with complex coefficients and complex variables. For the present case, the Laguerre's algebraic root-finding method [6] to solve the normalized form of polynomial equation has been used. The polynomial $P(\Omega')$ in the normalized form of the dispersion relation (1) in ion-beam frame is given below

$$P(\Omega')=a_0+a_1\Omega'+a_2\Omega'^2+a_3\Omega'^3+a_4\Omega'^4=0. \tag{6}$$

Here all the normalized notations used are usual, generic and defined by $\Omega'=\Omega/\omega_{pi}$, $\Omega_a'=\Omega_a/\omega_{pi}$ $k'=k\lambda_{De}$, $v'_{te}=v_{te}/c_s=\sqrt{m_i/m_e}$ and $M=v_0/c_s$. The expressions for the various coefficients in the polynomial $P(\Omega')$ are defined as follows

$$a_0=-(1+k'^2)(k'^2v_{te}'^2\Omega_a'^2),$$

$$a_1=0,$$

$$a_2=(|k'.M|)^2+(1+k'^2)(k'^2v_{te}'^2),$$

$$a_3=-2|k'.M|,\text{ and}$$

$$a_4 = 1.$$

It is found that out of four possible roots of $P(\Omega')$, only two roots are complex and these are the complex conjugates as a pair. For all the complex conjugated roots, only the complex root with positive imaginary part is useful, since this determines the growth rate of the instability. Real and imaginary parts of the corresponding complex roots are then plotted as shown in Figs. 2 and 3, respectively. Numerical characterization of the unstable mode of the instability clearly depicts the resonant character of the electron inertia-induced resonant acoustic instability [5].

3.4 Evaluation of wave energy

This is important to evaluate the wave energy in order to have a more complete picture of the basic source mechanism of the discussed instability. In presence of the beam, it is expected that one of the modes involved, has positive energy and the other has negative energy. The dispersion relation (1) can be put in the laboratory frame for a more clear identification and characterization of the positive and negative energy modes in the form of dispersion function $\varepsilon(\omega,k)$ as follows

$$\varepsilon(\omega,k) = 1 + \frac{1}{k^2\lambda_{De}^2}\left(1+\frac{{\omega_{pe}}^2}{k^2 v_{te}^2}\right) - \frac{\omega_{pi}^2}{(\omega - k.v_0)^2} = 0. \tag{7}$$

The average electric field energy stored in a propagating electrostatic (created by ambipolar effect) wave in a medium is given by the following relation [6]

$$W_\omega(\omega,k) = \frac{1}{2}\varepsilon_0\left\langle\left|\delta E(\omega,k)\right|^2\right\rangle\frac{\partial}{\partial\omega}\left[\omega\varepsilon(\omega,k)\right]. \tag{8}$$

Here ε_0 is the dielectric constant of free space, $\delta E(\omega,k)$ is the electric field amplitude of the ion acoustic fluctuations and $W_E = 1/2\left|\delta E(\omega,k)\right|^2$ is the corresponding counterpart of electric energy of the acoustic fluctuations through free space. Applying the equations (7) and (8), the following can explicitly be derived

$$\frac{W_\omega}{W_E} = \omega\frac{\partial\varepsilon(\omega,k)}{\partial\omega} = \left[\frac{2\omega^2}{k^2\lambda_{De}^2 k^2 v_{te}^2} + \frac{2\omega\omega_{pi}^2}{(\omega - k.v_0)^3}\right]. \tag{9}$$

Now, clearly, it is evident that the second term of equation (9) contributes negative energy value to the defined wave-plasma system. This occurs as because the sign of this term becomes negative for the values of $\omega < k.v_0$, which is the case for the reported instability. From a few typical plots in Fig.4, one can notice that the total wave energy suffers a sharp transition from negative to positive values at resonance frequency point of zero energy value. The resonance point lies in the domain of near-zero and non-zero frequencies in lab-frame. According to conventional definition and understanding, the wave energy expression in equation (9) classifies the near-zero frequency mode as the negative energy mode. Then immediately the non-zero frequency mode may be classified as the positive energy mode.

This is important to clarify that the theoretical concept of near-zero frequency mode is an outcome of the mathematical construct of weak but finite electron inertial response to the ion acoustic wave fluctuations. The blowing up character, as shown in Fig. 4, of the total wave energy in opposite directions suggests referring the discussed instability to as an 'explosive instability' in accordance with the law of conservation of energy. It signifies the transonic plasma condition with the resonant mode-mode coupling of the positive and negative energy modes. The time average of the hydrodynamic and wave potential energies of the considered wave-plasma system over the growth time scale is conserved during the energy exchange process between the unstable resonant eigen modes and the main source of ion flow dynamics. These two modes are clearly identified from equation (9) as the natural resonant modes of the defined plasma system that undergo linear resonant mode-mode coupling to produce the defined wave instability.

3.5 Estimation of quenching time

Under the cold ion approximation, even the small electrostatic potential will be able to distort the ion particle motion and associated trajectories, affecting the driving source flow velocity of the resonant instability under consideration. In wave frame, the streaming ion energy (E_i) can be expressed by the following relation

$$E_i = \frac{1}{2} m_i \left(v_0 - \frac{\omega}{k} \right)^2 . \tag{10}$$

For $v_o >> \omega/k \sim |(c_s - v_0)|$, which is a valid case for the considered instability [5], the condition for ion orbit distortion becomes of the following form,

$$W_w \geq \frac{1}{2} m_i n_0 v_0^2 . \tag{11}$$

From this condition, the quenching time is estimated under the assumption that the wave amplitude grows sufficiently from thermal noise level to physically measurable level such that

$$W_E(t) = W_i e^{\gamma t} . \tag{12}$$

Here W_i is the initial energy of the acoustic wave amplitude, which is of the order of the thermal fluctuations, i.e., $W_i \sim T_e / \lambda_{De}^3$ and is the unnormalized linear growth rate. Using the resonance values of $\omega = k|c_s - v_0|$ and $\omega - k.v_0 \sim kc_s$ as derived in [5] for long wavelength case of resonant mode, equation (12) for the quenching time τ with the help of (9) can be rewritten as follows

$$\tau = \sqrt{\frac{m_e}{2m_i}} \frac{1}{k\lambda_{De}} \left(|1 - M| \right)^{-1/2} \ln \left[\frac{1}{4} \left(n_0 \lambda_{De}^3 \right) \left(k^2 \lambda_{De}^2 \right) \frac{M^2}{|1 - M|} \right] . \tag{13}$$

For some typical plasma parameters in hydrogen plasma, $\ln\left(n_0 \lambda_{De}^3\right) \sim 15 - 30$. For $k\lambda_{De} \sim 0.3, 0.1, 0.05$ near resonant M as in Figs. 2 and 3, equation (13) gives $\tau > 1$, i.e., $\tau_q > \tau_{pi}$. This physically means that the resonant growth time scale is greater than that of the plasma ion oscillation time scale. Thus the resonant nature of the instability is observable in the present analysis.

3.6 Physical consequences

Wave energy analyses are carried out to depict the graphical appearance of poles (Fig. 1), nature of real parts of the roots (Fig. 2), nature of imaginary parts of roots (Fig. 3) and positive-negative energy modes (Fig. 4).

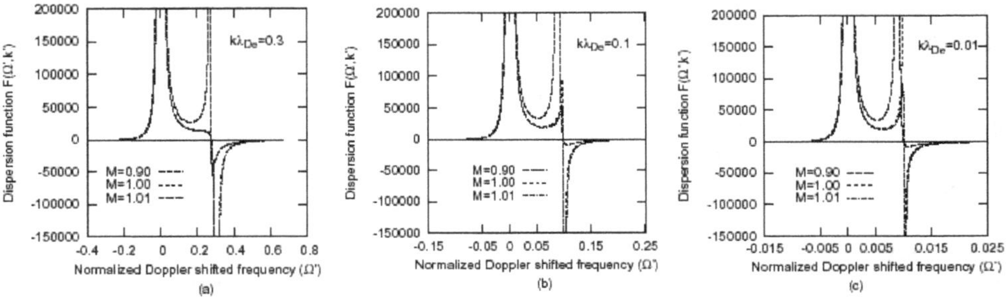

Fig. 1. Graphical appearance of resonance poles as a variation of the dispersion function $F(\Omega,k)$ with normalized Doppler-shifted frequency for dispersion scaling (a) $k\lambda_{De}=0.3$, (b) $k\lambda_{De}=0.1$, and (c) $k\lambda_{De}=0.01$

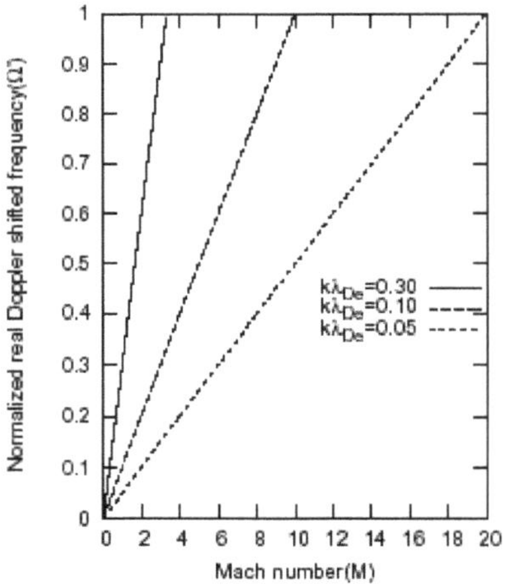

Fig. 2. Variation of the real part of the normalized Doppler shifted eigen mode frequency (Ω') with respect to Mach number (M) for different values of $k\lambda_{De}=0.30,0.10,0.05$

It is found that the instability arises out of linear resonance mode-mode intermixed coupling between the negative and positive energy modes. The total energy of the coupled mode-mode system comprising of hydrodynamical potential energy and wave kinetic energy, however, is in accordance with the law of conservation of energy in the observation time scale on the order of ion acoustic wave time scale. Identification and characterization of the resonance nature of the said instability through *transonic plasma* is presented in order to explore the acoustic richness in terms of collective waves, oscillations and fluctuations. This is an important point to be mentioned here that the same type of instability features are

expected to happen in plasma-wall interaction process and sheath-induced instability phenomena in other similar situations as well.
There are different sorts of analytical and numerical tools for studying the linear instabilities in a given plasma system. *Energy method,* based on energy minimization principle and the *normal mode analysis,* based on equilibrium perturbations are the two basic mathematical tools for analyzing the stability behavior of the given plasma systems. However, the latter is most popular and simple for common use in analyzing the threshold conditions of the instabilities and their growth rates. In the normal mode analysis, a linear dispersion relation is derived which can be put in the form of a polynomial with real or imaginary coefficients. The limitation of the analytical method depends upon the degree of the polynomial.

Computational technique broadly takes into account two ways of investigating instability. First, an unstable mode can be deduced by the derived dispersion relation. The obtained polynomial is then solved to delineate the complex roots having concern to the desired instability. Second, a more comprehensive computational method involves solving for the time dependent solution. Simulation technique used to solve the basic set of equations is supposed to give more complete picture of the space and time evolution of the wave phenomena. However, there is another very informative and simple method for analyzing the derived dispersion relation to predict for the unstable behavior of the plasma system under consideration. This is the graphical method in which the dispersion relation is graphically represented for different values of resonance characterization parameters. Source perturbation scale length $(k\lambda_{De})$ and deviation from sonic point $(1-M)$ are the characterization parameters for the defined acoustic resonance.

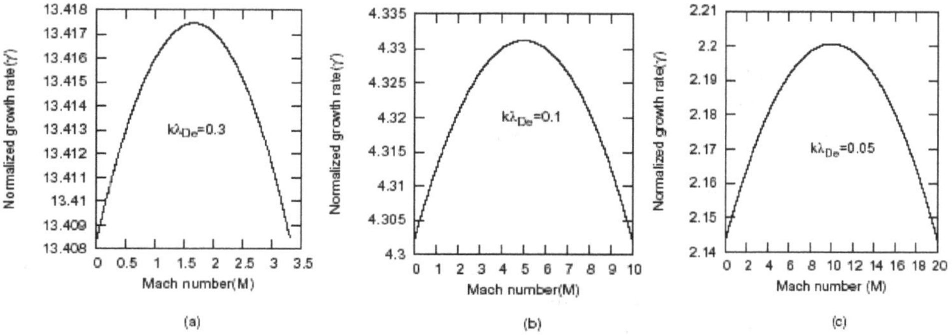

Fig. 3. Variation of the normalized growth rate of the electron inertia-induced resonant acoustic instability with Mach number for (a) $k\lambda_{De}=0.3$, (b) $k\lambda_{De}=0.1$ and (c) $k\lambda_{De}=0.05$ showing that transonic plasma is rich in wide range acoustic spectral components and hence, an unstable zone

This is quite natural and interesting to argue that the transonic plasma condition offers a unique example where the physical situation of localized hydrodynamic equilibrium of quasi-neutral plasma flow dynamics exists. Previous publication reports that the transonic plasma layer, assumed to have finite extension, can be considered as a good physical situation to study the acoustic instability, wave and turbulence driven by electron inertia-induced ion acoustic excitation physics.

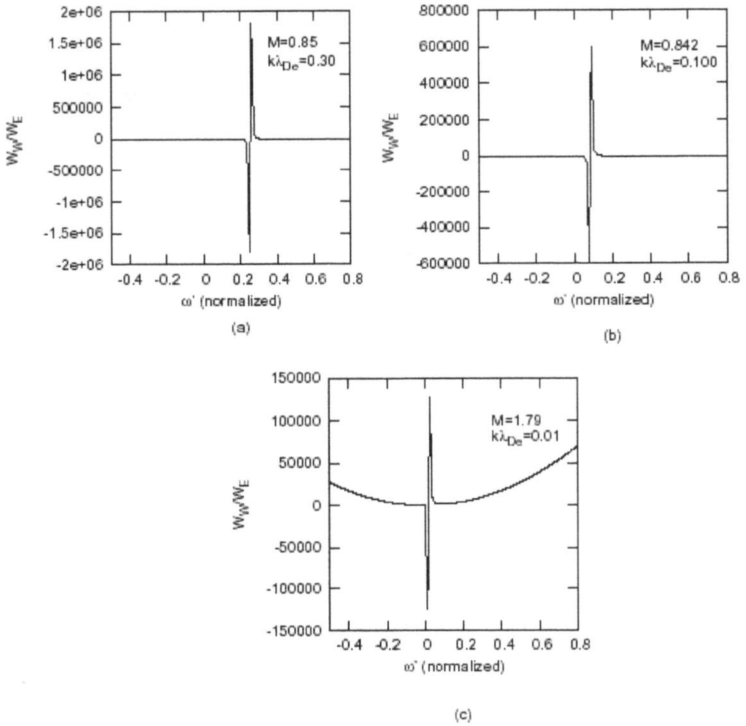

Fig. 4. Explosive nature of the electron inertia-induced ion acoustic wave instability as an outcome of an interplay for the linear resonant mode-mode coupling of positive and negative energy eigen modes. It shows how the normalized wave energy varies with normalized frequency under a set of fixed values of $M$and $k\lambda_{De}$ as (a) $M = 0.85, k\lambda_{De} = 0.30$; (b) $M = 0.842, k\lambda_{De} = 0.100$; and (c) $M = 1.79, k\lambda_{De} = 0.01$

In the present sections of the chapter, many features about the electron inertia-induced ion acoustic wave instability are observed. For example, we physically identify and demonstrate the following features of the instability obtained by theoretical and numerical means of analysis of the desired dispersion relation:

1. The transonic plasma layer is an unstable zone of hydrodynamic equilibrium of quasineutral plasma gas flow motion,
2. The instability is an outcome of the linear resonant mode-mode coupling of positive and negative energy modes,
3. The quenching time of the instability is estimated for some typical values of plasma and wave parameters as mentioned in the previous section. It is found to moderately exceed the ion plasma oscillation time scale, and
4. Lastly, this indicates that in lab frame observation the unstable mode of ion acoustic wave fluctuations at reduced frequencies may look like a purely growing mode. This is very likely to occur for almost entire unstable frequency domain of the frequency transformed ion acoustic waves.

In fact, the electron inertial responses naturally appear only at electron oscillation frequency. However, the *transonic plasma* condition creates a natural physical situation for the same to occur even at the ion acoustic wave frequency of the transformed reduced values. The linear process of resonant mode-mode coupling produces this and makes the coupled system of wave modes unstable.

We have identified and demonstrated the following features of the instability obtained by theoretical and numerical analysis of the dispersion relation: (i) The transonic plasma layer is indeed an unstable zone of hydrodynamic equilibrium of quasi-neutral plasma gas flow motion. (ii) The instability is an outcome of linear resonant mode-mode coupling of positive and negative energy modes. (iii) The normalized values of Doppler-shifted resonant frequencies of the unstable ion acoustic wave fluctuations in ion beam frame come out to be almost equal to 0.5. (iv) The estimated quenching time of the instability exceeds the ion plasma oscillation time scale moderately and hence, (v) In the lab-frame, the unstable modes of ion acoustic wave fluctuations at reduced frequencies may look physically like a purely growing mode.

This is further argued that the physical insights as listed above can be useful as theoretical, graphical and numerical recipes to (1) formulate and solve the problems of saturation mechanisms of the unstable ion acoustic wave fluctuations, (2) formulate and solve the problems of the ion acoustic wave turbulence, and (3) design and setup experiments to study the basic physics of linear and nonlinear ion acoustic wave activities in unique transonic plasma system. These investigations may be useful to improve the existing conceptual framework of physical and mathematical methods of two-scale theory of plasma sheath research to resolve the long-term mystery of the sheath edge singularity. These, in brief, are added to judge the didactic *vis-à-vis* the scientific qualities of the current research work too much specialized in the subject of ion acoustic wave physics.

3.7 Comments

The main conclusive comment here is that the graphical method successfully explains the unstable behavior of the fluid acoustic mode of the ion acoustic wave fluctuations in drifting plasmas with cold ions and hot electrons. A more vivid picture of linear resonant mode-mode coupling of positive and negative energy waves is obtained. This is important to note that simple formulae for wave energy and quenching time calculations [6] are used. This calculation further confirms the earlier results of stability analysis of drifting plasmas against the acoustic wave perturbations [5]. It is, therefore, reasonable to think of logical hypothesis of wave turbulence model approach to solve the sheath edge singularity problem [1, 4]. Actually, the local normal mode theory of the discussed instability implies that the entire transonic plasma zone should be rich in wide frequency range spectrum of the ion acoustic wave fluctuations. This leads to develop the conceptual framework of *situational definition* of the Debye sheath edge to behave as a turbulent zone with finite extension [12]. This hypothetical scenario of the transonic plasma condition can be examined by appropriate experiments of measuring wide range spectral components of the ion acoustic wave fluctuations.

This is a nontrivial problem to explicitly characterize the turbulent properties of the transonic region. The more realistic problem of wave turbulence analysis demands the self-consistent consideration of flow induced quasi-neutral plasma with inhomogeneity in equilibrium plasma background. Similar situations are likely to occur in stellar wind plasmas, where, the transonic behavior is brought about by *deLaval nozzle mechanism* [6-10] of gas flow through a tube of varying cross section. Recent experimental observation [12] in double plasma device (DPD) reports an instability even in a condition of symmetric bipotential ion-rich sheath case. Its frequency falls within zero frequency range and its source is believed to lie in presheath.

Finally, in a nutshell, it is concluded that the graphical method of analyzing the dispersion relation of the inertia-induced instability offers a simple and more informative method of practical importance in transonic plasma equilibrium. Moreover, the plasma environment of Debye sheath edge locality offers a realistic situation for self-excitation of the ion acoustic wave turbulence through resonant ion acoustic wave instability. This is induced by hydrodynamic tailoring of the ion acoustic wave-induced electron density fluctuations. Of course, no experimental observation of instability in transonic plasma has yet been reported to directly compare with the theoretical results. However it cannot be undermined in understanding wave turbulence phenomenon of flowing plasmas. This is informative to add that the frequency and amplitude transformation of the normal ion acoustic wave into unstable ion plasma wave at higher frequency is reported in high intense laser–plasma interaction processes [6-7] through the nonlinear ponderomotive action. This leads to the formation of soliton, double layers, etc. through the saturation mechanism of strong laser-plasma interaction processes due to non-zero average value of the spatially varying electric field associated with laser pulse.

4. Nonlinear normal acoustic mode analyses

4.1 Basic governing equations

A large amount of literature of theoretical and experimental investigations has been produced on the solitary wave propagation in plasmas since the theoretical discovery of ion acoustic soliton [4, 11-12, their references]. Varieties of physical situations of drifting ions of high energy with [5-12] and without [13-33] electron inertial correction have been considered in the ion acoustic wave dynamics. It is shown that the electron inertial motion becomes more important than the ion relativistic effect. Such situations exist in Earth's magnetosphere, stellar atmosphere and in Van Allen radiation belts [3]. Similar studies have been carried out in plasmas with additional ion beam fluid with full electron inertial response in motion [12 and references].

A number of experiments were performed in the unstable condition of beam plasma system in laboratory in order to observe soliton amplification [12]. There are many theoretical calculations and experiments on linear [7-8] and nonlinear [9-11] wave propagation properties of acoustic waves to see their behavior near the transonic point. For an assumed transonic region, it has been theoretically shown that the small amplitude acoustic wave fluctuations exhibit linear resonant growth of relaxation type under the consideration of weak but finite electron inertial delay effect [12-13]. In contrast to earlier claim [3] that the complex nature of coefficients in KdV equation prevents the soliton formation, we argue that their interpretation seems to be physically inappropriate. Instead, by global phase modification technique [12], we show that the usual soliton solution exists (even under the unstable condition), but only for infinitely long wavelength source perturbations. Otherwise, oscillatory shock-like solutions are more likely to exist.

Under fluid approximation, the self-consistently closed set of basic dynamical equations for transonic plasma system with all usual notations in normalized form is given as follows

Electron continuity equation:

$$\frac{\partial \phi}{\partial t}+v_e.\frac{\partial \phi}{\partial x}+\frac{\partial v_e}{\partial x}=0\text{, and} \tag{14}$$

Electron momentum equation:

$$\frac{m_e}{m_i}\left(\frac{\partial v_e}{\partial t}+v_e.\frac{\partial v_e}{\partial x}\right)=\frac{\partial \phi}{\partial x}-\frac{1}{n_e}\frac{\partial n_e}{\partial x}. \tag{15}$$

This is to remind the readers that equation (15) is obtained by substituting zero-order solution of Boltzmann electron density distribution into the normal electron continuity equation. In fact, in the asymptotic limit of $m_e/m_i \rightarrow 0$, electron continuity equation as such is redundant as because the left hand side (electron inertial effect) of (15) is ignorable. Equation (14) basically offers a scope to introduce the weak but finite role of electron to ion inertial mass ratio on the normal mode behavior of acoustic wave.

Ion continuity equation:

$$\frac{\partial n_i}{\partial t}+\frac{\partial}{\partial x}(n_i v_i)=0, \tag{16}$$

Ion momentum equation:

$$\frac{\partial v_i}{\partial t}+v_i\frac{\partial v_i}{\partial x}=-\frac{\partial \phi}{\partial x}\text{, and} \tag{17}$$

Poisson equation:

$$\frac{\partial^2 \phi}{\partial x^2}=n_e-n_i\,. \tag{18}$$

Following form of the derived *d*-KdV equation obtained from the above equations by the standard methodology of reductive perturbation [12] describes the nonlinear ion acoustic wave dynamics under transient limit (~soliton transit time scale) in a new space defined by the stretched coordinates (ξ,τ). This is to mention that $\phi(x,t)=\phi(\xi,\tau)e^{-\gamma\tau}$ and $\gamma\tau \rightarrow 0$ under the transient time action of the propagating ion acoustic soliton through transonic plasma

$$K_0\frac{\partial \phi}{\partial t}+M_0\phi\frac{\partial \phi}{\partial x}+\frac{1}{2}\frac{\partial^3 \phi}{\partial x^3}=\gamma K_0\phi\,. \tag{19}$$

Here the notations K_0 and M_0 termed as *complex response coefficients* [11-12, 26] and the linear resonant growth rate (γ) of the ion acoustic wave with complex Doppler-shifted Mach number $M_D=M_{Dr}+iM_{Di}$ and lab-frame Mach number $M=M_r+iM_i$ in transonic equilibrium appearing in equation (19) are as follows,

$$K_0=\left[A^2+B^2\right]^{1/2}\text{ where,}$$

$$A=\left(\frac{M_r}{\varepsilon_m}+\frac{M^3{}_{Dr}-3M_{Dr}M_i^2}{\left(M_{Dr}^2+M_i^2\right)^3}\right), \text{ and}$$

$$B=\left(\frac{M_i}{\varepsilon_m}+\frac{M_i^3-3M_{Dr}^2M_i}{\left(M_{Dr}^2+M_i^2\right)^3}\right),$$

$$M_0=\left[C^2+D^2\right]^{1/2} \text{ where,}$$

$$C=\frac{1}{2}\left[\frac{3\left\{\left(M_{Dr}^2-M_i^2\right)^2-4M_{Dr}^2M_i^2\right\}}{\left(M_{Dr}^2+M_i^2\right)^4}-\frac{\left(M_r^2-M_i^2\right)^2-4M_r^2M_i^2}{\varepsilon_m^2}-1\right], \text{ and}$$

$$D=-\frac{1}{2}\left\{\frac{12\left(M_{Dr}^2-M_i^2\right)M_{Dr}M_i}{\left(M_{Dr}^2+M_i^2\right)^4}+\frac{4\left(M_r^2-M_i^2\right)M_rM_i}{\varepsilon_m^2}\right\},$$

$$\gamma=\sqrt{2\left(\frac{m_i}{m_e}\right)}k\lambda_{De}\left|\left(1-v_{i0}\right)^{1/2}\right|.$$

The notations are usual and generic as discussed earlier [12]. In the system, plasma ions are self consistently drifting or streaming through a negative neutralizing background of hot electrons having relatively zero inertia. The time response of the electron fluid here is normally ignored. As a result, the unique role of weak but finite electron inertia to destabilize the plasma ion sound wave in transonic plasma equilibrium even within fluid model approach of normal mode description is masked.

4.2 Physical consequences

Now equation (19) after being transformed into an equivalent stationary ODE form by the Galilean transformations is numerically solved as an initial value problem. Some very small simultaneous values of ϕ, $\partial\phi/\partial\xi$ and $\partial^2\phi/\partial\xi^2$ are required for the numerical programme to proceed. A few numerical plots for the desired nonlinear evolutions are shown in Figs. 5-6. This is to note that the calculated amplitudes (as shown in Figs. 5a-6a) are the solutions of the present d-KdV equation (19) with bounded and unbounded phase potraits (as shown in Figs. 5b-6b). Now, the actual amplitudes of the resulting solutions can be deduced by multiplying the numerically obtained values with $\varepsilon\sim\left(k\lambda_{DE}\right)^2\approx10^{-2}$ [12]. In principle, the parameter ε is an arbitrary smallness parameter proportional to the dispersion strength or the amplitude of the weakly dispersive and weakly nonlinear plasma wave.

The unique motivation here is to characterize the possible nonlinear normal mode structure of ion acoustic fluctuations under unstable condition of the ion drifts [8-9,12]. By this very specific example, we show that the complex nature of the coefficients of the derived KdV equation in the unstable zone of transonic plasma doesn't prevent the existence of localized nonlinear solutions including usual soliton solution, too. The concept of global phase modification technique (DPMT) [11-12, 26] results into a *d*-KdV equation [8-9, 12] with variable nonlinear and dispersion coefficients.

Two distinct classes of solutions are obtained: *soliton* and *oscillatory shock-like* structures. Amplification and damping of the driven KdV soliton over the usual KdV soliton is noted for extremely large wavelength (dc) acoustic driving in source term as shown in Figs. 5-6. The amplification near resonance is associated with considerable reduction in nonlinear coefficient than unity as confirmed by numerical calculation. In other cases of shorter acoustic driving in source term as shown in Figs. 5-6, nonlinear solutions of oscillatory shock-like nature are obtained depending on the small deviation from resonant values. It is clearly seen that the peaks of oscillatory shock-like solutions are of either sinusoidal or non-sinusoidal nature with continuous elevation of the initial values of the successive peaks beyond the main nonlinear acoustic peak.

Most of the experimental results in Double Plasma Device (DPD) are reported to show that the obtained theoretical results may have practical relevance to understand the basic physics of ion acoustic wave activities in the transonic region [12] as in Fig. 7. The experiment is performed in a DPD of 90 *cm* in length and 50 *cm* in diameter equipped with multi-dipole magnets for surface plasma confinement [12]. The chamber is divided into source and target by a mesh grid of 85% transparency kept electrically floating. It is evacuated down to a pressure of $(5-6)\times10^{-5}$ Pa with a turbomolecular pump backed by a rotary pump. *Ar*-gas is bled into the system at a pressure $(3-5)\times10^{-2}$ Pa under continuous pumping condition. The source and target plasmas are produced by dc discharge between the tungsten filament of 0.1 *mm* diameter and magnetic cages.

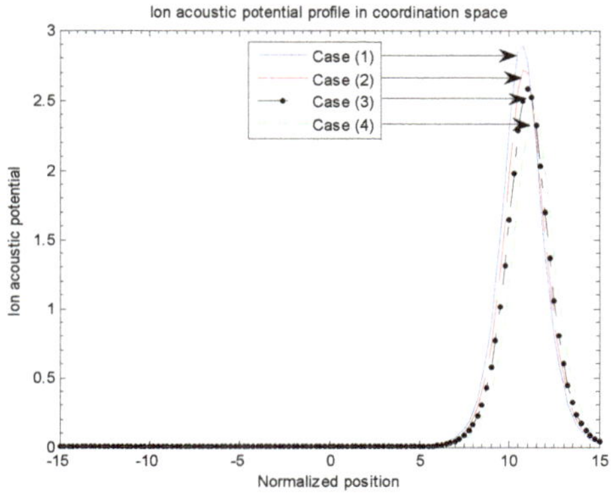

(a)

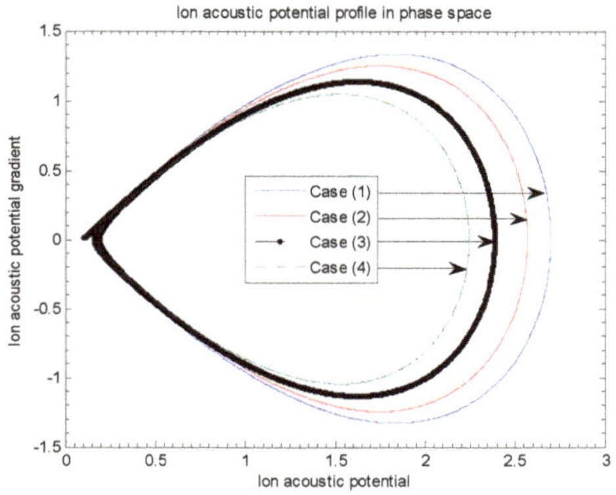

(b)

Fig. 5. Profile of (a) ion acoustic potential (ϕ) with normalized space variable (ξ), and (b) phase space geometry of ion acoustic potential in a phase space described by ϕ and $(\phi)_\xi$ with $k\lambda_{De} = 2.5\times10^{-8}$ (fixed) for Case (1): $\delta = 1.0\times10^{-7}$, Case (2): $\delta = 2.5\times10^{-7}$, Case (3): $\delta = 5.0\times10^{-7}$, and Case (4): $\delta = 7.5\times10^{-7}$

The plasma parameters are measured with the help of a plane Langmuir probe of 5 *mm* diameter and Retarding Potential Analyzer (RPA) of 2.2 *cm* in diameter. The probe and the analyzer are movable axially by a motor driving system so as to take data at any desired position. The plasma parameters are: electron density $n_0 = 10^8 - 10^9 cm^3$, electron temperature $T_e = 1.0 - 1.5eV$ and ion temperature $T_i = 0.1eV$. An ion-acoustic wave is excited with a positive ramp voltage of which the rise time is controllable and is applied to the source anode of the system. Propagating signals are detected by an axially movable Langmuir probe which is biased to $+4V$ with respect to the plasma potential in order to detect the perturbation in the electron current saturation region. The current is then converted into voltage by a resistance of 100Ω and the resultant signals are fed to the oscilloscope. The probe surface is repeatedly cleaned with ion bombardment by applying $-100V$ to it for a short time scale.

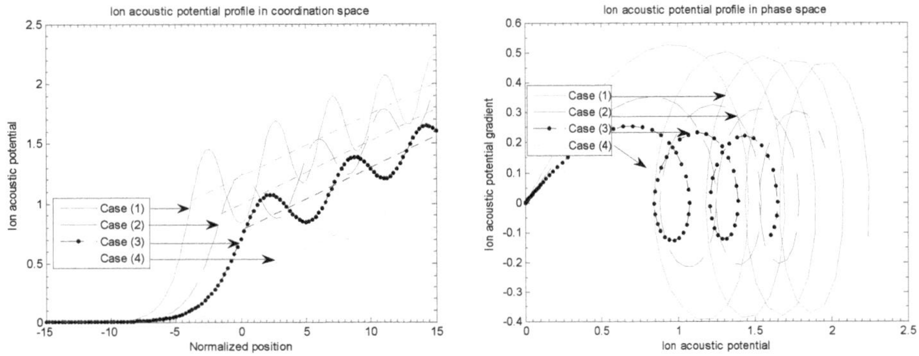

(a) (b)

Fig. 6. Same as Fig. 5 but with $k\lambda_{De} = 1.0 \times 10^{-1}$ (fixed) for Case (1): $\delta = 1.0 \times 10^{-5}$, Case (2): $\delta = 2.0 \times 10^{-5}$, Case (3): $\delta = 3.0 \times 10^{-5}$, and Case (4): $\delta = 5.0 \times 10^{-5}$

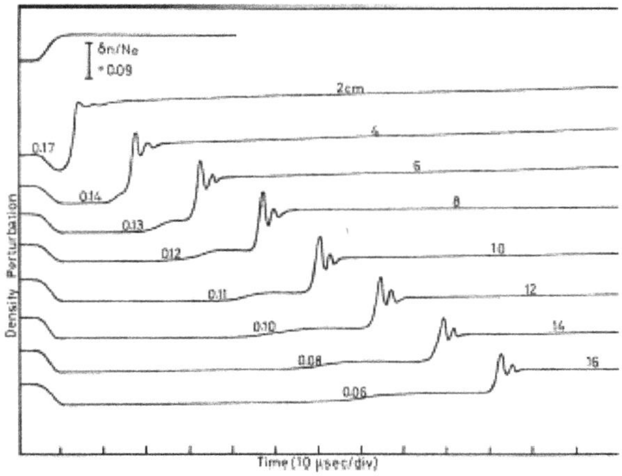

Fig. 7. Experimental profiles of variation of plasma density perturbation (δn) against time (t) at different position of the probe from the grid is shown. Along the x-axis, each division represents $10\mu s$ and along the y-axis, the density perturbation scale is given as $\delta n/n_e = 0.09$

It is also seen that the value of $\delta = 1 - M_{Dr}$ where resonance occurs remains invariant to spectral variation in source term even by orders of $1.0\times10^{-1} - 1.0\times10^{-10}$. The nonlinear and dispersive coefficients exhibit very sensitive role on even slight variation of δ from its resonance. It is noted that as the value of $\delta = 0$, the usual KdV soliton is recovered, irrespective of any wave number value in the source term. The source term plays an effective role only when finite $k\lambda_{De}$ and δ-values are assigned simultaneously

4.3 Comments

As per experimental observations, oscillatory shocks are reported to emerge from the transonic zone in the target plasma as shown in Fig. 7. One can qualitatively argue that as soon as the solitary wave passes through the unstable transonic zone, it may experience the transient phase modifications leading thereby to the formation of oscillatory shock. The observed damping of the oscillatory shock may be correlated to the non-resonant type of dissipation through phase incoherence among ion acoustic spectral components of the usual solitary wave. It seems to be more plausible to argue that the input energy to the usual soliton due to transonic plasma equilibrium may be shared among different spectral components through adiabatic energy exchange processes. This is concluded here that the complex coefficients of the KdV equation should, in principle, not become the criterion for the non-existence of localized nonlinear solutions including usual solilton, too. But the usual soliton solution exists only for infinitely long wavelength source perturbation. This conclusion is derived subject to the validity condition of our arguments of global temporal phase modification of usual soliton amplitude under unstable condition of the plasma medium. The unstable condition of the medium may cause structural deformation of the non-driven KdV solution. Such deformations may result into sinusoidal (linear) or non-sinusoidal (nonlinear) peaks of oscillatory shock-like solution depending on the wavelength of the source perturbations [8-9, 12].

Applying the wave packet model for a moving soliton leaving behind an acoustic tail of dispersive waves known as precursor or acoustic wind (in soliton frame), the asymmetry can be associated with elevation of the bottom potential by a finite dc value superposed with periodic repetition of linear or nonlinear peaks. The amplification or suppression of a single soliton can be possible only for infinitely long wavelength (dc) source. For shorter wavelength source driving, the transition from usual soliton solution to oscillatory shock-like solutions is more likely to occur. It is, in brief, concluded that the present mathematical study of *d*-KdV equation offers a significant contribution of analytical supports to our numerical prediction of structural transformation of the traveling nonlinear ion acoustic waves in transonic plasma equilibrium of desired quality. It clearly shows that the actual solution of *d*-KdV equation is a resultant of linear mixing (superposition) of soliton and shock both.

Dominating features of the individual nonlinear modes is decided by an appropriate choice of the specific values of unstable wave number (or wavelength) for a given value of the ion flow Mach number. It is obvious to note that in zero growth limit of *d*-KdV equation, the shock-term disappears and only soliton remains. This limit is correlated with dc range of the chosen unstable wave number of quite weaker dispersion strength. As the dispersion strength becomes significant to influence the original soliton strength of weak nonlineariy and weak dispersion in the defined transonic plasma of finite extension, structural modification of the usual KdV soliton profile occurs.

We further argue that the linear and nonlinear normal mode behaviors of the ion acoustic waves in transonic plasma condition differ qualitatively from those derived for static and dynamic equilibriums without electron inertial correction. The finite but weak hydrodynamic tailoring of the electron fluid motion on ion acoustic space and time scales brings about this difference. It is then argued that the plasma flows in transonic equilibrium should exhibit rich spectrum of linear and nonlinear ion acoustic waves and oscillations. Of course, under Vlasov model the hot electrons with streaming velocity comparable to the phase speed of the ion sound wave may destabilize the ion sound mode through wave-particle resonance effect [8 and references] too. However, our excitation mechanism of ion sound wave differs from the other known mechanisms [8] to excite the same ion sound wave on many grounds [8]. This kind of theoretical scenario of *transonic plasmas* offers a unique scope of *acoustic spectroscopy* to describe the internal state of transonic equilibrium of plasma flows [28].

These calculations have potential applications [26] extensively to understand plasma acoustic dynamics in colloidal plasmas too, but under transonic equilibrium configuration. A generalized statement thereby is reported that all possible sound modes in multi-species colloidal plasmas with drift motions (of inertial ionic species) could be destabilized by the inertial delay effect of the corresponding plasma thermal species that carry out thermal screening of acoustic potential developed due to respective inertial ionic species. Of course, threshold values may differ depending on the choice of the plasma sound mode under consideration. In technological application point of view, one may argue that the proposed theoretical model for inertia-induced acoustic instability mechanism may be utilized to make a plasma-based micro device for *acoustic amplifier* [26]. The amplified acoustic signals (developed due to respective inertial ionic species) from the amplifier could be detected, received and analyzed for the diagnosis and characterization of hydrodynamic flow of plasmas with embedded inertial dust contaminations. These analyses may have potential applications in different ion acoustic wave turbulence-related situations like aerodynamics, solar wind and space plasmas, fusion plasmas, industrial plasmas and plasma flows in astrophysical context, etc.

5. Astrophysical normal acoustics

A plasma-based Gravito-Electorstatic Sheath (GES) model is proposed to discuss the fundamental issues of the solar interior plasma (SIP) and solar wind plasma (SWP). Basic concepts of plasma-wall interaction physics are invoked. Here the wall is defined by a continuous variation of gravity associated with the SIP mass. The neutral gas approximation of the inertially confined SIP is relaxed, and as such the scope of quasi-neutral plasma sheath formation is allowed to arise near the self-consistently defined solar surface boundary (SSB). Analytical and numerical results are obtained to define the SSB and discuss the physics of the surface properties of the Sun, and hence, those of the SWP.

5.1 Physical model description

The SIP system can be idealized as a self-gravitationally bounded quasi-neutral plasma with a spherically symmetric surface boundary of nonrigid and nonphysical nature. The self-gravitational potential barrier of the solar plasma mass distribution acts as an enclosure to confine this quasi-neutral plasma. An estimated typical value $\sim 10^{-20}$ of the ratio of the solar plasma Debye length and Jeans length of the total solar mass justifies the quasi-neutral behavior of the solar plasma on both the bounded and unbounded scales. Here the zeroth-order boundary surface can be defined by the exact hydrostatic condition of gravito-electrostatic force balancing of the enclosed plasma mass at some arbitrary radial position from the center of the mean solar gravitational mass. With this much background in mind, let us now formulate the problem of the physical and mathematical descriptions GES formation around the SSB. For simplicity, we consider spherical symmetry of the inertially confined SIP mass, which helps to reduce the three dimensional problem of describing the GES into a simplified one dimensional problem in the radial direction. Thus, only a single radial degree of freedom is required for description of the dynamical behavior of the SWP under the assumed spherically symmetric self-gravitating solar plasma mass distribution.

The idea of the GES formation can be appreciated with quantitative estimates of the gravito-thermal coupling constants for the SIP electrons and hydrogen ions. Henceforth, "ions" and "hydrogen ions" will be used in the sense of the same ionic species. These parameters [10] can be defined and estimated as follows: The gravitothermal coupling constant for electrons can be estimated as $\Gamma_e = k_B T_e / m_e g_\Theta R_\Theta \approx 10$, for a mean electron temperature of $T_e \sim 10^5$ K and as $\Gamma_e \approx 800$ for mean $T_e = 10^6$ K. The notation k_B (=1.3806×10^{-23} JK^{-1}) denotes the Boltzmann constant. Similarly, the gravito-thermal coupling constant for ions can be estimated as $\Gamma_i = (T_i m_e / T_e m_i)\Gamma_e << 1$ for mean $T_e \sim 10^5$ K, and $\Gamma_i \approx 1$ for mean $T_e \sim 10^6 K$. Here $g_\Theta = GM_\Theta / R_\Theta^2$ denotes the value of the solar surface gravity. The values of the other constant quantities are taken to be $G = 6.6726\times10^{-8}$ dyn cm^2 g^{-2}, $M_\Theta = 1.90\times10^{33}$ g, and $R_\Theta = 6.97\times10^{10}$ cm.

These estimates are based on the condition of an isothermal SIP, where T_e and T_i respectively denote the electron and ion temperatures. It is now easy to see that the electrons can very well overcome the gravitational potential barrier at the SSB in the standard solar model, whereas the ions cannot. This is the reason why a surface polarization-induced space charge (electrostatic) field is likely to appear, due to thermal leakage of the electrons from the SSB in the radially outward direction. Moreover, the neutral gas approximation for the SIP may not be a good one for describing the properties of the SSB. Similar realizations have already occurred to previous authors [5, 9, 11, 14] for the SWP as well. We take the SIP to be an ideal nonisothermal plasma gas with relatively cold ions. The mean electron temperature $T_e > 10^6$ K for the SIP emerges as a more suitable choice for our theoretical consideration.

According to our GES-model analyses, the GES divides into two scales: one bounded, and the other unbounded. The former includes the steady state equilibrium description of the SIP dynamics bounded by the solar self-gravity. This extends from the solar center to the self-consistently defined and specified SSB. The unbounded scale encompasses the SWP dynamics extending from the SSB to infinity. The SIP electrons can easily escape from the defined SSB. On the other hand, the SIP ions cannot cross the gravitational potential barrier of the solar mass on their thermal energy alone. However, surface leakage of the SIP electrons is bound to produce an electrostatic field by virtue of surface charge polarization. This, in turn, provides an additional source to act on the SIP ions to further energize and encourage them cross over the solar self-gravitational potential barrier.

5.2 Basic governing equations

In order to describe the plasma-based GES physics of our model system, we adopt a collisionless unmagnetized plasma fluid for simplicity in mathematical development to obtain some physical insight into the solar wind physics. The role of magnetic field is also ignored (just for mathematical simplicity) in discussing the collisionless SIP and SWP dynamics. Applying the spherical capacitor charging model [3], the coulomb charge on the SSB comes out to be $Q_{SSB} \sim 120C$. The mean rotational frequency of the SSB about the centre of the SIP system is is determined to be $f_{SSB} \sim 1.59\times10^{-12}\,Hz$ [42]. Applying the electrical model [42] of the Sun, the mean value of the strength of the solar magnetic field at the SSB in our model analysis is estimated as $\langle|B_{SSB}|\rangle = 4\pi^2 Q_{SSB}\, f_{SSB} \sim 7.53\times10^{-11}\,T$, which is negligibly small for producing any significant effects on the dynamics of the solar plasma particles. Thus the effects of the magnetic field are not realized by the solar plasma particles due to the weak Lorentz force, which is now estimated to be $F_L = e(v\times B) \approx 3.61\times10^{-33}\,N$ corresponding to a subsonic flow speed $v \sim 3.00\,cm\,s^{-1}$ with the input data available [2, 42] with us and hence, neglected. Therefore our unmagnetized plasma approximation is well justified in our model configuration. In addition, the effects of solar rotation, viscosity, non-thermal energy. For further simplification, the electrons are assumed to obey a Maxwellian velocity distribution. Although these approximations may not be realistic, but they may be considered working hypotheses to begin with an ideal situation. Deviations indeed exist from a Maxwellian velocity distribution. We however use it as a working hypothesis for our model considerations. As a result, the usual form of the Boltzmann density distribution for plasma thermal electrons with all usual notations is given as

$$N_e = e^{\theta}. \tag{20}$$

Here $N_e = n_e/n_0$ denotes the normalized electron density. The generic notation $\theta = e\phi/T_e$ denotes the normalized value of the plasma potential associated with the GES on the bounded scale and with the SWP on unbounded scale. The general notation n_e stands for the nonnormalized electron density and $n_0 = \rho_\Theta/m_i$ defines the average bulk density of the equilibrium SIP. The notation $\rho_\Theta = 1.43$ g cm^{-3} stands for the average but constant solar plasma mass density and $m_i = 1.67\times10^{-24}$ g for the ionic (protonic) mass. Again e represents the electronic charge unit and ϕ, the nonnormalized plasma potential associated with both the GES and SWP.

The hydrogen ions are described by their full inertial response dynamics. This includes the ion momentum equation as well as the ion continuity equation. The first describes the change in ion momentum under the action of central gravito-electrostatic fields of potential gradient and forces induced by thermal gas pressure gradients. The latter equation is considered a gas dynamic analog of plasma flowing through a spherical chamber of radially varying surface area. In normalized forms, the ion momentum equation is

$$M\frac{dM}{d\xi} = -\frac{d\theta}{d\xi} - \varepsilon_T \frac{1}{N_i}\frac{dN_i}{d\xi} - \frac{d\eta}{d\xi}. \tag{21}$$

Here the minus sign in the gravitational potential term indicates the radially inward direction of the solar self-gravity. The deviation from the conventional neutral gas treatment of the SIP is introduced through the electric space charge-induced force (first term on right-hand side) effect. The normalized expression for conservation of ion flux density is

$$\frac{1}{N_i}\frac{dN_i}{d\xi} + \frac{1}{M}\frac{dM}{d\xi} + \frac{2}{\xi} = 0. \tag{22}$$

The normalizations are defined as follows:

$$\theta = \frac{e\phi}{T_e},\ \eta = \frac{\psi}{C_s^2},\ N_e = \frac{n_e}{n_0},\ N_i = \frac{n_i}{n_0},$$

$$M = \frac{v_i}{c_s},\ \xi = \frac{r}{\lambda_J},\ \lambda_J = \frac{c_s}{\omega_J},\ c_s = \left(\frac{T_e}{m_i}\right)^{1/2},$$

$$\omega_J = \left(4\pi\rho_\odot G\right)^{1/2},\ \varepsilon_T = \frac{T_i}{T_e}.$$

The notations ϕ and ψ respectively stand for the dimensional (*unnormalized*) values of the plasma electrostatic potential and the self-solar gravitational potential as variables associated with the GES. The dimensional values of the electron and ion population density variables are respectively denoted by n_e and n_i. Likewise, the dimensional ion fluid velocity variable is represented by the symbol v_i. The notation η stands for the normalized variable of the self-solar gravitation potential. The notation N_i denotes the normalized value of the ion particle population density variable. Notation M stands for the ion flow Mach number.

The notations r and ξ stand for the nonnormalized and normalized radial distance respectively from the heliocenter in spherical co-ordinates. The other notations λ_J, c_s and ω_J defined as above stand for the Jeans length, sound speed and Jeans frequency respectively. Finally, the notation ε_T as defined above stands for the ratio of ion to electron temperature. The ion flux density conservation (eq. 22) contains a term that includes the effect of geometry on the ion flow dynamics of the SIP mass, self-gravitationally confined in a spherical region, whose size is to be determined from our own model calculations. Equations (21) and (22) can be combined to yield a single expression representing the well-known steady state hydrodynamic flow,

$$\left(M^2 - \varepsilon_T\right)\frac{1}{M}\frac{dM}{d\xi} = -\frac{d\theta}{d\xi} + \varepsilon_T\frac{2}{\xi} - \frac{d\eta}{d\xi}. \tag{23}$$

There is an obvious difference in the above equation from the corresponding momentum equation under the neutral gas approximation for the SIP. The difference appears, as discussed above, in the form of a space charge effect originating from the Coulomb force on a collective scale (first term on the right-hand side of eq. (21)).

The gravito-electrostatic Poisson equations complement the steady dynamical equation (23) for a complete description of the gravito-electrostatic sheath structure, which is formed inside the non-rigid SSB. This is important to emphasize that in the case of a real physical boundary, the plasma sheath is always formed both inside and outside the boundary surface in its close vicinity [12]. The normalized forms of the gravitational and electrostatic Poisson equations for the SWP description are respectively given by

$$\frac{d^2\eta}{d\xi^2} + \frac{2}{\xi}\frac{d\eta}{d\xi} = N_i, \text{ and} \tag{24}$$

$$\left(\frac{\lambda_{De}}{\lambda_J}\right)^2\left[\frac{d^2\theta}{d\xi^2}+\frac{2}{\xi}\frac{d\theta}{d\xi}\right]=N_e-N_i\,. \tag{25}$$

Here $\lambda_{De}=\left(T_e/4\pi n_0 e^2\right)^{1/2}$ denotes the plasma electron Debye length of the defined SIP system. The other quantities are as defined above as usual. Equations (21)-(25) constitute a completely closed set of basic governing equations with which to discuss the basic physics of the GES-potential distribution on the bounded scale. Of course, the discussion also includes the associated ambipolar radial flow variation of the SIP towards an unknown SSB which we have to determine self-consistently in this problem with GES-based theory. For a typical value $T_e=10^6$ K, one can estimate that $\lambda_{De}/\lambda_J\approx 10^{-20}$ which implies that the Debye length is quite a bit smaller than the Jeans scale length of the solar plasma mass. Thus, on the typical gravitational scale length of the inertially bounded plasma, the limit $\lambda_{De}/\lambda_J\to 0$ represents a realistic (physical) approximation. By virtue of this limiting condition, the entire SIP extending up to the solar boundary and beyond obeys the plasma approximation. Thus, the quasi-neutrality condition as given below holds good

$$N_e=N_i=N=e^{\theta}. \tag{26}$$

This is to note that equation (26) does not mean that the plasma ions are Boltzmannian in thermal character, but inertial species. Equation (26) can be differentiated once in space and further rewritten as,

$$\frac{1}{N}\frac{dN}{d\xi}=\frac{d\theta}{d\xi}. \tag{27}$$

By virtue of the plasma approximation, one can justify that the GES of the SIP origin should behave as a quasi-neutral space charge sheath on the Jeans scale size order. The formation mechanism of the defined GES, however, is the same as in the case of plasma-wall interaction process in laboratory confined plasmas. From equations (26)-(27), it is clear that for the electrostatic potential and its gradient being negative, causes the exponential decrease of the plasma density. Finally, the reduced form of the basic set of autonomous closed system of coupled nonlinear dynamical evolution equations under quasi-neutral plasma approximation is enlisted as follows

$$\left(M^2-\varepsilon_T\right)\frac{1}{M}\frac{dM}{d\xi}=-\frac{d\theta}{d\xi}+\varepsilon_T\frac{2}{\xi}-\frac{d\eta}{d\xi}, \tag{28}$$

$$\frac{d\theta}{d\xi}+\frac{1}{M}\frac{dM}{d\xi}+\frac{2}{\xi}=0,\ \text{and} \tag{29}$$

$$\frac{d^2\eta}{d\xi^2}+\frac{2}{\xi}\frac{d\eta}{d\xi}=e^{\theta}. \tag{30}$$

This set of differential evolution equations constitutes a closed dynamical system of governing hydrodynamic equations that will be used to determine the existence of a bounded GES structure on the order of the Jeans scale length (λ_J) in our GES-model of the subsonic origin of the SWP of current interest. Thus the solar parameters $M(\xi)$, $g_s(\xi)$ and $\theta(\xi)$ representing the equilibrium Mach number, solar self-gravity and electrostatic potential, respectively, will characterize the gravito-electrostatic acoustics in our approach.

5.3 Theoretical analysis of solar surface boundary

5.3.1 Analytical calculations

We first wish to specify the overall condition for the existence of the SSB. Such existence demands the possibility of a self-consistent bounded solution for the solar self-gravity. The boundary will correspond to a maximum value of the solar self-gravity at some radial distance from the heliocenter. This defines a self-consistent location of the SSB. Before we proceed further, let us argue that the radially outward pulling bulk force effect of the GES-associated potential term in equation (28) demands a negative electrostatic potential gradient, that is, $d\theta/d\xi < 0$. This makes some physical sense because the ion fluid has to overcome the gravitational barrier to create a global-scale flow of the SIP in a quasi-hydrostatic way.

Now, if we invoke the concept of exact hydrostatic formation under gravito-electrostatic force balancing $(d\theta/d\xi \approx d\eta/d\xi)$, the surface potential can be solved to get $\theta - \theta_\Theta \approx \eta - \eta_\Theta$. Here the unknown boundary values of $\theta = \theta_\Theta$, $\eta = \eta_\Theta$ and $M = M_{SSB}$ are to be self-consistently specified numerically. The notation (M_{SSB}) stands for the Mach value associated with the SIP flow at the SSB. Now, by the exact hydrostatic equilibrium condition in the set of equations (28)-(30), one can get the following set of equations for the SSB description:

$$\left(M^2 - \varepsilon_T\right)\frac{1}{M}\frac{dM}{d\xi} = \varepsilon_T \frac{2}{\xi}, \tag{31}$$

$$-\frac{d\eta}{d\xi} + \frac{1}{M}\frac{dM}{d\xi} + \frac{2}{\xi} = 0\text{ , and} \tag{32}$$

$$\frac{d^2\eta}{d\xi^2} + \frac{2}{\xi}\frac{d\eta}{d\xi} = e^\theta . \tag{33}$$

For purpose of the GES analysis, we define the solar self-gravitational acceleration as $g_s = d\eta/d\xi$. Equation (34) thus reads

$$\frac{dg_s}{d\xi} + \frac{2}{\xi} g_s = e^\theta . \tag{34}$$

Finally, the SIP and hence, the SSB are described and specified in terms of the relevant solar plasma parameters $M(\xi)$, $g_s(\xi)$ and $\theta(\xi)$ representing respectively the equilibrium Mach number, solar self-gravity and electrostatic potential as a coupled dynamical system of the closed set of equations recast as the following

Solar self-gravity equation:

$$\frac{dg_s}{d\xi} + \frac{2}{\xi} g_s = e^\theta , \tag{34a}$$

Ion continuity equation:

$$\frac{d\theta}{d\xi}+\frac{1}{M}\frac{dM}{d\xi}+\frac{2}{\xi}=0, \text{ and} \tag{34b}$$

Ion momentum equation:

$$\left(M^2-\alpha\right)\frac{1}{M}\frac{dM}{d\xi}=\alpha\frac{2}{\xi}-g_s, \tag{34c}$$

where $\alpha=1+\in_T=1+\left(T_i/T_e\right)$, T_e is the electron temperature and T_i is the inertial ion temperature for the bounded solar plasma on the SIP-scale as already mentioned.

Let us now denote the maximum value (g_Θ) of solar gravity at some radial position $\xi=\xi_\Theta$ where $\theta=\theta_\Theta$ and apply the necessary condition for the maximization of g_s at a spatial coordinate $\xi=\xi_\Theta$. This condition $\left(dg_s/d\xi\right)_{\xi=\xi_\Theta}=0$ when used in equation (34) yields $\xi_\Theta=2g_\Theta e^{-\theta_\Theta}$. However, it is not sufficient to justify the occurrence of the maximum value of g_s until and unless the second derivative of g_s is shown to have negative value. To derive the sufficient condition for the maximum value of g_s at $\xi=\xi_\Theta$, let us once spatially differentiate equation (34) to yield

$$\frac{d^2g_s}{d\xi^2}-\frac{2}{\xi^2}g_s+\frac{2}{\xi}\frac{dg_s}{d\xi}=e^\theta\frac{d\theta}{d\xi}. \tag{35}$$

Now the condition for the maximization of g_s at the location $\xi=\xi_\Theta$ can be discussed by considering $d\theta/d\xi<0$ in equation (35) under the exact hydrostatic equilibrium approximation $\left(|d\theta/d\xi|\approx|d\eta/d\xi|=g_\Theta\right)$ near the solar surface to yield the following inequality

$$\left.\frac{d^2g_s}{d\xi^2}\right|_{\xi=\xi_\Theta}=\frac{2}{\xi_\Theta^2}g_\Theta-e^{\theta_\Theta}g_\Theta=g_\Theta\left(\frac{2}{\xi_\Theta^2}-e^{\theta_\Theta}\right)<0. \tag{36}$$

From these analytical arguments one can infer that the maximization of g_s indeed occurs at some arbitrary radial position that satisfies the inequality: $\xi_\Theta>\sqrt{2}e^{-\theta_\Theta/2}(=2.33)$ for $\theta_\Theta\sim-1$ (Figs. 8b, 9b, and 10b). Numerically the location of the SSB is found to lie at $\xi_\Theta\sim3.5$ that matches with $\xi_\Theta=2g_\Theta e^{-\theta_\Theta}$ for $\theta_\Theta=1.07$ and $g_\Theta=0.6$. It satisfies the analytically derived inequality (36) too. Now the other two equations (32)-(33) can be simultaneously satisfied in the SSB only for a subsonic solar plasma ion flow speed if Mach number gradient acquires some appropriate negative minimum near zero $\left(M\sim|(dM/d\xi)|\sim10^{-6}\right)$.

It is indeed seen numerically that near the maximum solar self-gravity of the SIP mass, the first and third terms in equation (32) are almost equal and hence the Mach number gradient term, which is negative in the close vicinity of the SSB, should be smaller than the other two terms so as to satisfy equations (31) and (32), simultaneously. Actually, the three equations (31)-(32) and (34) are solved numerically to describe the SSB of the maximum self-gravitational potential barrier properly where g_s associated with the self SIP mass is maximized.

5.3.2 Numerical calculations

Determination of the autonomous set of the initial values of the defined physical variables is a prerequisite to solve the nonlinear dynamical evolution equations (34a)-(34c) in general as an initial value problem. The initial values of the physical variables like $M(\xi), g_s(\xi)$ and $\theta(\xi)$ are defined inside the solar interior and are determined on the basis of extreme condition of the *nonlinear stability analysis* [4]. The self-consistent choice of the initial values is obtained by putting $dM/d\xi|_{\xi_i} = -e^{\theta_i/2}$, $dg_s/d\xi|_{\xi_i} = 0$ and $d\theta/d\xi|_{\xi_i} = 0$ in these three equations (34a)-(34c). But the realistic SWP model demands that $d\theta/d\xi|_{\xi_i} \neq 0$. Finally, we determine the expressions for a physically valid set of the initial values of the given physical variables as follows,

$$M_i = \frac{1}{2}\xi_i e^{\theta_i/2} \tag{37}$$

$$g_{si} = \frac{1}{2}\xi_i e^{\theta_i} \tag{38}$$

This is to note that the initial values of θ_i and ξ_i are chosen arbitrarily. As discussed later, we find that the SSB acquires a negative potential bias ($\theta_s \sim -1$) of about -1 kV. It also acquires the maximum value of solar interior gravity ($g_\Theta \sim 0.6$) and minimum value of non-zero SIP flow speed ($M_{SSB} \sim 10^{-7}$) at the SSB. The value of the electrostatic potential gradient at the surface comes out to be ~ -0.6. This means that the strength of the GES-associated solar surface gravity and electrostatic potential gradient is almost equal. As a result the SSB is defined by some constant values of the physical variables (g_s, θ, M). The SSB values of these parameters are determined through spatial evolution of the coupled system of equations (34a)-(34c) from the given initial values (37)-(38) inside the SIP zone.

We have used the well-known fourth-order Runge-Kutta method (RK 4) for numerical solutions of equations (34a)-(34c). By numerical analysis (Figs 8-11), we find that the solar radius is equal to thrice of the Jeans length (λ_J) for mean solar mass density $(\rho_\Theta = 1.41 g.cm^{-3})$. From this observation one can easily estimate that $\lambda_J = R_\Theta / 3.5$. Now comparing our own theoretical value of the solar mass self-gravity with that of the standard value, we arrive at the following relationship between the solar plasma sound speed (c_s) and the Jeans length (λ_J)

$$0.6\left(c_s^2/\lambda_J\right) = 2.74\times10^4 \ cm/sec^2. \tag{39}$$

By substituting the value of the Jeans length expressed in terms of the solar radius, one can determine and specify the mean value of the electron temperature, which is $T_e \sim 10^7$ K. The sound speed in the SIP under the cold ion model approximation is thus obtained as $c_s \sim 3\times10^7$ cm s^{-1}. Note that the SWP speed at 1 AU is fixed by the sound speed of the SWP medium, which is determined and specified by the requirement that a transonic transition solution occurs on the unbounded scale of the SWP dynamics description.

The gravito-acoustic coupling coefficient could be estimated as $\Gamma_{g-a} = m_i g_\Theta R_\Theta / T_e \sim 2.0$. In absence of the gravitational force, the bulk SIP ion fluid will acquire the flow speed corresponding to $M \sim 1.41$ for a negative potential drop of the order of $(-T_e/e)$ over a distance equal to that of the solar radius. If one estimates the value of gravito-acoustic coupling coefficient at this velocity defined by $M \sim 1.41$, it comes out to be unity. Thus the quasi-hydrostatic type of equilibrium gravitational surface confinement of the SIP is ensured. The GES-potential induced outward flow of the SIP is also ensured. Due to comparable strength of the solar surface gravitational effect of deceleration, the net SIP ion fluid flow is highly suppressed to some minimum value (~1.0-3.0 *cm/sec*) corresponding to $M \sim 10^{-7}$ at the SSB.

An interesting point to note here is that near the defined SSB, the electrostatic potential gradient terminates into an almost linear type of profile. The value of its gradient value will provide an estimate of the second order derivative's contribution into the electrostatic potential which measures the level of local charge imbalance near the solar surface. From our computational plots (Figs. 8b, 9b and 10b), this local charge imbalance comes out to be of the order -0.17, which is equivalent to 17% ion excess charge distributed over a region of size on the order of the plasma Debye sheath scale length. However, the same level of the electrostatic local charge imbalance on the Jeans scale length does not require the inclusion of the role of the Poisson term for the evolution of the electrostatic potential's profile under the GES-model. Hence, in this sense the GES is practically equivalent to a quasi-neutral plasma sheath with its potential profile tailored and shaped by the potential barrier of the self-gravity of the SIP mass distribution.

5.3.3 Properties of solar surface boundary

Table I lists the defined initial values of the physical variables (g_s, θ, M) as already discussed and their corresponding boundary values numerically obtained for the description of the desired SSB. The initial values of g_s, θ, and M are associated with the normalized mean SIP mass density, enclosed within a tiny spherical globule having normalized radius equal to an arbitrarily chosen value of ξ_i.

Parameter	At the Initial Radial Point (ξ_i)	At the Solar Surface Boundary (ξ_Θ)	Initial Values
Potential θ	$\frac{d\theta}{d\xi}=0$	$\frac{d\theta}{d\xi}\sim -0.62$, $\theta_\Theta \sim -1.00$	θ_i, arbitrarily chosen
Gravity g_s	$\frac{dg_s}{d\xi}=0$	$\frac{dg_s}{d\xi}=0$, $g_\Theta \sim 0.60$	$g_{si}=\frac{1}{2}\xi_i e^{\theta_i}$, derived
Mach number M	$\frac{dM}{d\xi}=-e^{\theta_i/2}$	$\frac{dM}{d\xi}=0$, $M_{SWP}\sim 10^{-7}$	$M_i=\frac{1}{2}\xi_i e^{\theta_i/2}$, derived

Table 1. Initial and Boundary Values of Relevant Solar Parameters

From the numerical plots shown in Figs. 8-10, we find that the minimum Mach number (M_{SSB}) at the specifically defined SSB comes out to be of order 10^{-7}. For this value of Mach number, equation (31) can be simplified to show that near the boundary, $dM/d\xi \approx -M/\xi_\Theta = -3\times10^{-8} \sim 0$. This corresponds to a quasi-hydrostatic type of the SSB equilibrium. It arises from gravito-electrostatic balancing with an outward flow of the SIP having a minimum speed of about 1-3 cm s^{-1}. With these inferences one can argue that the

SWP originates by virtue of the interaction of the SIP with the SSB. Hence an interconnection between the Sun and the SWP can be observed by applying the GES model. Here the boundary is not sharp but distributed over the entire region of the solar interior volume. The basic principles of the GES coupling govern the solar surface emission process of the subsonic SWP.

As depicted in table I, the time-independent solar g_s-profile associated with the SIP mass distribution terminates into a diffuse surface boundary. This is characterized and defined by the quasi-hydrostatic equilibrium $g_s = g_{\Theta} \sim |d\theta/d\xi|$, which occurs at about $\xi = \xi_{\Theta} \sim 3.5$ (see Figs. 8-10). As such, the basic physics of the subsonic origin of the SWP from the SSB is correlated with the bulk SIP dynamics. We note that the precise definition of the SSB influences the SWP velocity at 1 AU. Other models report similar observations too [3, 31-41]. The dependence on the ion to electron temperature ratio is quite visible in Fig. 11a. Let us now discuss the numerical results in the figures individually.

Figure 8 depicts the time-independent profiles of (g_s, θ, M) and their variations with the ion-to-electron temperature ratio ε_t for fixed values of the initial point (ξ_i=0.01) and plasma sheath potential ($\theta_i = -0.001$). As shown in Fig. 8a, the location of the SSB remains the same but its maximum value changes, and a most suitable choice of ε_t = 0.4 is identified for which the quasi-hydrostatic condition is fulfilled. The *E*-field profile is invariant for all chosen values ε_t =0-0.5. Again, as shown in Fig. 8b, the electrostatic potential corresponding to $\varepsilon_t \sim$ 0.4 comes out to be $\theta = \theta_{\Theta} \sim -1$ (i.e. ~1 kV). Similarly, Fig. 8c depicts the minimum Mach value of $M_{SSB} \sim 10^{-7}$ for ε_t ~0.4 varying by a factor of 2 for other values of ε_t.

Figure 9 depicts the time-independent profiles of g_s, θ, and M and their variations with initial position for fixed values of ε_t =0-0.4 and $\theta_i = -0.001$. It can be seen that the most suitable choice of the initial position for our fixed values θ_i and ε_T comes out to be $\xi_i \sim 0.01$, which is consistent with the earlier value shown in Fig. 8a. Moreover, the minimum value of $M \sim 10^{-7}$ (Fig. 9c) is also consistent with the earlier value shown in Fig. 8c.

Figure 10 depicts the time-independent profiles of g_s, θ, and M and their variations with electrostatic potential for fixed values of ξ_i=0.01 and ε_T =0.4. It is observed fascinatingly that the most suitable choice of the initial value of the normalized electrostatic potential for our fixed values of ξ_i and ε_T comes out to be $\theta_i = -0.001$.

It is notable that high initial drop of *M*-profile occurs as shown in Fig.8c, Fig. 9c and Fig. 10c. This indicates the over dominance of the solar interior gravity up to about $\xi \sim 1.5$, and thereafter, the *E*-field becomes comparable, balancing at about $\xi \sim 3.5$. Thus the normalized width of the *gravito-electrostatic sheath* could be estimated and denoted by $\xi_{G-E} \sim 2$. This is a quasi-neutral space charge region with positive charge (ion) excess near the defined SSB wall. Thus a self-consistent bounded solution of nonlinearly coupled gravito-electrostatic potential profiles exists. It forms a quasi-hydrostatic equilibrium at the SSB for the choice of the appropriate set of the initial parameter values $\theta_i = -0.001$ & $\xi_i = 0.01$ for $\varepsilon_T = 0.4$. This is not a rigid boundary at all. As a result the SSB is capable to exhibit many kinds of global oscillation dynamics governed by the nonlinear coupling of the gravito-electrostatic sheath potentials.

For a laboratory hydrogen plasma, the normalized floating potential can be estimated as $\theta_f = -3.76$ under the flat surface approximation. Now, if we consider the numerically calculated minimum value of M for our solar surface characterization, the estimated value of

θ_f is about -20 using the flat surface approximation. This is a crude estimate because the solar surface potential drop occurs over a distance of the order of the Jeans scale where the effect of curvature should not be ignored. The numerically obtained solar surface potential is quite a bit smaller than the floating potential. Simply put, this means that the defined SSB of the GES-model draws a finite amount of electron-dominated electric current that flows toward the heliocenter.

Let us invoke a generalized concept of the plasma sheath, which is traditionally associated with a localized electrostatic potential only in the plasma physics community. We argue that any localized nonneutral space charge layer (in our case, on the order of the Jeans length) is the result of a self-consistent nonlinear coupling of gravito-electrostatic force field variations. This is what we mean by the GES, which of course, obeys the global quasi-neutrality condition because of the smallness of the ratio λ_{De}/λ_J for the SIP parameters.

5.4 Acceleration of solar interior plasma

5.4.1 Basic equations for SWP descriptions

We have already argued that the subsonic origin of the SWP from the SSB is an outcome of the condition of quasi-hydrostatic equilibrium at the boundary. This is a result of the comparable, but competiting strengths of the gravitational deceleration and the electrostatic acceleration of the SIP near the SSB. Now we will try to look at the problem of solar wind acceleration from subsonic to supersonic speed. This is referred to as the *transonic transition* behavior of the outward-moving SIP in the form of the SWP. Let us now argue that the radial variation of Mach number and electrostatic potential beyond the defined SSB should be described by the following autonomous set of coupled nonlinear differential equations

$$\left(M^2-\varepsilon_T\right)\frac{1}{M}\frac{dM}{d\xi}=-\frac{d\theta}{d\xi}+\varepsilon_T\frac{2}{\xi}-\frac{1}{\xi^2}\frac{GM_\Theta}{C_s^2\lambda_J}\text{, and} \tag{40}$$

$$\frac{d\theta}{d\xi}+\frac{1}{M}\frac{dM}{d\xi}+\frac{2}{\xi}=0\ . \tag{41}$$

Let us note that the constant SIP mass acts as an external object to offer a source of gravity for tailoring and monitoring the outgoing SIP flow with the initially subsonic speed specified at the defined SSB. The Poisson equation for gravity is now redundant. It is important to comment that the electrostatic force field is not imposed from outside to control the solar wind's motion. In fact, the required electric field for the SWP acceleration is of internal origin. Equations (40) and (41) can be combined to yield a single coupled form as given below

$$\left[M^2-\left(1+\varepsilon_T\right)\right]\frac{1}{M}\frac{dM}{d\xi}=\left(1+\varepsilon_T\right)\frac{2}{\xi}-\frac{1}{\xi^2}\frac{GM_\Theta}{C_s^2\lambda_J}. \tag{42}$$

The quantity $a_0=GM_\Theta/c_s^2\lambda_J$ (*normalization coefficient*) is treated as a free parameter, which eventually provides a way to estimate the SWP electron temperature. The value of this parameter is determined by the condition that a transonic solution for the SWP exists for a given set of initial values of the required physical variables. The above equation can now be rewritten as

$$\left[M^2-(1+\varepsilon_T)\right]\frac{1}{M}\frac{dM}{d\xi}=(1+\varepsilon_T)\frac{2}{\xi}-\frac{a_0}{\xi^2}. \tag{43}$$

5.4.2 Numerical results

Equations (41) and (43) can be solved by numerical methods (by Runge-Kutta IV method) to determine the time independent $M-$ and $\theta-$ profiles associated with the SWP for some arbitrary values of a_0. However, we choose the minimum value of a_0 that yields transonic transition solutions. It is obvious from equation (43) that the critical distance will exist at $\xi=\xi_c=a_0/2(1+\varepsilon_T)$. This critical distance corresponds to $\sim 14R_\Theta$ from the defined solar surface. As shown in Fig. 10, the critical point for transonic transition, indeed, exists for a_0 = 95 for narrow range variation of $\varepsilon_T=0.0-0.1$, for the already derived solar surface values of $M_{SSB}=10^{-7}$ & $\theta_\Theta=-1.0$ as a set of initial values. The M-values at a distance of 1AU from the defined SSB, i.e., at $\xi\sim 750$ are about 3.3 and 3.5 for $\varepsilon_T=0.0$ & 0.1, respectively, as shown in Fig. 11a. The corresponding values of the electrostatic potential at the same distance are found to be $\theta=-31\,\&-30$ for $\varepsilon_T=0.0\,\&\,0.1$, respectively as shown in Fig. 11b. This is to note that for higher values of ε_T, solar breeze solutions are obtained.

Substitution of $\lambda_J=R_\Theta/3.5$ in the defined expression of a_0 = 95, we can estimate c_s ~100 *km/sec* and T_e ~100 eV for the SWP. The critical distance for transition behavior to occur for M_{SSB} ~10^{-7} (Fig. 11a) as an initial value for Mach number exists at about $14R_\Theta$ distance apart from the defined solar surface. This is to note that if we consider M_{SSB} ~10^{-6} as an initial value for the numerical solution of equations (41) and (43), the transonic transition occurs for a_0 = 71 that yields almost the same values of c_s ~100 *km/sec* and T_e ~100 eV for the SWP. But the critical transition location point exists at about $10R_\Theta$ distance apart from the defined solar surface. This implies the initial value of M_{SSB} plays an important role in the proper fixation of a_0 that determines the exact location of transonic transition point and the SWP-property. Accordingly, the speed of the SWP at 1 AU comes out to be 330-350 km s $^{-1}$, as shown in Fig. 11a (dotted vertical line).

Let us now look at Fig. 11b which the electrostatic potential's profile for a predetermined set of initial values of $M_{SSB}=10^{-7}$, and $\theta_\Theta\sim-1$ at the SSB, as in the case of Fig. 11a. It can be seen that the normalized value of the SWP-associated electrostatic potential at 1 AU is about -30 to -31 for $\varepsilon_T=0.0-0.1$. With some simple calculations, as illustrated in the next subsection, we can argue that beyond the transonic transition, the SWP seems to obey the zero-electric current approximation, but not before. This is inferred from the floating surface condition, which is defined by the equalization of escaping flux of the SWP particles in accordance with the law of conservation of particle flux.

5.4.3 Theoretical estimation of floating potential

In absence of any particle source and/or sink of a stellar origin under spherical geometry approximation, we get an expression for the steady state mass conservation of the SWP flow

$$r^2 n_{iT} v_i = r_{SSB}^2 n_0 v_{SSB}. \tag{44}$$

In normalized form the above expression (44) for $N_i=n_i/n_0=1$, can be written as

$$M = \left(\xi_{SSB}^2 / \xi^2\right) M_{SSB}. \tag{45}$$

Now using normal practice for floating potential estimation under net zero-electric current approximation, i.e. $J_e = J_i$,, one gets

$$e^{\theta_f} \sqrt{(m_i/m_e)} = M = \left(\xi_{SSB}^2 / \xi^2\right) M_{SSB}. \tag{46}$$

Now, from equation (47) the normalized floating potential at any normalized radial position from the SSB can be expressed as

$$\theta_f = \log\left[\sqrt{\frac{m_e}{m_i}} \left(\frac{\xi_{SSB}}{\xi}\right)^2 M_{SSB}\right]. \tag{47}$$

By simple calculations, one can generate the following comparative data of theoretical estimation of the SWP floating potential (using above expression (47)) at different distances from the obtained SSB as follows.

ξ	θ_f
3.50 (at ξ_Θ)	-19.57
47.50 (at ξ_c)	-24.78
100	-26.27
200	-27.66
300	-28.47
400	-29.04
500	-29.49
600	-29.86
700	-30.16
750 (1 AU)	-30.30

Table 2. Values of the Floating Potential

It looks as if the SSB was in non-floating condition as because it does not acquire floating potential during evolution of the GES-potential distribution of the SIP. However, beyond the critical distance and up to a distance of 1 *AU*, the calculated values of the floating potential almost match with those of the SWP obtained numerically (Fig. 11b). This implies that a finite divergence-free electric current exists at the SSB up to the transonic transition region! Beyond the transonic point, zero electric current approximation seems to hold good.

It is commented that the zero-electric current approximation at the SSB assumed in previous model calculations [3, 11, 31-41] for the qualitative description of the SWP properties seems to be physically unjustified. Furthermore, our model calculation does not require outside imposition of the electric field to ensure the validity of the zero-electric current approximation at the SSB. Probably the imposition of the zero-electric current approximation is not suitable for proper description of the SWP properties. Now the natural question may arise, "What happens to the SWP current after the transonic transition?" It seems the current dissipates mainly through a channel of inertial resistance of the plasma ions due to solar gravity.

5.5 Physical consequences

5.5.1 Description of numerical results

The proposed GES-model predicts that the GES formation (of the SIP origin) drives the subsonic SWP at the solar surface. The quasi-hydrostatic equilibrium defines the solar boundary and ensures the GES formation. Numerically $\theta_{\Theta} \sim -1$, $M_{SSB} \sim 10^{-7}$, and $g_{\Theta} \approx |d\theta/d\xi| \sim 0.60$ prescribe the defined solar boundary (Table I). It requires specific initial values $\theta_i = -0.001$ & $\xi_i = 0.01$ in the solar interior for $\varepsilon_T = 0.4$.

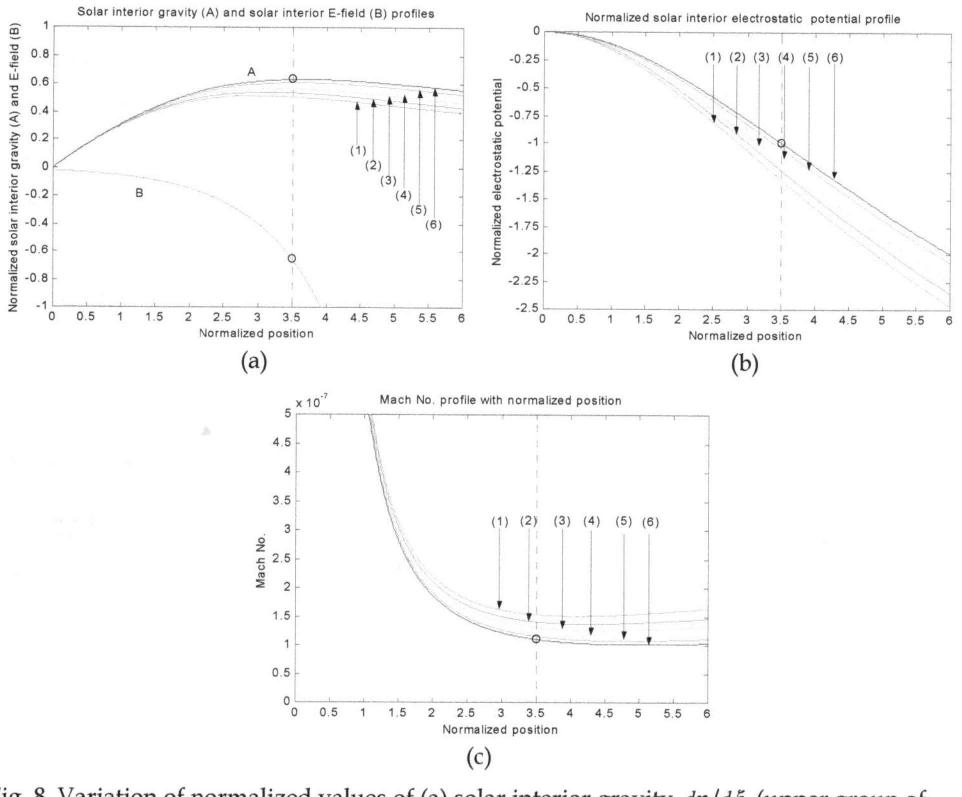

Fig. 8. Variation of normalized values of (a) solar interior gravity $d\eta/d\xi$ (upper group of curves) and electric field $d\theta/d\xi$ (lower curve), (b) electrostatic potential θ, and (c) speed M associated with solar interior plasma flow dynamics with normalized position (ξ) from the heliocenter $(\xi = 0)$. The values of initial position $\xi_i = 0.01$ and initial electrostatic potential $\theta_i = -0.001$ are held fixed. The lines correspond to the cases ε_T = 0.0 (graph 1), 0.1 (graph 2), 0.2 (graph 3), 0.3 (graph 4), 0.4 (graph 5), and 0.5 (graph 6) respectively. The defined solar surface boundary lies at a radial position $\xi_{\Theta} \sim 3.5$ (implying $R_{\Theta} \sim 3.5\lambda_J$) with circled points corresponding to the solar surface values

Item	Parker's model	GES Model
1.	Deals with an unbounded solution of steady state hydrodynamic equilibrium of the solar wind (SW)	Deals with bounded (SIP) and unbounded (SWP) solutions of a continuum steady state hydrodynamic equilibrium
2.	Considers a single neutral fluid (gas) model approximation for the SW gas flow dynamics	Considers a two-fluid ideal plasma (gas) model for the SIP and SWP gas flow
3.	Predicts an unbounded solution of supersonic expansion of the SW provided that a sub-sonic flow pre-exists at the SSB	Predicts a bounded solution of the SIP mass distribution with its subsonic outflow at the SSB
4.	Genesis of the subsonic solar surface origin of the SW is not precisely known: discusses the acceleration of the SW by analogy with the de Laval nozzle	Discusses the genesis of the subsonic SSB origin of the SIP in terms of the basic principles of the GES acceleration of ions: the transonic acceleration mechanism of the SWP is the same as Parker's
5.	Does not specify precisely the SW-base definition and prescription for the self-consistent SSB	Offers a precise definition of and prescription for a self-consistent SSB
6.	Standard solar surface is electrically uncharged and unbiased	SSB acquires a negative electrostatic potential (~1 kV) at the cost of thermal loss of the electrons
7.	Does not consider plasma-boundary wall interaction, plasma sheath formation and spontaneous thermal leakage through squeezing mechanism	Considers it
8.	Concept of floating surface (at which no net electric current) is not involved	It is involved
9.	Considers one-scale (SW) theoretical description	Considers two-scale (SIP and SWP) theoretical description
10.	Extensive research has already been done on the SW acceleration and heating	Opens a new chapter of the GES-based theory for interior (bounded) and exterior (unbounded) solar plasma flow dynamics

Table 3. Parker versus GES Model

The GES-formation occurs due to solar surface leakage of the thermal electrons of solar interior plasma outgoing radially outwards. It causes an appreciable space charge polarization effect near the boundary. The depth of the electrostatic potential well for the plasma ions, so formed, is such as to allow the incoming ions from the solar interior bulk plasma to acquire the kinetic energy of ion motion to overcome the maximum gravitational potential barrier height near the boundary. The SIP ions come out of the solar gravitational barrier with a minimum speed M_{SSB} ~10^{-7}. From the floating potential calculation with no net current flow, we infer that the solar surface boundary drives out some finite electric current in the outward flow. That is, it seems a finite electric current loss of the SIP occurs through its leakage process near the SSB! It can be shown that the total surface charge in the boundary, however, comes out to be about 10^{20} times the electronic charge. Table III gives a glimpse of distinction between Parker's model and GES-model of the subsonic SWP origin and its acceleration from subsonic to supersonic flow speed.

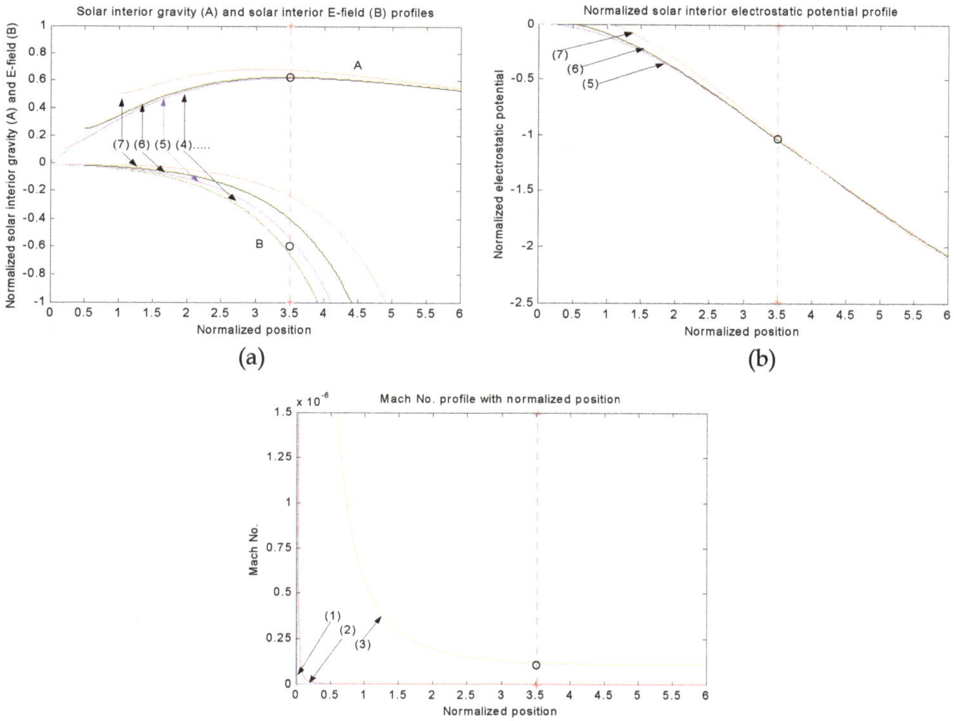

(a) (b)

(c)

Fig. 9. Same as Fig. 8 but with the ion-to-electron temperature ratio $\varepsilon_T = 0.40$ and electrostatic potential $\theta_i = -0.001$ held fixed. The lines correspond to the cases with initial positions $\xi_i = 10^{-4}$ (graph 1), 10^{-3} (graph 2), 10^{-2} (graph 3), 10^{-1} (graph 4), 0.2 (graph 5), 0.5 (graph 6), and 1.0 (graph 7), respectively. The circled points indicate the most suitable choice of the solar surface values

This is to clarify that the GES-model is a quite simplified one in the sense that it does not include any role of magnetic forces, interplanetary medium or any other complications like rotations, viscosity, etc. It opens a new chapter for further study on the coupled system of the solar interior and exterior plasma flow dynamics.

According to GES-model, the normal acoustic modes of the global solar surface oscillations can be analyzed in terms of the local and global gravito-electrostatic plasma sheath oscillations governed by the basic principles of linear and nonlinear nonlocal theory of the Jeans collapse model [24-25] of charged dust clouds in plasma environment.

The magnified view of the Mach number variation in *transonic transition zone* of the SWP (Fig. 11c) indicates the existence of an extended region having almost uniform sonic flow speed. It can be deduced from Fig. 11c that the transonic point does not always coincide with the critical point. We define the critical point as a radial point (away from that defined solar surface) where the net force on the SWP ions, due to GES-induced *E*-field and external gravity due to total solar interior plasma mass, becomes almost zero.

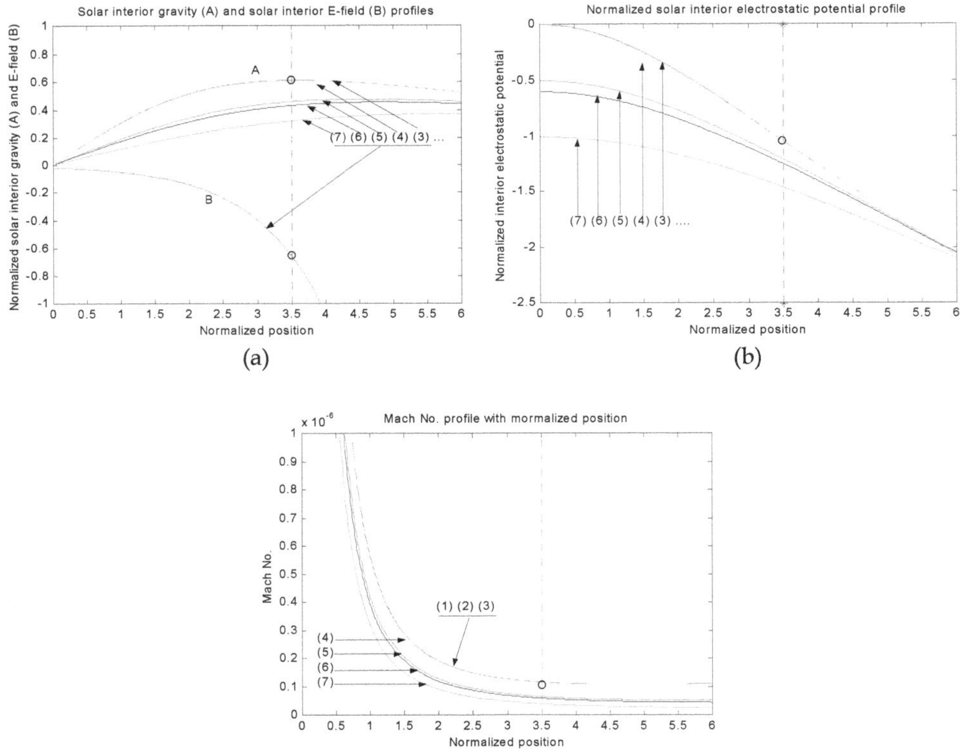

(a) (b)

(c)

Fig. 10. Same as Fig. 8 but with the initial position $\xi_i = 0.01$ and ion-to-electron temperatue ratio $\varepsilon_T = 0.40$ held fixed. The lines correspond to the cases of $\theta_i = 0.0$ (graph 1), -0.001 (graph 2), -0.01 (graph 3), -0.1 (graph 4), -0.5 (graph 5), -0.6 (graph 6), and –1.0 (graph 7), respectively. The circled points indicate the most suitable choice of the solar surface values

Numerical solution in the GES-model reproduces the Parker model values of the SWP speed at 1 AU (Fig. 11a) for the numerically predetermined set of initial values of M_{SSB} ~10^{-7} and SWP ion-to-electron temperature ratio $\varepsilon_T = 0.0 - 0.1$. The estimated critical point for the transonic transition to occur $(r_c \sim 14R_\Theta)$ differs from that $(r_c \sim 10R_\Theta)$ in Parker's model. We find that the latter can be obtained with a choice of M_{SSB} ~10^{-6} (or larger than this by orders of magnitude) as an initial Mach value at the solar surface.

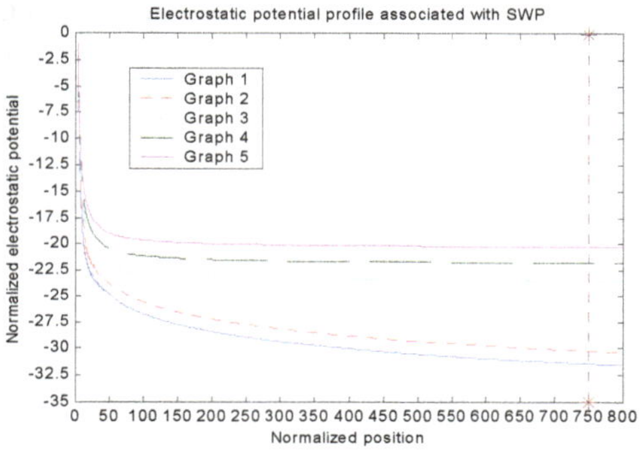

(a)

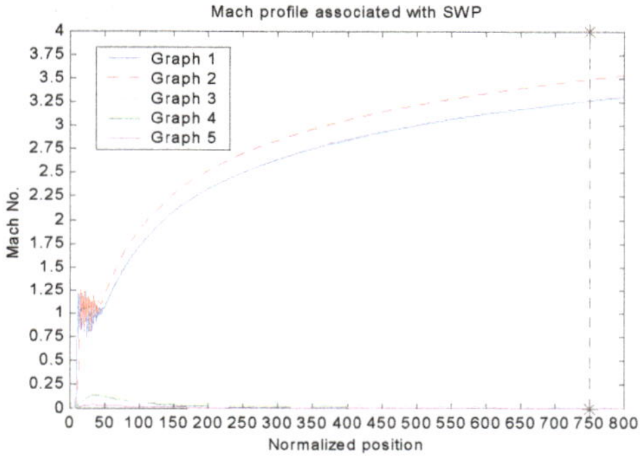

(b)

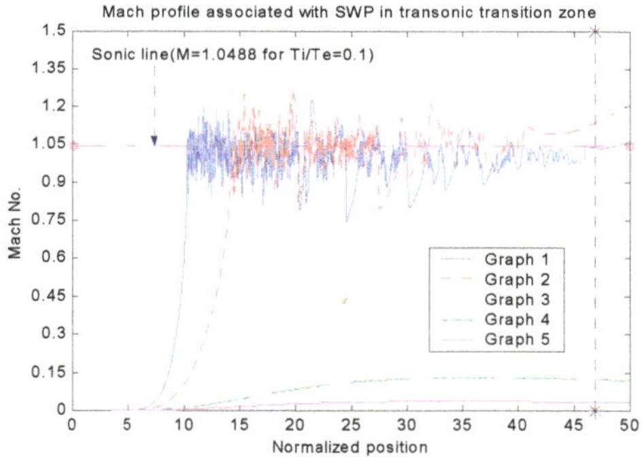

(c)

Fig. 11. Variation of normalized values of (a) speed M, (b) electrostatic potential θ, and (c) speed M in the transonic transition zone associated with SWP flow dynamics with respect to normalized position (ξ) from the solar surface boundary $(\xi_\Theta = 3.5)$ in magnified form. The predetermined solar surface boundary parameter values of $M_{SSB} \sim 10^{-7}$, $\theta_\Theta \sim -1.0$ and $a_0 = GM_\Theta / c_s^2 \lambda_J = 95$ are considered as the set of initial values. The lines correspond to the cases of ε_T = 0.0 (graph 1), 0.1 (graph 2), 0.2 (graph 3), 0.3 (graph 4), and 0.4 (graph 5), respectively. The critical distance lies at $\xi_c \cong 47.5$, which corresponds to a radial position of $r \sim 14R_\Theta$ from the solar surface boundary

This is to clarify that the GES-model is a quite simplified one in the sense that it does not include any role of magnetic forces, interplanetary medium or any other complications like rotations, viscosity, etc. It opens a new chapter for further study on the coupled system of the solar interior and exterior plasma flow dynamics.

According to GES-model, the normal acoustic modes of the global solar surface oscillations can be analyzed in terms of the local and global gravito-electrostatic plasma sheath oscillations governed by the basic principles of linear and nonlinear nonlocal theory of the Jeans collapse model [24-25] of charged dust clouds in plasma environment.

The magnified view of the Mach number variation in *transonic transition zone* of the SWP (Fig. 11c) indicates the existence of an extended region having almost uniform sonic flow speed. It can be deduced from Fig. 11c that the transonic point does not always coincide with the critical point. We define the critical point as a radial point (away from that defined solar surface) where the net force on the SWP ions, due to GES-induced *E*-field and external gravity due to the total SIP mass, becomes almost zero.

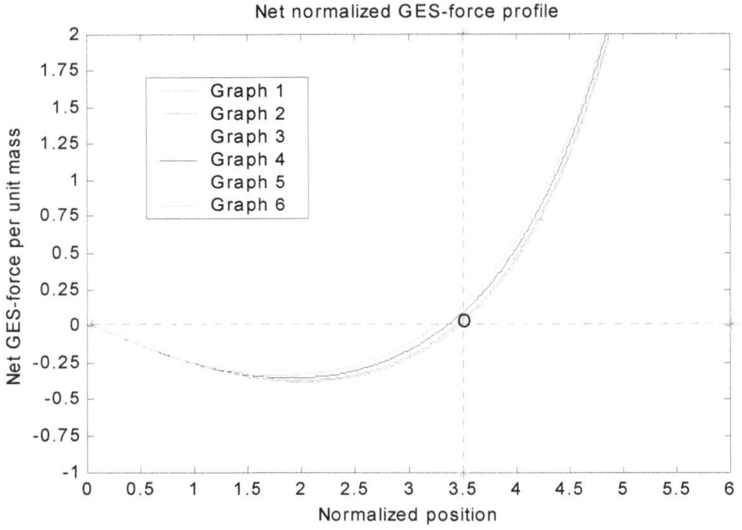

Fig. 12. Variation of net normalized GES force per unit mass associated with solar interior plasma flow dynamics with respect to normalized position (ξ) from the heliocentre $(\xi = 0)$. The initial position $\xi_i = 0.01$ and initial electrostatic potential $\theta_i = -0.001$ are held fixed. The lines correspond to the cases of ε_T = 0.5 (graph 1), 0.4 (graph 2), 0.3 (graph 3), 0.2 (graph 4), 0.1 (graph 5), and 0.0 (graph 6), respectively. The defined solar surface boundary is found to lie at a radial position $\xi_\Theta \sim 3.5$ for a more suitable choice of ε_T = 0.4-0.5 variations

Now, one can see that the sonic point for $\varepsilon_T = 0.0$ (graph 1 in Fig. 11c) falls around $\xi \cong 10$ whereas the transonic point for the same value of $\varepsilon_T = 0.0$ falls around the critical point $\xi_c \cong 47.5$. Similarly, one can see that the sonic point for $\varepsilon_T = 0.1$ (graph 2 in Fig. 11c) falls around $\xi \cong 15$ whereas the transonic point for the same value of $\varepsilon_T = 0.1$ falls around the non-critical point $\xi \cong 50$. From these numerical observations of the transonic transition region, one can clearly notice that an extended zone of about $40\lambda_J$ exists having almost a uniform sonic flow speed of the SWP between sonic and transonic points.

In this region, the inertia-induced acoustic excitation theory [5-12] may have potential applications provided it is improved with a proper inclusion of the solar gravity under nonlocal normal-mode analysis. Similar situations are likely to arise in laboratory plasmas of gravitationally sensitive multi-ion colloidal plasma systems [14-15, 24-25].
This is to point out that the intense acoustic fluctuations appearing in the Mach-profile (Fig. 11c) are merely the results of a numerical instability arising due to the mathematically indeterminate situations localized mainly near the sonic speed. These fluctuations, however, are found to disappear beyond the critical distance. Physically, however, the physical appearance of such indeterminate structures in the graphical plots is because of some chaotic interference and intermittency of acoustic background fluctuations in the emitted SWP. It may produce some dissipative effects in course of the electrodynamical process of the electrodynamical interaction of the SWP particles with background particles of ambient stellar atmosphere.

5.6 Comparison with exospheric model

Exospheric model [38-39] is a simple kinetic model for the solar coronal plasma expansion. This model assumes that beyond a given altitude termed as the '*exobase*' from the SSB, binary collisions between the SWP-particles are negligible. The coronal plasma expansion is believed to occur due to thermal evaporation of the hot protons that have velocity exceeding the escape velocity so as to cross over the barrier of the external solar gravitational field. The generalized model [19-23] considers the non-Maxwellian velocity distribution function for the coronal plasma electrons.
Of course, this model has succeeded to explain the observation of the high speed SWP. This model indeed explains the high speed SWP without requiring any additional source to heat the coronal plasma electrons. In this model, the exobase is defined by the condition that the density scale length equals the mean free path of the SWP particles. Due to the complexities of coronal based physics and multiplicities of plasma species different exobases are likely to coexist. Moreover, the appropriate electrostatic potential is determined by applying the approximations of both local quasi-neutrality and zero current. Of course, this model has succeeded in explaining the observation of the high-speed SWP. But according to our model calculations, the zero-current approximation of the exospheric model seems to be valid only on large scales but not near and above the SSB.
By our GES-model analyses, a finite electron-dominated [10-11] current with a positive finite divergence exists on the solar interior scale for $d\theta/d\xi < 0$. Immediately after the SSB, i.e., on the unbounded scale of the SWP, a divergence-free current exists. This seems to exhibit a discontinuous behavior. How to resolve this? In reality electron temperature has variable profiles on both the bounded and unbounded scales. Probably a self-consistent profile of two distinct electron temperatures on two regions of bounded and unbounded scales separately may resolve the interfacial transition problem of the proposed two-scale theory of the GES-associated solar plasma current system.
According to our GES theory and model calculations, the zero-electric current approximation of exospheric model calculations [38-39] requires further review. The appropriate electrostatic potential estimate from numerical analysis emphasizes that the zero electric current approximation is an outcome of the GES model on the large scale of the SWP. Now the question may naturally arise, "What happens to the SWP current after the transonic transition?" It seems that the electron-dominated electric current dissipates mainly through a channel of inertial resistance of the plasma ions due to solar gravity as a barrier.

The other dissipation channels of the electric current may be through the SWP heating generation of fluctuations in thermal noise level, etc. The uniform flow region of the SWP is, in addition, found to have a large number of conservation rules [11] under the lowest order inertial correction of the thermal electrons in the solar plasma system approximately from applied mathematical point of view. The details of the associated physical mechanism and fluctuations will be communicated to somewhere else.

5.7 Comments

Before we conclude with any physical comment, we must admit that the neglect of collisional dissipation and deviation from a Maxwellian velocity distribution of the plasma particles is not quite realistic. But our GES model under these simple approximations may provide quite interesting results. For example, it provides deep physical insight into the interconnection between the Sun and the SWP. The violation of the zero-current approximation is indeed noted in the neighborhood of the SSB and above. Of course, the zero current approximation seems to be satisfied beyond the transonic region. This conclusion is based on the well-known condition of the floating surface boundary in basic plasma sheath physics.

An estimated value $\lambda_{De}/\lambda_J \sim 10^{-20}$ of the ratio of the solar plasma Debye length and the Jeans length of the total solar mass justifies the quasi-neutral behavior of the solar plasma on both the bounded and unbounded scales.

Applying the spherical capacitor charging model, the coulomb charge on the SWP at a distance of ~ 1 AU comes out to be . For rotation frequency of the solar plasma system corresponding to the mean angular frequency about the centre of the system (Gunn 1931), the mean value of the strength of the solar magnetic field associated with the SWP in our model analysis is estimated as $\langle|B_{SWP}|\rangle = 4\pi^2 Q_{SWP}\, f_{SWP} \sim 3.01\times10^{11}\, T$. This is obviously considerably higher for producing any significant effects on the dynamics of the SWP particles. Thus the effects of the magnetic field are not ignorable for the SWP particles due to the significantly strong Lorentz force, which is now estimated to be $F_L = e(v_0 \times B_{SWP}) \approx 1.64\times10^{-2}\, N$ corresponding to a supersonic flow speed $v_0 \sim 340.00\, km\, s^{-1}$. Thus the Lorentz force may have some remarkable effect on the SWP particles and hence, may not justifiably neglected for the unbounded scale description. It justifies the convective and circulation dynamics to be considered in that context. Therefore our unmagnetized plasma approximation may not prove well justified in our GES model configuration for the SWP flow dynamics description. Although collision processes are dominant in the realistic solar interior [2, 39-44], collisionless models [2, 39] are also equally useful for the solar plasma description. Thus our collisionless model approximation for mathematical simplicity may be justified here. In our GES model, the calculated values of the mean free paths for the solar plasma electrons, $\lambda_e \sim 1.50\times10^{198}\, m$ and for ions, $\lambda_i \sim 3.05\times10^{132}\, m$ justify the collisionless model approximation. This approximation holds good justifiably under the fulfillment of the validity condition $\lambda_e, \lambda_i >> \lambda_J$.

One can note that the SIP electron temperature, specified by T_{e1}, differs from (exceeds) the SWP electron temperature specified T_{e2} by one order of magnitude. This is discussed already discussed above. It simply means the SWP has been relatively cooled. It is quite natural for expanding plasma gas to be cool. This is to further comment that these two different electron temperatures are considered constant over their respective scales.

Actually, a discontinuity exists at the interface of the bounded and unbounded scales. This is an open problem to resolve.

Let us clarify once again that equalizing the maximum value of the numerically determined solar self-gravity with the standard value specifies the SIP electron temperature. Similarly, the appropriate choice of the defined constant a_0 specifies the SWP electron temperature, which ensures that a transonic solution of the SWP dynamics exists. Now, with this simplified treatment our theoretical model calculations yield the following conclusions.

1. Contrary to the general belief that the SWP emerges from the SSB, our theory provokes us to argue that the genesis of the subsonic origin of the SWP at the SSB in fact lies in the SIP dynamics. It is governed by the basic principles of the GES formation near the SSB and beyond. The surface boundary is located at a radial distance defined by $\xi \sim 3.5$ (Figs. 8–10) from the heliocentric origin. This specific location in the plots (Figs. 8a, 9a, and 10a) is marked by a vertical line with small circles.
2. Thereafter, the outward moving SIP forms the SWP with a highly subsonic speed at the SSB. Initially the outward acceleration of the SWP is quite rapid allowing a *transonic transition* solution to exist for a specific choice of $\varepsilon_T = 0.0 - 0.1$ (Fig. 11a). This occurs as a consequence of the predominantly self-consistent electric field associated with the SWP (Fig. 11b). It produces a *transonic transition region* of sufficient length scale with the critical point lying at about $14R_{\Theta}$ (Fig. 11c) from the SSB.
3. It is noted that initially the gravitational potential barrier decelerates the SIP dynamics rapidly. As soon as the *E*-field of the SIP origin gathers sufficient strength, an outward flow occurs with a reduced minimum speed of M_{SSB} ~10^{-7} (Figs. 8c, 9c, 10c) at the SSB defined by the quasi-hydrostatic equilibrium condition at a point of the maximum solar gravity, as clearly depicted in Fig. 12. This figure clearly shows the strong solar self-gravity up to the solar boundary and relatively weaker strength of the solar external gravity beyond the boundary.
4. According to our model calculations, the SSB behaves as a negatively biased grid with a bias potential of about 1 kV. The surface draws a finite current dominated by the thermal electrons and flows towards the surface. As a result, the solar surface oscillations may naturally be attributed to the resulting consequences of the GES oscillations. Under the neutral ideal gas approximation of the SIP, this property cannot be deduced.
5. We therefore conclude that our GES-based model may be useful to study the properties of the SSB and the properties of the slow speed SWP. Of course, the properties of the high speed SWP description under our model will require a kinetic treatment as already reported by previous workers in the case of the exospheric model.

A few more reminders are in order:

1. The exact location of the SSB and that of 1 AU distance as specified in Figs. 8-12 on the normalized scale are estimated for the normalization factor, which is, decided by the SIP parameters.
2. In the absence of magnetic field in our model approach, the Lorenz force term is absent, but it will be needed for further improvements under the fluid and/or kinetic regime to see the realistic dynamics of the solar plasma system. However, the estimated mean value of the solar magnetic field $\langle |B_{SIP}| \rangle \sim 7.53 \times 10^{-11}\, T$ in the SIP justifies and supports our unmagnetized plasma approximation in the present context.

3. The GES-model is a useful theoretical construct with which to study SWP dynamics in terms of solar interior dynamical behavior (generator of the SWP) through active dynamical coupling processes of solar exterior regions in the light of localized electric space charge effects.

Finally, it is important to comment that the further improvements and modifications to the model will be needed to make it more realistic for actual SWP conditions. These form the basic problems of future research on the GES model. The genesis of the SWP is now found to be associated with the coupling of the SIP potential and self-gravitational potential of the SIP mass. We finally argue that the lines of communications should be kept open between theorists and observers and solar and stellar physicists, and more importantly also between the solar and plasma physics communities, in order to further the study of stellar wind plasmas. Ours is a first step, albeit very simplified and external-field free and ideal, in this particular direction. We have tried to provide an integrated theoretical outlook on the SIP dynamics on the bounded scale, and SWP on the unbounded scale. This model could further be useful to study the properties of the helio-seismic dynamics of the Sun and other stars [36-37] too.

5.8 Overall summary

The presented chapter reviews the latest findings of normal acoustic mode analyses through different types of transonic plasma equilibrium models [5-12] under the lowest order inertial correction of plasma thermal species. Different types of acoustic resonances are observed in transonic plasma equilibria depending on different plasma inertial ions. The linear analyses show the graphical nature of the associated resonance poles. This implies that transonic plasma is an unstable zone, which is rich in wide range spectra of acoustic wave fluctuations. The acoustic wave kinetics in the nonlinear normal mode analyses in different types of plasmas [8-9, 12, 26] is describable by a linear source driven KdV (*d*-KdV) equation. After integration, it shows two distinct classes of soluations, i. e., solitons and oscillatory shocks. The fundamental condition to observe inertia-induced (ion) acoustic wave resonant excitation is that the ion flow speed must be uniform. Accordingly, the same applies to the solar wind dynamics [10-11, 35-41] in self-gravitating plasma systems as well. A large number of conservation laws of applied mathematical significance associated with the *d*-KdV flow dynamics are also pointed out [9] in transonic plasma domain in different situations including solar plasma. Of course, convective and circulation dynamics which are the primary sources of magnetic field [41], are neglected throughout for simplicity. Similar observations of acoustic kinetics of the formation of soliton-type structures are also found in self-gravitating dust molecular clouds in presence of partially ionized dust grains through the active mechanism of gravito-acoustic coupling processes [27]. Some future scopes including realistic sources of acoustic perturbation of the presented analyses are also pointed out in brief.

Very similar to Geoseismology dealing with the Earth's interior through the various seismic (acoustic) waves produced during the earthquakes, Helioseismology is the study of the various linear and nonlinear surface waves and oscillations of solar origin (like *p-mode*, and *f-mode*) to measure the internal structure and dynamics of the Sun [36-44]. The acoustic dynamics in the Sun (or Star) is understandable by considering it to have a resonant cavity like an organ pipe in which acoustic waves are trapped (by reflections or refractions) [41]. One of the earliest studies of solar oscillations and fluctuations established that the power spectrum of the Sun's full disk contained a multitude of Doppler shift peaks between 2.5 mHz - 4.5 mHz [36-37 and references therein]. The Global Oscillation Network Group

(GONG), Stellar Observations Network Group (SONG), Helio- and Asteroseismology (HELAS) Network, and Birmingham Solar Oscillations Network (BiSON) are examples of recent studies being undertaken to measure these surface oscillations through space and ground based remote-sensing observations [36-37, 44]. Michelson Doppler Imager (MDI) onboard Solar and Heliospheric Observatory (SOHO) and recently, Helioseismic Magnetic Imager (HMI) onboard Solar Dynamics Observatory (SDO) also measure these oscillations from space [36-37, 41, 44]. Significant power has been observed at frequencies ranging from 1.4 mHz to 5.6 mHz, corresponding to periods of 3 to 12 minutes. They are called '*5 minute oscillations*' due to their dominant mean period [44]. Besides, the behavior of the solar intermediate-degree modes (during extended minimum) is also investigated to explore the time-varying solar interior dynamics with the help of contemporaneous helioseismic GONG and MDI data [44]. The basic physics behind these helioseismic and helioacoustic observations (*in situ*) reported in the literature, however, needs to be more clearly understood in a broader horizon. Moreover, there are many more experimental observations [3, 35-44] on seismic activities that will require self-gravitating plasma wave theory for further development of our stability analyses and seismic diagnostics. In conclusion, we strongly believe that the presented mathematical strategies and techniques of linear and nonlinear acoustic mode analyses amidst more realistic plasma-boundary interaction processes may have some potential applications in such future helio- and astero-seismic directions.

6. Acknowledgements

I am thankful to Ms. Sandra Bakic, Publishing Process Manager, InTech - Open Access Publisher, for the invitation, chapter proposal and continuous cooperation. I am also thankful to each and everybody of the InTech family whoever involved for extending cooperation. Lastly, I gratefully recognize the Intech Editorial Board for giving me this rare opportunity to publish this chapter in the book "Acoustic Waves - From Microdevices to Helioseismology" without any article processing charge.

7. References

[1] B. W. Tillery, *Physical Sciences* (8th Edition, McGraw-Hill 2009), Ch. 5, pp. 115-134.
[2] F. F. Chen, *Introduction to Plasma Physics and Controlled Fusion* (Plenum Press, New York and London, 1988), Ch.1, pp. 1-17.
[3] S. R. Cranmer, *Space. Sci. Rev.* 101, 229 (2002).
[4] M. Lakshamanan and S. Rajasekhar, *Nonlinear Dynamics: Integrability, Chaos, and Patterns* (Springer-Verlag Heidelberg 2003), Ch. 11-14, pp. 341-454.
[5] C. B. Dwivedi and R. Prakash, *J. Appl. Phys.* 90, 3200 (2001).
[6] P. K. Karmakar, U. Deka and C. B. Dwivedi, *Phys. Plasmas* 12, 032105 (2005).
[7] P. K. Karmakar, U. Deka and C. B. Dwivedi, *Phys. Plasmas* 13, 104702 (1) (2006).
[8] P. K. Karmakar and C.B. Dwivedi, *J. Math. Phys.* 47, 032901(1) (2006).
[9] P. K. Karmakar, *J. Phys.: Conf. Ser.* 208, 012059(1) (2010).
[10] C. B. Dwivedi, P.K. Karmakar and S.C. Tripathy, *Astrophys. J.* 663 (2), 1340 (2007).
[11] P. K. Karmakar, *J. Phys.: Conf. Ser.* 208, 012072(1) (2010).
[12] U. Deka, A. Sarma, R. Prakash, P. K. Karmakar and C. B. Dwivedi, *Phys. Scr.* 69, 303 (2004).
[13] P. Mora, *Phys. Rev. Lett.*, 90, 185002-1 (2003).

[14] C. B. Dwivedi, *Pramana-J. Phys.* 55, 843 (2000).
[15] C. B. Dwivedi, *Phys. Plasmas* 6, 31 (1999).
[16] M. S. Sodha and S. Guha, *Advances in Plasma Physics,* edited by A. Simon and W. B. Thompson (Wiley, New York, 1971) vol. 4.
[17] C. K. Goertz, *Rev. Geophys.* 27, 271 (1989).
[18] F. Verheest, *Space Sci. Rev.* 77, 267 (1996).
[19] H. Thomas, G. E. Morfill, V. Demmel, J. Goerce, B. Feuerbacher and D. Mohlmann, *Phys. Rev. Lett.* 73, 652 (1994).
[20] G. S. Selwyn, J. E. Heidenreich and K. L. Haller, *Appl. Phys. Lett.* 57, 1876 (1990).
[21] L. Boufendi, A. Bouchoule, P. K. Porteous, J. Ph. Blondeau, A. Plain and C. Laure, *J. Appl. Phys.* 73, 2160 (1993).
[22] A. Gondhalekar, P. C. Stangeby and J. D. Elder, *Nucl. Fusion* 34, 247 (1994).
[23] B. N. Kolbasov, A. B. Kukushkin, V. A. Rantsev Kartinov, and P. V. Romanov, *Phys. Lett. A* 269, 363 (2000).
[24] N. N. Rao, P. K. Shukla and M. Y. Yu, *Planet. Space Sci.* 38, 345 (1990).
[25] P. K. Shukla, *Waves in dusty, solar wind and space plasmas,* (AIP Conference proceedings, Leuven, Belgium) 537, 3 (2000).
[26] P. K. Karmakar, *Pramana- J. Phys.* 68, 631 (2007).
[27] P. K. Karmakar, *Pramana- J. Phys.* 76 (6), 945 (2011).
[28] J. P. Goedbloed, R. Keppens and S. Poedts, *Space Sci. Rev.* 107, 63 (2003).
[29] S. P. Kuo and D. Bivolaru, *Phys. Plasmas* 8 (7), 3258 (2001).
[30] K. Itoh, K. Hallatschek and S-I Itoh, *Plasma Phys. Control. Fusion* 47, 451 (2005).
[31] I. Ballai, *PADEU* 15, 73 (2005).
[32] I. Ballai, E. Forgacs and A. Marcu, *Astron. Nachr.* 328 (8), 734 (2007).
[33] P. K. Shukla and A. A. Mamun, *New J. Phys.* 5, 17.1 (2003).
[34] M. Khan, S. Ghosh, S. Sarkar and M. R. Gupta, *Phys. Scr.* T116, 53 (2005).
[35] M. S. Ruderman, *Phi. Trans. R. Soc. A*, 364, 485 (2006).
[36] W. J. Chaplin, *Astron. Nachr.* 331 (9), 1090 (2010).
[37] J. C. Dalsgaard, *Astron. Nachr.* 331 (9), 866 (2010).
[38] J. Lemaire and V. Pierrard, *Astrophys. Space Sci.* 277, 169 (2001).
[39] E. Marsch, *Liv. Rev. Solar Phys.* 3, 1 (2006) (http: //livingreviews.org/lrsp-2006-1).
[40] U. Narain and P. Ulmschneider, *Space Sci. Rev.* 75, 453 (1996).
[41] A. Nordlund, R. F. Stein and M. Asplund, *Living Rev. Solar Phys.* 6, 2 (2009) (http: //livingreviews.org/lrsp-2009-2).
[42] R. Gunn, *Phys. Rev.* 37, 983 (1931).
[43] V. M. Nakariakov and E. Verwichte, *Living Rev. Solar Phys.* 2, 3 (2005) (http: //livingreviews.org/lrsp-2005-3).
[44] S. C. Tripathy, F. Hill, K. Jain and J. W. Leibacher, *Astrophys. J.* 711, L84 (2010).

4

An Operational Approach to the Acoustic Analogy Equations

Dorel Homentcovschi and Ronald Miles
Department of Mechanical Engineering
State University of New York at Binghamton
USA

1. Introduction

Great progress has been made in the last sixty years in the study of the important problem of noise generated by the interaction of flow with stationary or mobile bodies such as occurs in jets, rotating blade propulsion machinery (propellers, turbofans helicopter rotors) and last but not least in aircraft at all ranges of flight and speed. An important part of this progress was based on a rigorous theory known as the Acoustic Analogy initiated by Sir James Lighthill in (Lighthill, 1952) and (Lighthill, 1954). Lighthill considered a free flow, as for example with a jet engine, and the nonstationary fluctuations of the stream represented by a distribution of quadrupole sources in the same volume. The flow parameters such as the surface pressure and the Lighthill tensor T_{ij} are assumed known from solving the aerodynamic problem in the region of sound generation or furnished by measurements. For the first time, this revealed a clear distinction between Aerodynamic Theory, meant to determine mainly the aerodynamic parameters as the lift and damping on the moving object (and also supplying the data for the noise determination) and the Aeroacoustic Theory needed for studying the noise produced, generally at large distances, by the flying (or moving) objects. A primary aim of the following is to show that by using an operational calculus based on the multidimensional Fourier Transform all the theory involved in obtaining the Ffowcs Williams-Hawkings formula (Ffowcs Williams and Hawkings, 1969) can be performed using only classical mathematical analysis.

Curle's contribution (Curle, 1955) is a formal solution of the Acoustic Analogy which takes stationary hard surfaces into consideration. The theory developed by Ffowcs Williams and Hawkings (FW-H) (Ffowcs Williams and Hawkings, 1969) is valid for aeroacoustic sources in relative motion with respect to a hard surface, as is the case in many technical applications for example in the automotive industry or in air travel. The calculation involves quadrupole, dipole and monopole terms. An important point is that FW-H theory, developed in (Ffowcs Williams and Hawkings, 1969) assumed that the boundary surface coincides with the physical body surface and is impenetrable. In both Aerodynamic and Aeroacoustic theories the domain was the same: the infinite air domain external to the moving body.

When the Aeroacoustic Theory was developed by Lighthill, Curle, Ffowcs Williams and Hawkings there were not a lot of experimental or theoretical data to be used as input to their aeroacoustic theoretical work. For this reason, they derived mainly qualitative results which were quite useful in guiding many significant acoustic experiments and in designing low noise propulsion machinery.

The situation has changed dramatically in the last 20 years. The rapid growth in high speed digital computer technology, the availability of turbulent flow simulation codes as well as high quality measured fluid dynamic data and advances in the theory of partial differential equations, resulted in obtaining the needed data for many important problems of aeroacoustics. However, the development of reliable Computational Fluid Dynamics (CFD) methods made them also useful in the evaluation of the near-field aerodynamic parameters. Unfortunately, a fully CFD-based computational aeroacoustic methodology is so far too inefficient and beyond the capability of supercomputers of today.

To avoid these computational difficulties, the philosophy of approaching the Aeroacoustic problem has changed by introducing a surface S as a "permeable" control surface. The surface S is assumed to include inside, in the volume V, all the nonlinear flow effects and noise sources. This splitting of the problem into a linear problem for an infinite domain and a nonlinear setting for a bounded region allows the use of the most appropriate numerical methodology for each of them. In the bounded domain V the CFD methods or advanced measurement techniques will be used for obtaining the aerodynamic near-field and providing the data on the surface S needed for the external, infinite domain modelling. The analysis of the flow information inside V is, in general, expensive either using experiments or CFD. Therefore, it is advantageous to make the volume V as small as possible.

The FH-W equation involving a permeable surface is the proper model for determining the far-field pressure in the infinite domain. The case of permeable surfaces was analyzed by Ffowcs Williams in (Dowling and Ffowcs Williams, 1982) and (Crighton, et al.), by Francescantonio in (Francescantonio, 1997) who called it the KFWH (Kirchhoff FW-H) formula, by Pilon and Lyrintzis in (Pilon and Lyrintzis, 1997) calling it an improved Kirchhoff method and by Brentner and Farassat in (Brentner & Farassat, 1998). Besides the accessibility to the surface data the advantage of the methods using a permeable control surface is that the surface integrals and the first derivative needed can be evaluated more easily than the volume integrals and the second derivatives necessary for the calculation of the quadrupole terms when the traditional Acoustic Analogy is used.

The Acoustic Analogy approach and especially the theory based on the FW-H equation is the most widely used tool for deterministic noise sources. The beauty (and the power) of this model is that all the manipulations are completely rigorous without any *ad hoc* reasoning.

Besides the approach based on generalized functions, used in most papers approaching the FW-H formula, we note the work by Goldstein (Goldstein, 1976) where all the formulas are obtained starting with a generalized Green's formula. A similar approach was used in (Wu and Akay, 1992) . However, the algebra involved in their construction is substantial.

In the second of the two reports by Farassat (Farassat, 1994),(Farassat, 1996), which covers the details of the mathematics used for the wave equation with sources on a moving surface, the author correctly claims that the Ffowcs Williams-Hawkings famous paper published in 1969 used a level of mathematical sophistication including multidimensional generalized functions (distributions) and differential geometry unfamiliar to most researchers and designers working in the field. Many people use Dirac's δ (generalized) function starting with its integral definition. On the other hand, to learn about more complicated operations such as the derivative of a generalized function (distribution) we need a change of paradigm in the way we look at ordinary functions. For some people involved in practical applications this is not a simple task. The power of the theory of generalized functions stems from its operational properties. Thus, for example, discontinuous solutions of linear equations using the Green's function are easily obtained by posing the problem in generalized function spaces. In the following we show that the theory connected with the FW-H formula can be made much simpler, without manipulating multidimensional generalized functions, by using an operational calculus based on the multidimensional Fourier Transform. The approach based on using the Fourier Transform preserves all the good operational properties of the generalized functions without the need to introduce a new sophisticated mathematical tool. The method was used previously by us (Homentcovschi and Singler, 1999) for a direct introduction to the Boundary Element Method.

In the case of permeable surfaces an alternate method for solving the Aeroacoustic problem for the infinite external domain is based on the Kirchhoff formula for the wave equation. Due to its use in Aeroacoustic theory we included also in Section 6 the Kirchhoff formulation for the solution of the wave equation in the case of mobile surfaces. The proof is based again on 3-D Fourier Transform of discontinuous functions and the final formula includes the volume sources and the surface sources as well. As a comparison of the two approaches (that based on FW-K equation and that using Kirchhoff's equation) we notice that Kirchhoff's method requires less memory because fewer quantities on the control surface needed to be stored. On the other hand, Brentner and Farassat in (Brentner & Farassat, 1998) have shown that the FW-H equation is superior to the Kirchhoff formula for aeroacoustic problems because it is based on conservation laws of fluid mechanics rather than on the wave equation. Thus, the FW-H equation is valid even if the integration surface is in the nonlinear region being therefore more robust with the choice of control surface. Another advantage of the FW-H method is that it does not require computation of the normal derivatives on the permeable surface.

A comprehensive review of the use of Kirchhoff's method in computational aeroacoustics was given by Lyrintzis in (Lyrintzis, 1994). The same author reviewed the advances in the use of surface integral methods in aeroacoustics, including Kirchhoff's method and permeable Ffowcs-Williams Hawkings methods in (Lyrintzis, 2003). Morino in (Morino, 2003) addresses commonalities and differences between aeroacoustics and aerodynamics. A discussion about the acoustic analogy and alternative theories for jet noise prediction is the subject of (Morris and Farassat, 2002). Finally we note some interesting work about this subject included in the book edited by (Raman, 2009).

It is our hope that the elementary derivations included in this chapter will make the application of the FW-H equation more clear, avoiding in the future comments such as those generated by (Zinoviev and Bies, 2004) (See (Farassat, 2005), (Zinoviev and Bies, 2005), (Farassat and Myers, 2006), (Zinoviev and Bies, 2006)).

2. The equations of the acoustic analogy and their operational form

In order to apply the 3-D Fourier Transform to a certain physical variable this has to be defined in the whole space. Thus, to utilize the Fourier Transform to examine the sound field in a finite physical domain, it is necessary to imbed this domain within an infinite space. For example, in the case of studying the air motion around a finite body, the interior of the body is considered as an air-filled domain separated from the exterior, infinite domain by an impermeable surface, S_b enclosing the body. This introduces a first class of discontinuity surfaces of the motion between two air-filled regions. The second class contains the natural discontinuity surfaces S_s inside the flow domain as shock fronts, wakes, etc. Finally, to aid computations, it is often helpful to also define a virtual permeable surface, S_p, enclosing body along with the portion of the air domain where viscous effects and the nonlinear terms in the Navier-Stokes equations are important. It is convenient to evaluate the aerodynamic field in this region numerically using CFD, due to the difficulties imposed by viscosity and nonlinear effects. These calculations furnish data describing the hydrodynamic state on the virtual permeable surface, S_p. In the domain exterior to the permeable surface the acoustical analogy is applied to predict the sound field. In other words, the field inside the virtual permeable surface is assumed to be strongly influenced by hydrodynamic effects, while the field in the external, infinite domain can be modeled according to the usual assumptions of linear acoustics.

2.1 The equations of the acoustic analogy

In the case where the whole space $\mathbb{R}^3$ is filled by a compressible viscous fluid, containing several discontinuity surfaces, the flow inside each domain is governed by the following equations:
the continuity equation (mass conservation)

$$\frac{\partial \rho}{\partial t} + \frac{\partial}{\partial x_j}\left(\rho u_j\right) = 0, \tag{1}$$

and the conservation of momentum in the form written by Lighthill in Refs. (Lighthill, 1952) and (Lighthill, 1954)

$$\frac{\partial\left(\rho u_i\right)}{\partial t} + c^2 \frac{\partial \rho}{\partial x_i} = -\frac{\partial T_{ij}}{\partial x_j}. \tag{2}$$

Here ρ is the density, c is the velocity of sound in the uniform medium, u_j is the component of fluid velocity in the direction x_j $(j = 1,2,3)$, and a repeated index implies a summation over these values, and,

$$T_{ij} = P_{ij} + \rho u_i u_j - c^2 \rho\, \delta_{ij},$$

is Lighthill's stress tensor. Also, we denoted by

$$P_{ij} = p\delta_{ij} + \mu\left(-\frac{\partial u_j}{\partial x_j} - \frac{\partial u_j}{\partial x_i} + \frac{2}{3}\left(\frac{\partial u_\kappa}{\partial x_\kappa}\right)\delta_{ij}\right),$$

the compressive stress tensor, δ_{ij} is the Kronicker delta function, p is the pressure and μ is the viscosity.

Remark 1. *The acoustical analogy equations (1) and (2) are in fact the exact fluid flow equations in the form written by Lighthill. This means that if one solves these equations correctly for a problem satisfying the Lighthill assumptions, then one will get the correct answer to the aerodynamic and aeroacoustic problems simultaneously.*

In the case where inside the flow domain there are discontinuity surfaces of type S_s (as shock fronts) the conservation laws for mass and momentum yield the Rankine-Hugoniot type junction conditions

$$\left[\rho\left(u_n - v_n\right)\right] = 0 \tag{3}$$

$$\left[\rho u_\kappa\left(u_n - v_n\right) + P_{j\kappa} n_j\right] = 0. \tag{4}$$

We denote by square brackets the jump of its content across the discontinuity surface (see also formula (54) in Appendix A). u_n is the velocity projection on the normal to the surface, $\mathbf{n}$, and v_n is the normal velocity of the surface.

Finally, on solid (deformable) impermeable surfaces S_b inside the flow the obvious non-penetration condition proves true

$$u_n - v_n = 0. \tag{5}$$

2.2 The operational form of the acoustic analogy equations

Since the density ρ_0 and pressure p_0 at large distances (in the unperturbed fluid) are different from zero we introduce the perturbation of the density $\rho' = \rho - \rho_0$ and the perturbation of the stress tensor as $P'_{j\kappa} = P_{j\kappa} - p_0\delta_{j\kappa}$ (where $\delta_{j\kappa}$ is the Kronecker delta) and will write equations (1) and (2) as

$$\frac{\partial \rho'}{\partial t} + \frac{\partial}{\partial x_j}\left(\rho u_j\right) = 0 \tag{6}$$

$$\frac{\partial\left(\rho u_j\right)}{\partial t} + c^2\frac{\partial \rho'}{\partial x_j} = -\frac{\partial T'_{j\kappa}}{\partial x_\kappa}, \tag{7}$$

where

$$T'_{j\kappa} = P'_{j\kappa} + \rho u_j u_\kappa - c^2\rho'\delta_{j\kappa}$$

According to the previous discussion, these equations are valid within the flow domains in the whole space. Assuming, as in Appendix A, that there is a discontinuity surface S, separating the inner domain $D^{(i)}$ and the external domain $D^{(e)}$, by applying the Fourier transform to equations (6) and (7) and using the formulas (53) and (68) there results *the operational equations*

$$\frac{d\widetilde{\rho'}}{dt} + ik_j\widetilde{\rho u_j} = \int_S \left[\rho\left(u_n - v_n\right)\right] e^{-i\mathbf{k}\cdot\mathbf{y}} dS \tag{8}$$

$$\frac{d\widetilde{\rho u_j}}{dt} + ik_j c^2\widetilde{\rho'} = -ik_\kappa \widetilde{T'}_{j\kappa} + \int_S \left[\rho u_j\left(u_n - v_n\right) + P_{\kappa j}n_\kappa\right] e^{-i\mathbf{k}\cdot\mathbf{y}} dS \tag{9}$$

Here the overhead tilde denotes a Fourier Transform (see Appendix A), $\mathbf{k}(k_1, k_2, k_3)$ is the wave vector and $\mathbf{y}(y_1, y_2, y_3)$ is the position vector of the integration (source) point. The *operational* form of the conservation equations contains in the left-hand sides the Fourier transform of corresponding terms in (1) and (2) and in the right-hand sides the integrals accounting for the influence of the discontinuity surfaces. As was noted in Remark 1 in Appendix A, in the general case, in the right-hand side of equations (8) and (9) the sum of the contributions of all discontinuity surfaces will appear.

Remark 2. *The square brackets in equations (8) and (9) can also be written as* $\left[\rho_0 u_n + \rho'\left(u_n - v_n\right)\right]$ *and* $\left[\rho u_\kappa\left(u_n - v_n\right) + P'_{\kappa j}n_j\right]$ *respectively.*

2.3 Some particular cases

2.3.1 Shock-type discontinuity surfaces (S_s)

In the case of a discontinuity surface S_s (of shock-type) inside the fluid domain the junction conditions (3) and (4) will cancel out the integral terms in equations (8) and (9). Consequently the shock-type discontinuity surfaces are not introducing any supplementary terms in the operational equations.

2.3.2 Impermeable solid deformable surfaces (S_b)

In the case of a solid with a deformable and impermeable boundary surface the condition (5) will cancel out the integral term in the operational form of the continuity equation (8) and a part of the integral in the momentum equation (9). In this case, the Fourier transform of the traction $\mathbf{P}=(P_1, P_2, P_3)$ of the surface on the fluid enters in the integral term of the momentum equations as $P_j = P_{\kappa j} n_\kappa$. The operational momentum equation now contains in the right-hand side the action of the solid surface S_b on the fluid flow.

3. Solution of the operational form of the acoustical analogy equations.

3.1 The case of a permeable surface (S_p)

Let S_p be a discontinuity surface of the flow variables inside the external fluid flow such that in the domain $D^{(i)}$ we have $p^{(i)} = p_0, \rho^{(i)} = \rho_0, \mathbf{v}^{(i)}= \mathbf{0}$.
We write the system of equations (8), (9) as

$$\frac{d\widetilde{\mathbf{w}}}{dt} = \widetilde{\mathbb{A}}\widetilde{\mathbf{w}}+\widetilde{\mathbf{f}} \tag{10}$$

where

$$\widetilde{\mathbf{w}}^T = (\widetilde{\rho'}, \widetilde{\rho u_1}, \widetilde{\rho u_2}, \widetilde{\rho u_3}), \tag{11}$$

$$\widetilde{\mathbb{A}}= \begin{bmatrix} 0 & -i k_1 & -i k_2 & -i k_3 \\ -c^2 i k_1 & 0 & 0 & 0 \\ -c^2 i k_2 & 0 & 0 & 0 \\ -c^2 i k_3 & 0 & 0 & 0 \end{bmatrix}, \tag{12}$$

$$\begin{aligned} \widetilde{f}_1 &= \textstyle\int_S \left[\rho_0 u_n + \rho' \left(u_n - v_n\right)\right] e^{-i\mathbf{k}\cdot\mathbf{y}} dS \\ \widetilde{f}_{j+1} &= -ik_\kappa \widetilde{T}_{j\kappa} + \textstyle\int_S \left[\rho u_j \left(u_n - v_n\right) + P_j\right] e^{-i\mathbf{k}\cdot\mathbf{y}} dS, \quad j = 1,2,3. \end{aligned} \tag{13}$$

The solution of equation (10) can be obtained by using the exponential $\widetilde{\mathbb{H}}\,(t)$ of the matrix $\widetilde{\mathbb{A}}$ as

$$\widetilde{\mathbf{w}}= \int_{-\infty}^{t} \widetilde{\mathbb{H}}\left(t - t'\right) \widetilde{\mathbf{f}}\left(t'\right) dt' \tag{14}$$

The exponential of the matrix $\widetilde{\mathbb{A}}$ can be written as

$$\widetilde{\mathbb{H}}= \left[\widetilde{h}_{j\kappa}\right] \tag{15}$$

$$\begin{aligned} \widetilde{h}_{1,1} &= \cos\left(ckt\right), \quad \widetilde{h}_{1,\kappa+1} = -\frac{i\, k_\kappa \sin\left(ckt\right)}{ck}, \ \kappa = 1,2,3 \\ \widetilde{h}_{j+1,1} &= -\frac{i\, k_j c \sin\left(ckt\right)}{k}, \quad j = 1,2,3 \\ \widetilde{h}_{j+1,\kappa+1} &= \delta_{j\kappa} + \frac{k_j k_\kappa \left(\cos\left(ckt\right) - 1\right)}{k^2}, \quad j,\kappa = 1,2,3 \end{aligned} \tag{16}$$

where $k^2 = k_1^2 + k_2^2 + k_3^2 = |\mathbf{k}|^2$. By introducing the function

$$\begin{aligned} \widetilde{F}(\mathbf{k}, t) = {} & \frac{d}{dt} \int_S \left\{\rho_0 u_n + \rho' \left(u_n - v_n\right)\right\} e^{-i\mathbf{k}\cdot\mathbf{y}} dS \\ & -ik_j \int_S \left\{\rho u_j \left(u_n - v_n\right) + P_j'\right\} e^{-i\mathbf{k}\cdot\mathbf{y}} dS + ik_j ik_\kappa \widetilde{T}'_{j\kappa} \end{aligned} \tag{17}$$

the components of the vector $\widetilde{\mathbf{w}}$ can be written as

$$\widetilde{\rho'}(\mathbf{k},t) = \int_{-\infty}^{t} \widetilde{F}(\mathbf{k},\tau)\frac{\sin(ck(t-\tau))}{ck}d\tau \qquad (18)$$

$$\widetilde{\rho u_j}(\mathbf{k},t) = -\int_{-\infty}^{t} ik_\kappa \widetilde{T}_{j\kappa}(\mathbf{k},\tau)d\tau + \int_{-\infty}^{t} d\tau \int_{S(\tau)} \left\{\rho u_j(u_n - v_n) + P'_j\right\} e^{-i\mathbf{k}\cdot\mathbf{y}}dS \qquad (19)$$
$$+ik_j \int_{-\infty}^{t} \widetilde{F}(\mathbf{k},\tau)\frac{\cos(ck(t-\tau))-1}{k^2}d\tau$$

The method given in this section has the advantage of furnishing the integral representations for the operational density and operational velocity as well. A simpler deduction of the representation of the operational density is given in the next section.

3.2 The equation for the operational density

By eliminating the Fourier transform of the momentum $\widetilde{\rho u_j}$ between equations (8) and (9) the operational equation satisfied by the density perturbation becomes

$$\frac{d^2\widetilde{\rho'}}{dt^2} + c^2k^2\widetilde{\rho'} = \widetilde{F}(\mathbf{k},t). \qquad (20)$$

The relationship (20) is the operational form of the nonhomogeneous wave equation. The general solution of the homogeneous wave equation in operational form can be written as

$$\widetilde{\rho'}_h = A(k)\cos(ckt) + B(k)\sin(ckt))$$

and by using Lagrange's method of variation of parameters there results the same representation formula (18) for the operational density as the solution of equation (20).

3.3 The case of an impermeable surface (S_b)

In the case where the surface S_b is impermeable, the condition (5) cancels out some terms in formula (17). In this case the nonhomogeneous term is

$$\widetilde{F}_b(\mathbf{k},t) = \frac{d}{dt}\int_S \rho_0 u_n e^{-i\mathbf{k}\cdot\mathbf{y}}dS - ik_j\int_S P'_j e^{-i\mathbf{k}\cdot\mathbf{y}}dS + ik_j ik_\kappa \widetilde{T}'_{j\kappa} \qquad (21)$$

which coincides with the nonhomogeneous term in the operational form of the FW-H equation.

By introducing the new variables suggested by Francescantonio (Francescantonio, 1997), in the form modified in (Brentner & Farassat, 1998)

$$U_j = \left(1 - \frac{\rho}{\rho_0}\right)v_j + \frac{\rho u_j}{\rho_0} \qquad (22)$$

$$L_j = P'_j + \rho u_j(u_n - v_n) \qquad (23)$$

the nonhomogeneous term of the operational wave equation in the case of a permeable surface coincides with that corresponding to an impermeable case

$$\widetilde{F}(\mathbf{k},t) = \frac{d}{dt}\int_S \rho_0 U_n e^{-i\mathbf{k}\cdot\mathbf{y}}dS - ik_j\int_S L_j e^{-i\mathbf{k}\cdot\mathbf{y}}dS + ik_j ik_\kappa \widetilde{T}_{j\kappa} \qquad (24)$$

The terms U_j and L_j can be interpreted respectively as a modified velocity and a modified traction, which take into account the flow across S.

4. Determination of velocity

4.1 The lift component of velocity

The lift component of velocity is given by inverse Fourier transform of the term

$$\left(\widetilde{\rho u_j}\right)_L = ik_j ik_r \int_{-\infty}^{t} d\tau \int_{S_\tau} P'(\mathbf{y},\tau)\, n_r'(\mathbf{y},\tau)\, e^{-i\mathbf{k}\cdot\mathbf{y}} \frac{\cos(ck(t-\tau))-1}{k^2} dS$$

Therefore, by using formula (73) we obtain

$$\left(\rho u_j\right)_L = \frac{\partial^2}{\partial x_j \partial x_r} \int_{-\infty}^{t} d\tau \int_{S_\tau} P' n_r' \frac{H(t-\tau-r/c)}{4\pi r} dS$$

where

$$r = |\mathbf{x}-\mathbf{y}| \equiv \sqrt{(x_1-y_1)^2+(x_2-y_2)^2+(x_3-y_3)^2}.$$

$\mathbf{x}(x_1,x_2,x_3)$ being the position vector of the observation point.

5. Determination of density

Since the general permeable case can be studied by means of an equivalent impermeable case we shall determine the density in the case where the nonhomogeneous term is (21).

5.1 The case of sources on a surface

Consider the simpler case when the noise sources are on a rigid surface having only translation and rotation motions. Then,

$$\widetilde{F}_S(\mathbf{k},t) = \int_S Q(\mathbf{y},t)\, e^{-i\mathbf{k}\cdot\mathbf{y}} dS \tag{25}$$

where $Q(\mathbf{y},t)$ is the surface intensity. In this case we have

$$\widetilde{\rho}_S'(\mathbf{k},t) = \int_{-\infty}^{t} d\tau \int_{S_\tau} Q(\mathbf{y},\tau)\, e^{-i\mathbf{k}\cdot\mathbf{y}} \frac{\sin(ck(t-\tau))}{ck} dS \tag{26}$$

Hence,

$$\begin{aligned}
\rho_S'(\mathbf{x},t) &= \frac{1}{(2\pi)^3}\int e^{i\mathbf{k}\cdot\mathbf{x}} d\mathbf{k} \int_{-\infty}^{t} d\tau \int_{S_\tau} Q(\mathbf{y},\tau)\, e^{-i\mathbf{k}\cdot\mathbf{y}} \frac{\sin(ck(t-\tau))}{ck} dS \\
&= \int_{-\infty}^{t} d\tau \int_{S_\tau} Q(\mathbf{y},\tau)\, dS \int \frac{\sin(ck(t-\tau))}{(2\pi)^3 ck} e^{i\mathbf{k}\cdot(\mathbf{x}-\mathbf{y})} d\mathbf{k} \\
&= \int_{-\infty}^{t} d\tau \int_{S_\tau} Q(\mathbf{y},\tau)\, \frac{\delta(\tau-(t-r/c))}{4\pi c^2 r} dS
\end{aligned} \tag{27}$$

Here the relationship (71) and the property $\delta(-t) = \delta(t)$ have been used.
The only contribution in the last integral in formula (27) comes from the time τ_e which is the solution of the equation

$$g(\tau,\mathbf{y},t,\mathbf{x}) = 0 \tag{28}$$

where

$$g(\tau,\mathbf{y},t,\mathbf{x}) = \tau - t + \frac{|\mathbf{x}-\mathbf{y}|}{c} \tag{29}$$

In other worlds, the value of the density at the observation point $\mathbf{x}$ at the moment t is determined by the noise sources at the *emission (radiating) time* τ_e on the *emission (radiating) surface* $S_e \equiv S_{\tau_e}$.

It is now necessay to consider the coordinate systems. Let a fixed point P_0 reside on a material surface such as an airplane wing or a blade etc. We consider two coordinate frames:

a) A frame η fixed relative to the moving material surface. This is the frame used by a designer to describe the structure geometry for purposes of fabrication. The variable η will be called the Lagrangian variable of the point P_0 and in the case supposed here (nondeformable material surface) is independent of time.

b) A coordinate frame fixed with respect to the undisturbed medium having the origin O (see Fig.1). The position of the observation point is given by the position vector $\mathbf{x}$. The position of the point P_0 is described by the position vector $\mathbf{y}\,(\eta,\tau)$. The position vectors $\mathbf{x}$ and $\mathbf{y}$ will give the Eulerian coordinates of the observation and source terms, respectively. The formula

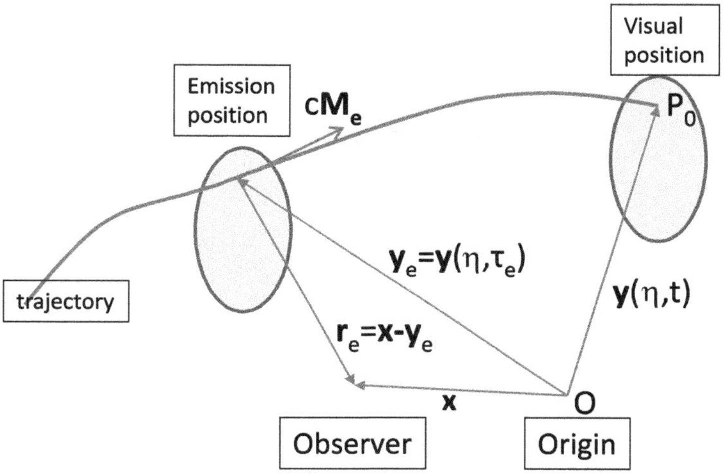

Fig. 1. Coordinate frame

$$\mathbf{y} = \eta + \int^{\tau} c\mathbf{M}\,(\eta,\tau')\,d\tau' \tag{30}$$

gives the connection between the Lagrangian and Eulerian coordinates of the point P_0. For η fixed the equation $\mathbf{y} = \mathbf{y}\,(\eta,\tau)$ gives the trajectory of the point P_0 and $c\mathbf{M} = d\mathbf{y}/d\tau$ is the velocity of the source point with respect to the undisturbed medium (source convection velocity). Formula (30) can be viewed as a transformation between the Lagrangian and Eulerian coordinates of the point P_0. The inverse transformation will be denoted by $\eta = \eta\,(\mathbf{y},\tau)$. In the case where the transformation involves only translations and rotations we have $\det\,(\partial\mathbf{y}/\partial\eta) = \det\,(\partial\eta/\partial\mathbf{y}) = 1$. Fig.1 shows the observer's position $\mathbf{x}$ at time t, the emission (radiation) position of the material surface ($S_e \equiv S_{\tau_e}$), the position of the same surface at the observation time (S_t), the position vector $\mathbf{y}_e$ at the emission time and at the observation time $\mathbf{y}\,(\eta,t)$, the trajectory of the point P_0, the convection velocity of the source at the emission moment $c\mathbf{M}_e$, the radiation vector $\mathbf{r}_e = \mathbf{x} - \mathbf{y}_e$, and the emission distance $r_e = |\mathbf{r}_e|$.

To change the integration variable in the last integral in formula (27) from τ to g, we calculate

$$\frac{dg}{d\tau} = 1 - \frac{1}{c}\nabla r \cdot \frac{\partial \mathbf{y}}{\partial \tau} = 1 - \frac{\mathbf{r}}{r} \cdot \frac{\mathbf{v}}{c} = 1 - M_r \tag{31}$$

where M_r is the Mach number at the point η in the radiation direction at the time τ. The density perturbation can therefore be written as

$$\rho'_S(\mathbf{x}, t) = \int_{S_e} \left[\frac{Q(\mathbf{y}, \tau)}{4\pi c^2 r \left|1 - M_r\right|} \right]_{ret} dS_\eta \equiv \int_{S^*(\tau_e)} \frac{Q^*(\eta, \tau_e)}{4\pi c^2 r_e \left|1 - M_{r_e}\right|} dS_\eta \tag{32}$$

Here, $|1 - M_{r_e}|$ is the *Doppler factor*. The square brackets $[]_{ret}$ imply that the contents are to be evaluated at the retarded (emission or radiating time) τ_e given implicitly by $g(\tau) = 0$. The emission position is $\mathbf{y}_e = \mathbf{y}(\eta, \tau_e)$, the emission distance r_e of the source point η to the observer position $\mathbf{x}$ is $r_e = |\mathbf{x} - \mathbf{y}(\eta_e, \tau_e)|$, and $Q^*(\eta, \tau_e) = Q(\mathbf{y}_e, \tau_e)$.

5.1.1 The thickness noise

The thickness noise is given by the term

$$\widetilde{F}_{thickness}(\mathbf{k}, t) = \frac{d}{dt} \int_S \rho_0 u_n e^{-i\mathbf{k}\cdot\mathbf{y}} dS$$

An analysis similar with that of the section 5.1 yields

$$\rho'_{thickness}(\mathbf{x}, t) = \frac{\partial}{\partial t} \int_S \left[\frac{\rho_0 u_n}{4\pi c^2 r \left|1 - M_r\right|} \right]_{ret} dS_\eta \tag{33}$$

5.1.2 The loading noise

The last term in relationship (21) describes the loading noise

$$\widetilde{F}_{loading}(\mathbf{k}, t) = -ik_j \int_S P'_j e^{-i\mathbf{k}\cdot\mathbf{y}} dS$$

Its contribution to the perturbed density ρ' is

$$\rho'_{loading}(\mathbf{x}, t) = -\frac{\partial}{\partial x_j} \int_S \left[\frac{P'_j}{4\pi c^2 r \left|1 - M_r\right|} \right]_{ret} dS_\eta \tag{34}$$

We have given here a very short presentation of formulas for thickness noise and loading noise. A more complete presentation about these formulas and their implementation can be found in (Farassat, 2007).

5.2 The quadrupole noise term

The last term in formula (21) corresponds to a quadrupole noise source:

$$\begin{aligned} \rho'_q(\mathbf{x}, t) &= \mathcal{F}^{-1}\left\{ ik_j ik_\kappa \widetilde{T}_{j\kappa} \right\} \\ &= \int_{-\infty}^{t} d\tau \int ik_j ik_\kappa \widetilde{T}_{j\kappa} \frac{\sin(ck(t-\tau))}{(2\pi)^3 ck} e^{i\mathbf{k}\cdot\mathbf{x}} d\mathbf{k} \\ &= \frac{\partial^2}{\partial x_j \partial x_\kappa} \int_{-\infty}^{t} d\tau \int_{D_\tau^{(e)}} T_{j\kappa}(\mathbf{y}, \tau) \, d\mathbf{y} \int \frac{\sin(ck(t-\tau))}{(2\pi)^3 ck} e^{i\mathbf{k}\cdot(\mathbf{x}-\mathbf{y})} d\mathbf{k}, \end{aligned} \tag{35}$$

where $D_{\tau}^{(e)}$ denotes the 3-D domain occupied by volume sources at the moment τ. The last integral in formula (35) was calculated in Appendix B. Introducing its expression given by formula (71) there results

$$\rho_q'(\mathbf{x},t) = \frac{\partial^2}{\partial x_j \partial x_\kappa} \int_{-\infty}^{t} d\tau \int_{D_{\tau}^{(e)}} T_{j\kappa}(\mathbf{y},\tau) \frac{\delta(\tau - (t - r/c))}{4\pi c^2 r} d\mathbf{y}, \tag{36}$$

the relationship (36) becomes

$$\rho_q'(\mathbf{x},t) = \frac{\partial^2}{\partial x_j \partial x_\kappa} \int_{D^{(e)}(\tau)} \left[\frac{T_{j\kappa}(\mathbf{y},\tau)}{4\pi c^2 r \left|1 - M_r\right|} \right]_{ret} d\eta. \tag{37}$$

where the effect of source convection is revealed by the Doppler factor; convection effectively increases the source strength by $\left|1 - M_r\right|^{-1}$. Further transformations of the formula (37) useful for its numerical implementation were made by Farassat and Brentner (Farassat and Brentner, 1988) and by Brentner in (Brentner, 1997). In the case where the discontinuity surface is permeable (of type S_p) this term is missing, the surface being usually chosen outside the space containing the quadruple sources. In the general case the sum of the solutions $\rho_q'(\mathbf{x},t)$, $\rho_{thickness}'$ and $\rho_{loading}'$ completely specify the density field.

6. The Kirchhoff method in Aeroacoustics

Besides the Acoustic Analogy approach for the solution of the Aeroacoustic noise, another widely used method is based on Kirchhoffs' solution of the wave equation. We start with the nonhomogeneous wave equation for the pressure perturbation

$$\left[\Delta - \frac{1}{c^2}\frac{\partial^2}{\partial t^2}\right] p'(\mathbf{x},t) = g(\mathbf{x},t) \tag{38}$$

where $g(\mathbf{x},t)$ represents the density of pressure sources. By applying the Fourier transform with respect to the spatial variables and using formulas (58) and (70) we obtain the operational form of the wave equation (38)

$$\frac{d^2\widetilde{p}'}{dt^2} + c^2k^2\widetilde{p}' = -c^2\widetilde{g}(\mathbf{k},t) + \widetilde{G}(\mathbf{k},t) \tag{39}$$

where

$$\widetilde{G}(\mathbf{k},t) = -\int_{S_t} \left(c^2 \left[\frac{\partial p'}{\partial n}\right] + v_n \left[\frac{\partial p'}{\partial t}\right] \right) e^{-i\mathbf{k}\cdot\mathbf{y}} dS \tag{40}$$
$$- \int_{S_t} c^2 i\mathbf{k}\cdot\mathbf{n} \left[p'\right] e^{-i\mathbf{k}\cdot\mathbf{y}} dS - \frac{d}{dt}\int_{S_t} v_n \left[p'\right] e^{-i\mathbf{k}\cdot\mathbf{y}} dS$$

Equation (39) is similar to equation (20). Consequently, its solution can be written as

$$\widetilde{p}'(\mathbf{k},t) = \int_{-\infty}^{t} \left(-c^2\widetilde{g}(\mathbf{k},\tau) + \widetilde{G}(\mathbf{k},\tau)\right) \frac{\sin(ck(t-\tau))}{ck} d\tau \tag{41}$$

Hence the contribution of the nonhomogeneous term in equation (38) can be written as

$$
\begin{aligned}
p'_g(\mathbf{x},t) &= -c^2 \int_{-\infty}^{t} d\tau \int \widetilde{g}(\mathbf{k},\tau) \frac{\sin(ck(t-\tau))}{(2\pi)^3 ck} e^{i\mathbf{k}\cdot\mathbf{x}} d\mathbf{k} \\
&= -\int_{-\infty}^{t} d\tau \int g(\mathbf{y},\tau) \frac{\delta(t-\tau-|\mathbf{x}-\mathbf{y}|/c)}{4\pi|\mathbf{x}-\mathbf{y}|} d\mathbf{y} \\
&= -\int g(\mathbf{y}, t-|\mathbf{x}-\mathbf{y}|/c) \frac{d\mathbf{y}}{4\pi|\mathbf{x}-\mathbf{y}|}
\end{aligned}
\tag{42}
$$

Finally, the contribution of the terms corresponding to the boundary conditions on the mobile surface S can be written as

$$
\begin{aligned}
p'_G(\mathbf{x},t) &= -\frac{\partial}{\partial x_i} \int_{S_t} \left[\frac{p' n_i}{4\pi r |1-M_r|}\right]_{ret} dS \\
&-\frac{\partial}{\partial t} \int_{S_t} \left[\frac{p' M_n}{4\pi r |1-M_r|}\right]_{ret} dS \\
&-\int_{S_t} \left[\left(\frac{\partial p'}{\partial n} + M_n \frac{\partial p'}{\partial \tau}\right) \frac{1}{4\pi r |1-M_r|}\right]_{ret} dS
\end{aligned}
\tag{43}
$$

which coincides with the relationship (5.3) given in (Ffowcs Williams and Hawkings, 1969).

7. Concluding remarks

Acoustic Analogy is one of the greatest contributions to the field of acoustics of the previous century. It is a major extension of acoustics made by Sir M. J. Lighthill (and other contributors) who formulated for the first time the science of how sound is created by fluid motion. This theory completes the previous work by famous researchers in the field of acoustics who had discovered how sound propagates through various media and across surrounding surfaces. In this chapter we have attempted to simplify the application of the Acoustic Analogy by showing how to apply it using only classical mathematical analysis tools.

8. Acknowledgment

This work has been supported by the National Institute on Deafness and Other Communication Disorders grant R01 DC009429 to RNM. The content is solely the responsibility of the authors and does not necessarily represent the official views of the National Institute on Deafness and Other Communication Disorders or the National Institutes of Health.

9. Appendix A: Fourier Transform of piecewise differentiable functions

Let $D^{(i)}(t)$ be a bounded mobile domain in $\mathbb{R}^3$ having a smooth boundary surface S_t. Denote by $D^{(e)}(t)$ the domain external to the surface S_t (See Fig.2): $D^{(i)} \cup S \cup D^{(e)} = \mathbb{R}^3$, $D^{(i)} \cap D^{(e)} = \phi$. Let also $\varphi^{(i)}(\mathbf{x},t)$ be a continuous differentiable function defined in the closed domain $\overline{D^{(i)}} \times \mathbb{R}$ and $\varphi^{(e)}(\mathbf{x},t)$ a continuous differentiable function defined in $\overline{D^{(e)}} \times \mathbb{R}$. We define also

$$
\varphi(\mathbf{x},t) = \begin{cases} \varphi^{(i)}(\mathbf{x},t), & \text{in } D^{(i)} \\ \varphi^{(e)}(\mathbf{x},t), & \text{in } D^{(e)} \end{cases}
\tag{44}
$$

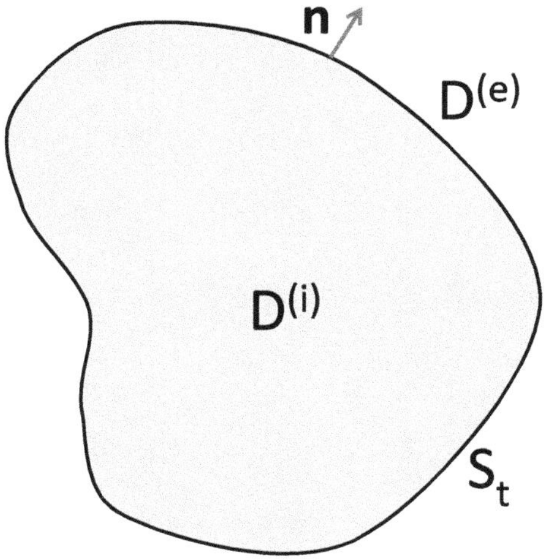

Fig. 2. Boundary Surface

Generally, the function $\varphi(\mathbf{x},t)$ is discontinuous across the surface S. We call it a piecewise differentiable function (pdf). Assuming that the function $\varphi^{(e)}(\mathbf{x},t)$ is decreasing sufficiently rapidly at infinity (for more precise conditions about the function $\varphi^{(e)}(\mathbf{x},t)$ see (Homentcovschi and Singler, 1999)) we can take the Fourier Transform of the function $\varphi(\mathbf{x},t)$ with respect to space variables

$$\widetilde{\varphi}(\mathbf{k},t) \equiv \mathcal{F}\{\varphi(\mathbf{x},t)\} = \int \varphi(\mathbf{x},t)\, e^{-i\mathbf{k}\cdot\mathbf{x}} d\mathbf{x} \tag{45}$$

Here $\mathbf{x} = (x_1,x_2,x_3)$, $\mathbf{k} = (k_1,k_2,k_3)$, $\mathbf{k}\cdot\mathbf{x} = k_1x_1 + k_2x_2 + k_3x_3$ is the inner product of the two vectors, $d\mathbf{x} = dx_1dx_2dx_3$ and the integral is extended over the whole $\mathbb{R}^3$ space. The inversion formula can be written as

$$\varphi(\mathbf{x},t) \equiv \mathcal{F}^{-1}\{\widetilde{\varphi}(\mathbf{k},t)\} = \frac{1}{(2\pi)^3}\int \widetilde{\varphi}(\mathbf{k},t)\, e^{i\mathbf{x}\cdot\mathbf{k}} d\mathbf{k} \tag{46}$$

where $d\mathbf{k} = dk_1dk_2dk_3$. Accounting for relationship (44) we can write

$$\mathcal{F}\{\varphi(\mathbf{x},t)\} = \int_{D^{(i)}} \varphi^{(i)}(\mathbf{x},t)\, e^{-i\mathbf{k}\cdot\mathbf{x}} d\mathbf{x} + \int_{D^{(e)}} \varphi^{(e)}(\mathbf{x},t)\, e^{-i\mathbf{k}\cdot\mathbf{x}} d\mathbf{x} \tag{47}$$

9.1 Fourier transform of the derivative with respect to a spatial variable of a piecewise differentiable function

9.1.1 The first basic formula

We write

$$\mathcal{F}\left\{\frac{\partial\varphi}{\partial x_1}\right\} = \int_{D^{(i)}} \frac{\partial\varphi^{(i)}(\mathbf{x},t)}{\partial x_1} e^{-i\mathbf{k}\cdot\mathbf{x}} d\mathbf{x} + \int_{D^{(e)}} \frac{\partial\varphi^{(e)}(\mathbf{x},t)}{\partial x_1} e^{-i\mathbf{k}\cdot\mathbf{x}} d\mathbf{x} \tag{48}$$

But

$$\int_{D^{(i)}} \frac{\partial\varphi^{(i)}}{\partial x_1} e^{-i\mathbf{k}\cdot\mathbf{x}} d\mathbf{x} = \int_{D^{(i)}} \frac{\partial}{\partial x_1}\left(\varphi^{(i)} e^{-i\mathbf{k}\cdot\mathbf{x}}\right) d\mathbf{x} + ik_1 \int_{D^{(i)}} \varphi^{(i)}(\mathbf{x},t)\, e^{-i\mathbf{k}\cdot\mathbf{x}} d\mathbf{x} \tag{49}$$

The first integral in the right-hand side of relationship (49) can be replaced, by using the divergence theorem by an integral over the boundary surface S_t.

$$\int_{D^{(i)}} \frac{\partial}{\partial x_1}\left(\varphi^{(i)} e^{-i\mathbf{k}\cdot\mathbf{x}}\right) d\mathbf{x} = \int_{S_t} n_1 \varphi^{(i)}(\mathbf{y},t)\, e^{-i\mathbf{k}\cdot\mathbf{y}} dS \tag{50}$$

where $\mathbf{n} = (n_1, n_2, n_3)$ is the external unit normal to S_t. Therefore,

$$\int_{D^{(i)}} \frac{\partial \varphi^{(i)}}{\partial x_1} e^{-i\mathbf{k}\cdot\mathbf{x}} d\mathbf{x} = ik_1 \int_{D^{(i)}} \varphi^{(i)}(\mathbf{x},t)\, e^{-i\mathbf{k}\cdot\mathbf{x}} d\mathbf{x} + \int_{S_t} n_1 \varphi^{(i)}(\mathbf{y},t)\, e^{-i\mathbf{k}\cdot\mathbf{y}} dS \tag{51}$$

Similarly,

$$\int_{D^{(e)}} \frac{\partial \varphi^{(e)}}{\partial x_1} e^{-i\mathbf{k}\cdot\mathbf{x}} d\mathbf{x} = ik_1 \int_{D^{(e)}} \varphi^{(e)}(\mathbf{x},t)\, e^{-i\mathbf{k}\cdot\mathbf{x}} d\mathbf{x} - \int_{S_t} n_1 \varphi^{(e)}(\mathbf{y},t)\, e^{-i\mathbf{k}\cdot\mathbf{y}} dS \tag{52}$$

Finally, the relationships (49), (51), and (52) give *the first basic formula*:

$$\mathcal{F}\left\{\frac{\partial \varphi}{\partial x_1}\right\} = ik_1\, \widetilde{\varphi}(\mathbf{k},t) - \int_S n_1 [\varphi(\mathbf{y},t)]\, e^{-i\mathbf{k}\cdot\mathbf{y}} dS \tag{53}$$

Here we denoted by square bracket the jump of the function $\varphi(\mathbf{x},t)$ across the surface S_t

$$[\varphi(\mathbf{y},t)] = \lim_{\mathbf{x}^{(e)}\to\mathbf{y}} \varphi^{(e)}\left(\mathbf{x}^{(e)},t\right) - \lim_{\mathbf{x}^{(i)}\to\mathbf{y}} \varphi^{(i)}\left(\mathbf{x}^{(i)},t\right) \tag{54}$$

for $\mathbf{y} \in S_t$, $\mathbf{x}^{(i)} \in D^{(i)}$ and $\mathbf{x}^{(e)} \in D^{(e)}$. Similar relationships to (53) can be proved for the derivatives in the directions x_2 and x_3.

Remark 3. *It is clear that the obtained relationships can be extended immediately to the case where there are more discontinuity surfaces of the given function. The resulting formulas will contain sums of integrals corresponding to each discontinuity surface.*

9.1.2 Other formulas

The relationship (53) gives also

$$\mathcal{F}\{\nabla\varphi\} \equiv \widetilde{\nabla\varphi} = i\mathbf{k}\, \widetilde{\varphi}(\mathbf{k},t) - \int_{S_t} \mathbf{n}\, [\varphi(\mathbf{y},t)]\, e^{-i\mathbf{k}\cdot\mathbf{y}} dS \tag{55}$$

In the case where we write $\mathbf{V}(\mathbf{x},t) = (V_1(\mathbf{x},t), V_2(\mathbf{x},t), V_3(\mathbf{x},t))$ where $V_j(\mathbf{x},t)$ is a piecewise differentiable function defined by a relationship similar to (44) we can write also the formulas

$$\mathcal{F}\{\nabla\cdot\mathbf{V}\} \equiv \widetilde{\nabla\cdot\mathbf{V}} = i\mathbf{k}\cdot\widetilde{\mathbf{V}}(\mathbf{k},t) - \int_{S_t} \mathbf{n}\cdot [\mathbf{V}(\mathbf{y},t)]\, e^{-i\mathbf{k}\cdot\mathbf{y}} dS \tag{56}$$

$$\mathcal{F}\{\nabla\times\mathbf{V}\} \equiv \widetilde{\nabla\times\mathbf{V}} = i\mathbf{k}\times\widetilde{\mathbf{V}}(\mathbf{k},t) - \int_{S_t} \mathbf{n}\times [\mathbf{V}(\mathbf{y},t)]\, e^{-i\mathbf{k}\cdot\mathbf{y}} dS \tag{57}$$

The formulas (55), (56) and (57) permit the calculation of the Fourier Transforms of a gradient of a scalar field of a divergence and a curl of a vector field in the case of piecewise differentiable scalar and vector fields.

Moreover, we can write

$$\mathcal{F}\{\Delta\varphi\} \equiv \mathcal{F}\{\nabla\cdot\nabla\varphi\} = i\mathbf{k}\cdot\widetilde{\nabla\varphi}(\mathbf{k},t) - \int_{S_t}\mathbf{n}\cdot[\nabla(\mathbf{y},t)\varphi]\,e^{-i\mathbf{k}\cdot\mathbf{y}}dS \tag{58}$$

$$= i\mathbf{k}\cdot\left(i\mathbf{k}\,\widetilde{\varphi}(\mathbf{k},t) - \int_{S_t}\mathbf{n}\,[\varphi(\mathbf{y},t)]\,e^{-i\mathbf{k}\cdot\mathbf{y}}dS\right) - \int_{S_t}\left[\frac{\partial\varphi}{\partial n}(\mathbf{y},t)\right]e^{-i\mathbf{k}\cdot\mathbf{y}}dS$$

Finally,

$$\widetilde{\Delta\varphi} = -k^2\widetilde{\varphi}(\mathbf{k},t) - \int_{S_t} i\mathbf{k}\cdot\mathbf{n}\,[\varphi(\mathbf{y},t)]\,e^{-i\mathbf{k}\cdot\mathbf{y}}dS - \int_{S_t}\left[\frac{\partial\varphi}{\partial n}(\mathbf{y},t)\right]e^{-i\mathbf{k}\cdot\mathbf{y}}dS \tag{59}$$

where $k^2 = k_1^2 + k_2^2 + k_3^2 = |\mathbf{k}|^2$.

9.2 Fourier transform of the time derivative of a piecewise differentiable function

9.2.1 The displacement velocity of a mobile surface

Let $S(y_1, y_2, y_3, t) = 0$ be the equation of the mobile surface S_t. Then, for $S(y_{01}, y_{02}, y_{03}, t_0)$ on S_{t_0} we can write

$$0 = S(y_1, y_2, y_3, t) - S(y_{01}, y_{02}, y_{03}, t_0) = \tag{60}$$
$$\left(\frac{\partial S}{\partial y_1}\frac{dy_1}{dt} + \frac{\partial S}{\partial y_2}\frac{dy_2}{dt} + \frac{\partial S}{\partial y_3}\frac{dy_3}{dt} + \frac{\partial S}{\partial t}\right)(t - t_0),$$

the partial derivatives of the function S being calculated at a certain point $\mathbf{x}'$ lying between the points $\mathbf{x}(t_0)$ and $\mathbf{x}(t)$. But,

$$n_1 = \frac{\partial S/\partial y_1}{|grad\, S|}, n_2 = \frac{\partial S/\partial y_2}{|grad\, S|}, n_3 = \frac{\partial S/\partial y_3}{|grad\, S|} \tag{61}$$

are the components of the external normal unit vector $\mathbf{n}$ and

$$\frac{dy_1}{dt} = v_1, \frac{dy_2}{dt} = v_2, \frac{dy_3}{dt} = v_3, \tag{62}$$

are the projections on the velocity vector of a point on the surface S_t on the three axes. The relation (60) yields

$$v_n = -\frac{\partial S/\partial t}{|grad\, S|} \tag{63}$$

which is the displacement velocity of the geometric surface S_t. We mention that the displacement velocity of a surface has the direction of the normal vector to this surface.

9.2.2 Reynolds' transport theorem

For calculating the Fourier Transform of a time derivative of a certain function we write

$$\int_{D^{(i)}(t)}\varphi^{(i)}(\mathbf{x},t)\,e^{-i\mathbf{k}\cdot\mathbf{x}}d\mathbf{x} - \int_{D^{(i)}(t_0)}\varphi^{(i)}(\mathbf{x},t_0)\,e^{-i\mathbf{k}\cdot\mathbf{x}}d\mathbf{x} = \tag{64}$$
$$\int_{D^{(i)}(t)}\varphi^{(i)}(\mathbf{x},t)\,e^{-i\mathbf{k}\cdot\mathbf{x}}d\mathbf{x} - \int_{D^{(i)}(t)}\varphi^{(i)}(\mathbf{x},t_0)\,e^{-i\mathbf{k}\cdot\mathbf{x}}d\mathbf{x} +$$
$$\int_{D^{(i)}(t)}\varphi^{(i)}(\mathbf{x},t_0)\,e^{-i\mathbf{k}\cdot\mathbf{x}}d\mathbf{x} - \int_{D^{(i)}(t_0)}\varphi^{(i)}(\mathbf{x},t_0)\,e^{-i\mathbf{k}\cdot\mathbf{x}}d\mathbf{x} =$$
$$\int_{D^{(i)}(t)}\left\{\varphi^{(i)}(\mathbf{x},t) - \varphi^{(i)}(\mathbf{x},t_0)\right\}e^{-i\mathbf{k}\cdot\mathbf{x}}d\mathbf{x} + \int_{D^{(i)}(t)-D^{(i)}(t_0)}\varphi^{(i)}(\mathbf{x},t_0)\,e^{-i\mathbf{k}\cdot\mathbf{x}}d\mathbf{x}$$

Now, dividing by $(t - t_0)$ and taking the limit for $t \to t_0$ the first term gives the Fourier transform of the time derivative of the function $\varphi^{(i)}(\mathbf{x}, t)$ while in the second integral we can write $d\mathbf{x} = v_n (t - t_0)\, dS$ (see (Jacob, 1959), (Currie, 2003)). Finally, we obtain the following form of the Reynolds' transport theorem

$$\frac{d}{dt}\int_{D^{(i)}(t)} \varphi^{(i)}(\mathbf{x}, t)\, e^{-i\mathbf{k}\cdot\mathbf{x}} d\mathbf{x} = \int_{D^{(i)}(t)} \frac{\partial \varphi^{(i)}(\mathbf{x}, t)}{\partial t} e^{-i\mathbf{k}\cdot\mathbf{x}} d\mathbf{x} + \int_{S_t} v_n(\mathbf{y}, t)\, \varphi^{(i)}(\mathbf{y}, t)\, e^{-i\mathbf{k}\cdot\mathbf{y}} dS \tag{65}$$

v_n being the displacement velocity of the surface $S(t)$.

9.2.3 The second basic formula

We calculate now

$$\mathcal{F}\left\{\frac{\partial \varphi(\mathbf{x}, t)}{\partial t}\right\} = \int_{D^{(i)}(t)} \frac{\partial \varphi^{(i)}(\mathbf{x}, t)}{\partial t} e^{-i\mathbf{k}\cdot\mathbf{x}} d\mathbf{x} + \int_{D^{(e)}(t)} \frac{\partial \varphi^{(e)}(\mathbf{x}, t)}{\partial t} e^{-i\mathbf{k}\cdot\mathbf{x}} d\mathbf{x}$$

By using formula (65) we get

$$\int_{D^{(i)}(t)} \frac{\partial \varphi^{(i)}(\mathbf{x}, t)}{\partial t} e^{-i\mathbf{k}\cdot\mathbf{x}} d\mathbf{x} = \frac{d}{dt}\int_{D^{(i)}(t)} \varphi^{(i)} e^{-i\mathbf{k}\cdot\mathbf{x}} d\mathbf{x} - \int_{S_t} v_n\, \varphi^{(i)}(\mathbf{y}, t)\, e^{-i\mathbf{k}\cdot\mathbf{y}} dS \tag{66}$$

$$\int_{D^{(e)}(t)} \frac{\partial \varphi^{(e)}(\mathbf{x}, t)}{\partial t} e^{-i\mathbf{k}\cdot\mathbf{x}} d\mathbf{x} = \frac{d}{dt}\int_{D^{(e)}(t)} \varphi^{(e)} e^{-i\mathbf{k}\cdot\mathbf{x}} d\mathbf{x} + \int_{S_t} v_n \varphi^{(e)}(\mathbf{y}, t)\, e^{-i\mathbf{k}\cdot\mathbf{y}} dS \tag{67}$$

The sum of formulas (66) and (67) gives the *second basic formula*

$$\mathcal{F}\left\{\frac{\partial \varphi(\mathbf{x}, t)}{\partial t}\right\} = \frac{d\,\widetilde{\varphi}(\mathbf{k}, t)}{dt} + \int_{S_t} v_n(\mathbf{y}, t)\, [\varphi(\mathbf{y}, t)]\, e^{-i\mathbf{k}\cdot\mathbf{y}} dS. \tag{68}$$

Formula (68) permits us to calculate the Fourier transform of the time derivative of a piecewise differentiable function. For the second time derivative we can write

$$\mathcal{F}\left\{\frac{\partial^2 \varphi(\mathbf{x}, t)}{\partial t^2}\right\} = \frac{d}{dt}\widetilde{\frac{d\varphi}{dt}} + \int_{S_t} v_n(\mathbf{y}, t) \left[\frac{\partial \varphi(\mathbf{y}, t)}{\partial t}\right] e^{-i\mathbf{k}\cdot\mathbf{y}} dS \tag{69}$$

By using again formula (68) we obtain finally,

$$\widetilde{\frac{\partial^2 \varphi}{\partial t^2}} = \frac{d^2\,\widetilde{\varphi}}{dt^2} + \frac{d}{dt}\int_{S_t} v_n\, [\varphi]\, e^{-i\mathbf{k}\cdot\mathbf{y}} dS + \int_{S_t} v_n \left[\frac{\partial \varphi}{\partial t}\right] e^{-i\mathbf{k}\cdot\mathbf{y}} dS \tag{70}$$

such that in the Fourier transform of second time derivative of the piecewise differentiable function φ enters the jump of the function φ across the discontinuity and the jump of the first time derivative of φ as well.

10. Appendix B: Determination of the Greens' function for the wave equation

By using formula 11, given at page 364 in (Gelfand and Shilov, 1964) we can write

$$\mathcal{F}^{-1}\left\{\frac{\sin(ckt)}{k}\right\} = \frac{\delta(t - r/c)}{4\pi c\, r} \tag{71}$$

where $r = |\mathbf{x}|$.
By taking the derivative with respect to parameter t in the both sides of relationship (71) there results

$$\mathcal{F}^{-1}\{\cos(ckt)\} = \frac{\delta'(t - r/c)}{4\pi c^2 r} \tag{72}$$

Similarly, integrating the formula (71) over the interval $(0, t)$ we obtain

$$\mathcal{F}^{-1}\left\{\frac{1 - \cos(ckt)}{k^2}\right\} = \frac{H(t - r/c)}{4\pi r} \tag{73}$$

$H(x)$ being the Heaviside's function.

11. References

Brentner, K. S., An efficient and robust method for predicting helicopter high-speed impulsive noise, Journal of Sound and Vibration, 1997, 203(1), 87-100.

Brentner, K. S. & Farassat, F. 1998 Analytical comparison of the acoustic analogy and Kirchhoff formulation for moving surfaces. AIAA J. 36(8), 1998, 1379–1386.

Crighton, D. G., Dowling, A. P., Ffowcs Williams, J. E., Heckl, M., and Leppington, F. G., Modern Methods in Analytical Acoustics: Lecture Notes, Springer–Verlag, London, 1992. Chap.11, Sec. 10.

Curle, N., The influence of solid boundaries upon aerodynamic sound, Proceedings of the Royal Society A 231 (1955) 505–514.

Currie, I. G., Fundamental Mechanics of Fluids, 3rd edition, Marcel Dekker, 2003, pg.12.

Dowling, A. P., and Ffowcs Williams, J.E., Sound and Sources of Sound, Wiley &Sons, New York, 1982. Chap. 9, Sec. 2.

Farassat, F., Introduction to generalized functions with applications in aerodynamics and aeroacoustics, Corrected Copy (April 1996), NASA Technical Paper 3428, 1994,

Farassat, F., The Kirchhoff formulas for moving surfaces in aeroacoustics—the subsonic and supersonic cases, NASA Technical Memorandum 110285, 1996.

Farassat, F., Comments on the paper by Zinoviev and Bies "On acoustic radiation by a rigid object in a fluid flow", Journal of Sound and Vibration 281 (2005) 1217–1223.

Farassat, F., Derivation of Formulations 1 and 1A of Farassat, NASA Technical Memorandum 214853, 2007.

Farassat, F., and Brentner, K. S., The uses and abuses of the acoustic analogy in helicopter rotor noise prediction, Journal of the American Helicopter Society, 1988, 33, 29-36

Farassat, F., Myers, M.K., Further comments on the paper by Zinoviev and Bies, "On acoustic radiation by a rigid object in a fluid flow", Journal of Sound and Vibration 290 (2006), 538-547.

Ffowcs Williams, J.E., and Hawkings, D.L., Sound generation by turbulence and surfaces in arbitrary motion, Philosophical Transactions of the Royal Society A 264 (1969) 321–342.

Francescantonio, P. Di,A new boundary integral formulation for the prediction of sound radiation, J. Sound Vibr. 202 (1997) (4), pp. 491–509.

Gelfand, I. M., and Shilov, G. E., Generalized Functions, Volume 1, Academic Press, New York and London, 1964.

Goldstein, M. E., Aeroacoustics, McGraw-Hill Book Co., 1976.

Homentcovschi, D., and Singler, T., An introduction to BEM by integral transforms, Engineering Analysis with Boundary Elements, 23 (1999) 603-609.

Jacob, C., Introduction Mathematique a la Mechanique des Fluides, Gauthier-Villlars, Paris, 1959.

Lighthill, M.J., On sound generated aerodynamically, Proceedings of the Royal Society of London, A 211 (1952) 564–586.

Lighthill, M.J., On sound generated aerodynamically: Turbulence as a source of sound, Proceedings of the Royal Society of London, A 222 (1954), 1-32.

Lyrintzis, A. (1994). Review: The use of Kirchhoff's method in computational aerodynamics, *Transactions of the ASME, Journal of Fluid Engineering* 116(4): 665–676.

Lyrintzis, A.S., Surface integral methods in computational aeroacoustics.-From the (CFD) near field to the (Acoustic) far-field, Aeroacoustics, 2 (2003) pp. 95-128.

Morino, L. (2003). Is there a difference between aeroacoustics and aerodynamics? An aeroelastician's viewpoint, *AIAA JOURNAL* 41(7): 1209–1223.

Morris, P.J. and Farassat, F., Acoustic analogy and alternative theories for jet noise prediction, AIAA Journal, 40 (2002), pp. 671-680.

Pilon, A. R., and Lyrintzis, A. S., "Integral Methods for Computational Aeroacoustics," AIAA paper No. 97-0020, presented at the 35th Aerospace Science Meeting, Reno, NV, Jan. 1997.

Raman, G. (editor), Computational Aeroacoustics, Multiscience Publishing Co. Ltd., 2009, 507pp.

Wu, X-F, and Akay A., Sound radiation from vibrating bodies in motion, J. Acoust. Soc. Am., 91 (1992) pp.2544-2555.

Zinoviev, A., Bies, D.A., On acoustic radiation by a rigid object in a fluid flow, Journal of Sound and Vibration 269 (2004) 535–548.

Zinoviev, A., Bies, D.A., Author's Reply to: F. Farassat, Comments on the paper by Zinoviev and Bies "On acoustic radiation by arigid object in a fluid flow", Journal of Sound and Vibration 281 (2005) 1224–1237.

Zinoviev, A., Bies, D.A., Author's Reply, Journal of Sound and Vibration 290 (2006) 548-554.

Exact Solutions Expressible in Hyperbolic and Jacobi Elliptic Functions of Some Important Equations of Ion-Acoustic Waves

A. H. Khater[1] and M. M. Hassan[2]
[1] *Mathematics Department, Faculty of Science, Beni-Suef University, Beni-Suef*
[2] *Mathematics Department, Faculty of Science, Minia University, El-Minia*
Egypt

1. Introduction

Many phenomena in physics and other fields are often described by nonlinear partial differential equations (NLPDEs). The investigation of exact and numerical solutions, in particular, traveling wave solutions, for NLPDEs plays an important role in the study of nonlinear physical phenomena. These exact solutions when they exist can help one to well understand the mechanism of the complicated physical phenomena and dynamical processes modeled by these nonlinear evolution equations (NLEEs). The ion-acoustic solitary wave is one of the fundamental nonlinear wave phenomena appearing in fluid dynamics [1] and plasma physics [2, 3]. It has recently became more interesting to obtain exact analytical solutions to NLPDEs by using appropriate techniques and symbolical computer programs such as Maple or Mathematica. The capability and power of these software have increased dramatically over the past decade. Hence, direct search for exact solutions is now much more viable. Several important direct methods have been developed for obtaining traveling wave solutions to NLEEs such as the inverse scattering method [3], the tanh-function method [4], the extended tanh-function method [5] and the homogeneous balance method [6]. We assume that the exact solution is expressed by a simple expansion $u(x,t) = U(\xi) = \sum_{i=0}^{N} A_i F^i(\xi)$ where A_i are constants to be determined and the function $F(\xi)$ is defined by the solution of an auxiliary ordinary differential equation (ODE). The tanh-function method is the well known method as a direct selection of the function $F(\xi) = tanh(\xi)$. Recently, many exact solutions expressed by various Jacobi elliptic functions (JEFs) of many NLEEs have been obtained by Jacobi elliptic function expansion method [7-10], mapping method [11, 12], F-expansion method [13], extended F-expansion method [14], the generalized Jacobi elliptic function method [15] and other methods [16-20]. Various exact solutions were obtained by using these methods, including the solitary wave solutions, shock wave solutions and periodic wave solutions.

The main steps of the F-expansion method [13] are outlined as follows:

Step 1. Use the transformation $u(x,t) = u(\xi)$; $\xi = k(x - \omega t) + \xi_0$, ξ_0 is an arbitrary constant, and reduce a given NLPDE, say in two independent variables,

$$F(u, u_t, u_x, u_{tt}, u_{xx}, ...) = 0, \tag{1.1}$$

to the (ODE)

$$G(u,u',u'',...) = 0, \qquad u' = \frac{du}{d\xi}. \tag{1.2}$$

In general, the left hand side of Eq. (1.1) is a polynomial in u and its various derivatives.
Step 2. The F-expansion method gives the solution of (1.1) in the form

$$u(x,t) = u(\xi) = \sum_{i=0}^{N} a_i F^i(\xi), \qquad a_N \neq 0, \tag{1.3}$$

where a_i $(i = 0,1,2,...,N)$ are constants to be determined and $F(\xi)$ satisfies the first order nonlinear ODE in the form

$$(F'(\xi))^2 = q_0 + q_2 F^2(\xi) + q_4 F^4(\xi), \tag{1.4}$$

where q_0, q_2 and q_4 are constants and N in Eq. (1.3) is a positive integer that can be determined by balancing the nonlinear term(s) and the highest order derivatives in Eq. (1.1).
Step 3. Substituting the F-expansion (1.3) into (1.2) and using (1.4); setting each coefficient of the polynomial to zero yields a system of algebraic equations involving a_0, a_1, ...a_N, k and ω.
Step 4. Solving these equations, probably with the aid of Mathematica or Maple, then a_0, a_1, ...a_N, k and ω can be expressed by q_0, q_2, q_4.
Step 5. Substituting these results into F-expansion (1.3), then a general form of traveling wave solution of the NLPDE (1.1) can be obtained. Many solutions of equation (1.4) have been reported in [13, 14]. Substituting the values of q_0, q_2, q_4 and the corresponding JEF solution $F(\xi)$ into the general form of solution, we may get several classes of exact solutions of equations (1.1) involving JEFs.
Also, we give a brief description of the mapping method to seek the traveling wave solutions of (1.1) in the form $u(x,t) = u(\eta)$, $\eta = kx - \omega t + \eta_0$, η_0 is an arbitrary constant. Thus, Eq. (1.1) reduces to Eq. (1.2), whose solution can be express in the form

$$u(\eta) = \sum_{i=0}^{n} A_i f^i(\eta), \tag{1.5}$$

where n is a balancing number, A_i are constants to be determined and $f(\eta)$ satisfies the nonlinear ODE

$$f'^2(\eta) = 2p\,f(\eta) + q\,f^2(\eta) + \frac{2}{3} r\,f^3(\eta). \tag{1.6}$$

Here p, q and r are constants. After substituting Eq. (1.5) into the ODE (1.2) and using Eq. (1.6), the constants A_i, k and ω may be determined. By using the solutions of auxiliary nonlinear equation (1.6), many JEF solutions of NLEEs have been obtained [19, 20].
The JEFs $\mathrm{sn}(\xi) = \mathrm{sn}(\xi, m)$, $\mathrm{cn}(\xi) = \mathrm{cn}(\xi, m)$ and $\mathrm{dn}(\xi) = \mathrm{dn}(\xi, m)$ are double periodic and have the following properties:

$$\mathrm{sn}^2(\xi) + \mathrm{cn}^2(\xi) = 1, \qquad \mathrm{dn}^2(\xi) + m^2\,\mathrm{sn}^2(\xi) = 1.$$

In the limit $m \longrightarrow 1$, the JEFs degenerate to the hyperbolic functions, i.e.,

$$\mathrm{sn}(\xi,1) \longrightarrow \tanh(\xi), \;\; \mathrm{cn}(\xi,1) \longrightarrow \mathrm{sech}(\xi), \;\; \mathrm{dn}(\xi,1) \longrightarrow \mathrm{sech}(\xi).$$

Detailed explanations about JEFs can be found in [21].
Some of the nonlinear models in fluids, plasma and dust plasma are described by canonical models and include the Korteweg-de Vries (KdV) and the modified KdV equations [22-25].

The evolution of small but finite-amplitude solitary waves, studied by means of the Korteweg-de Vries (KdV) equation, is of considerable interest in plasma dynamics. In the study of multidimensional version two type of nonlinear waves are well known, the so called Kadomtsev-Petviashvilli (KP) equation and Zakharov - Kuzentsov (ZK) equation. Employing the reductive perturbation technique on the system of equations for hydrodynamics and the dynamics of plasma waves to derive such equation.
We construct several classes of exact JEF solutions of some nonlinear evolution equations of plasma physics by using the mapping method and the F-expansion method. The rest of this chapter is organized as follows: in section 2, we present the JEF solutions to the KdV equation, combined KdV - modified KdV equation. In section 3, we apply the F-expansion method to the Schamel- KdV equation. Moreover, using the ansatz solution (1.5) and the solutions of nonlinear ODE (1.6), many exact solutions of Schamel equation, ZK equation and modified fifth order KdV equation are given in sections 4, 5, 6.

2. The KdV and modified KdV equations

The Korteweg de-Vries (KdV) equation

$$u_t + \alpha u u_x + u_{xxx} = 0,$$

models a variety of nonlinear phenomena, including ion acoustic waves in plasmas, dust acoustic solitary structures in magnetized dusty plasmas, and shallow water waves. On the other hand, the modified KdV equation (mKdV)

$$u_t + b u^2 u_x + u_{xxx} = 0,$$

models the dust-ion acoustic waves, electromagnetic waves in size-quantized films, ion acoustic solitons, traffic flow problems, and in other applications. The KdV equation and the modified KdV equation are completely integrable equations that have multiple-soliton solutions and possess infinite conservation quantities. The KdV equation is the earliest soliton equation that was firstly derived by Korteweg and de Vries to model the evolution of shallow water wave in 1895. In the study of the KdV equation, traveling wave solution leads to periodic solution which is called cnoidal wave solution [22, 23]. Exact solutions of KdV equation have been studied extensively since they were first found. Solitary wave solutions and periodic wave solutions were obtained for the KdV and modified KdV equations [3, 7, 22]. The JEF solutions to two kinds of KdV equations with variables coefficients have been constructed by using the method of the auxiliary equation [19]. The reductive perturbation method [24] has been employed to derive the KdV equation for small but finite amplitude electrostatic ion-acoustic waves [23, 25, 26]. The basic equations describing the system in dimensionless variables is studied by El-Labany [26] and the KdV equation for the first-order perturbed potential has been obtained using the reductive perturbation method.
We consider the combined KdV and mKdV equation [22, 27, 28]

$$u_t + \alpha u u_x + \beta u^2 u_x + \delta u_{xxx} = 0, \quad \beta \neq 0. \tag{2.1}$$

where α, β and δ are constants. Equation (2.1) is widely used in various fields such as quantum field theory, dust-acoustic waves, ion acoustic waves in plasmas with a negative ion, solid-state physics and fluid dynamics.
Let $u = u(\xi)$, equation (2.1) transformed to the reduced equation

$$-\omega u' + \alpha u u' + \beta u^2 u' + \delta k^2 u''' = 0. \tag{2.2}$$

Balancing u''' with u^2u' yields $N = 1$, so the F-expansion method gives

$$u(x,t) = a_0 + a_1 F(\xi). \tag{2.3}$$

Substituting (2.3) into (2.2) and equating the coefficients of like powers of $F(\xi)$ to zero, we obtain a set of algebraic equations. Solving these algebraic equations, we obtain the exact solutions of (2.1) as follows:
When $q_0 = 1,\ q_2 = -1 - m^2,\ q_4 = m^2$, solutions of Eq. (1.4) is $F(\xi) = \text{sn}\xi$, we have

$$u = -\frac{\alpha}{2\beta} \pm k\sqrt{\frac{-6m^2\delta}{\beta}}\,\text{sn}\Big(k(x + (\frac{\alpha^2 + 4\beta\delta k^2(m^2+1)}{4\beta})t) + \xi_0\Big), \tag{2.4}$$

If $q_0 = m^2 - 1,\ q_2 = 2 - m^2,\ q_4 = -1$, the solution of Eq (1.4) is $F(\xi) = \text{dn}\xi$. Thus, we obtain the periodic wave solutions of Eq. (2.1)

$$u = -\frac{\alpha}{2\beta} \pm k\sqrt{\frac{6\delta}{\beta}}\,\text{dn}(k(x + \frac{\alpha^2 - 4\beta\delta k^2(2-m^2)}{4\beta}t) + \xi_0), \tag{2.5}$$

Selecting the values of the q_0, q_2 and q_4 of equation (1.4) and the corresponding function F, we can construct various JEF solutions of (2.1). Other JEF solutions are omitted here for simplicity. If we put $\alpha = 0$ in (2.4), we get the periodic solution of the modified KdV equation which coincides with that given by Liu et al. [7]. Moreover, the solutions (2.5) to equation (2.1) given in [28] are recovered. With $m \longrightarrow 1$ in (2.4) , (2.5), the solitary wave solutions to (2.1) given in [7, 27, 28] are also recovered.
We notice that the solutions of the KdV equation cannot obtain from (2.4) and (2.5) as $\beta = 0$. In this case, the general form of cnoidal wave solutions of the KdV equation are given by

$$u(x,t) = -\frac{3\omega q_4}{\alpha\, q_2}F^2(\xi), \quad \xi = \sqrt{\frac{\omega}{4\delta q_2}}(x - \omega t) + \xi_0. \tag{2.6}$$

Thus we can obtain abundant cnoidal wave solutions of the KdV equation in terms of JEFs. Some periodic wave solutions of the KdV equation and modified KdV equation have been studied in [7,23, 28]. As $m \longrightarrow 1$, these solutions will degenerate into the corresponding solitary wave solutions.

3. The JEF solutions of Schamel- KdV equation

We consider the Schamel- KdV equation [29, 30]

$$u_t + (\alpha u^{1/2} + \beta u)u_x + \delta u_{xxx} = 0, \quad \beta \neq 0 \tag{3.1}$$

where α, β and δ are constants and u is the wave potential.
In order to find the periodic wave solution of (3.1), we use the transformations $u = v^2$, $v(x,t) = V(\xi);\ \xi = k(x - \omega t) + \xi_0$, then (2.7) becomes

$$-\omega V V' + (\alpha V^2 + \beta V^3)V' + \delta k^2[VV''' + 3V'V''] = 0. \tag{3.2}$$

The balancing procedure implies that $N = 1$. Therefore, the F-expansion method gives the solution

$$V(x,t) = V(\xi) = a_0 + a_1 F(\xi), \tag{3.3}$$

where a_0 and a_1 are constants to be determined and $F(\xi)$ is a solution of Eq. (1.4). Substituting Eq. (3.3) into Eq. (3.2) and equating the coefficients of the like powers of F to zero, yields a set of algebraic equations for a_0, a_1, k and ω:

$$\begin{aligned}
&[\beta a_1^2 + 12\delta k^2 q_4]a_1 = 0,\\
&[\alpha a_1^2 + 3\beta a_0 a_1^2 + 6\delta k^2 a_0 q_4]a_1 = 0,\\
&[-\omega + 2\alpha a_0 + 3\beta a_0^2 + 4\delta k^2 q_2]a_1 = 0,\\
&[-\omega + \alpha a_0 + \beta a_0^2 + \delta k^2 q_2]a_0 = 0.
\end{aligned} \tag{3.4}$$

Solving these algebraic equations, we gave a general form of traveling wave solutions of Eq. (3.1)

$$u = \frac{4\alpha^2}{25\beta^2}\left[1 \pm \sqrt{\frac{-2q_4}{q_2}}\,F(\xi)\right]^2. \tag{3.5}$$

Therefore, we obtained in [30] the JEF solutions of Eq. (3.1) as follows:
When $q_0 = 1,\ q_2 = -1 - m^2,\ q_4 = m^2$, solutions of Eq. (1.4) is $F(\xi) = \mathrm{sn}\xi$, we have

$$u_1 = \frac{4\alpha^2}{25\beta^2}\left[1 \pm \sqrt{\frac{2m^2}{m^2+1}}\mathrm{sn}\left(\frac{2\alpha}{5\sqrt{-6\delta\beta(m^2+1)}}(x + \frac{16\alpha^2}{75\beta}t) + \xi_0\right)\right]^2, \quad \beta\delta < 0, \tag{3.6}$$

If $q_0 = 1 - m^2,\ q_2 = 2m^2 - 1,\ q_4 = -m^2,\ F(\xi) = \mathrm{cn}\xi$, thus yields the exact solutions of Eq. (3.1)

$$u_2 = \frac{4\alpha^2}{25\beta^2}\left[1 \pm \sqrt{\frac{2m^2}{2m^2-1}}\,\mathrm{cn}\left(\frac{2\alpha}{5\sqrt{6\delta\beta(2m^2-1)}}(x + \frac{16\alpha^2}{75\beta}t) + \xi_0\right)\right]^2, \quad \beta\delta > 0, \tag{3.7}$$

If $q_0 = m^2 - 1,\ q_2 = 2 - m^2,\ q_4 = -1$, the solution of Eq (1.4) is $F(\xi) = \mathrm{dn}\xi$. So, we obtained the exact solutions of Eq. (3.1) in the form

$$u_3 = \frac{4\alpha^2}{25\beta^2}\left[1 \pm \sqrt{\frac{2}{2-m^2}}\,\mathrm{dn}\left(\frac{2\alpha}{5\sqrt{6\delta\beta(2-m^2)}}(x + \frac{16\alpha^2}{75\beta}t) + \xi_0\right)\right]^2, \quad \beta\delta > 0, \tag{3.8}$$

Many types of JEF solutions of Eq. (3.1) are given [30]. As $m \longrightarrow 1$, Eqs. (3.6)-(3.8) degenerate to

$$\begin{aligned}
u_4 &= \frac{4\alpha^2}{25\beta^2}\left[1 \pm \tanh\left(\frac{\alpha}{5\sqrt{-3\delta\beta}}(x + \frac{16\alpha^2}{75\beta}t) + \xi_0\right)\right]^2, \quad \beta\delta < 0,\\
u_5 &= \frac{4\alpha^2}{25\beta^2}\left[1 \pm \sqrt{2}\,\mathrm{sech}\left(\frac{2\alpha}{5\sqrt{6\delta\beta}}(x + \frac{16\alpha^2}{75\beta}t) + \xi_0\right)\right]^2, \quad \beta\delta > 0,
\end{aligned} \tag{3.9}$$

The solitary wave solutions (3.9) in terms of tanh are equivalent to the solutions given in [31]. The JEF solutions of (3.1) may be describe various features of waves and may be helpful in understanding the problems in ion acoustic waves.

4. Schamel equation and modified KP equation

The equation describing ion-acoustic waves in a cold-ion plasma where electrons do not behave isothermally during their passage of the wave is

$$u_t + u^{1/2} u_x + \delta u_{xxx} = 0. \tag{4.1}$$

Schamel [29] derived this equation and a simple solitary wave solution having a sech^4 profile was obtained. Therefore the Schamel equation (4.1) containing a square root nonlinearity is very attractive model for the study of ion-acoustic waves in plasmas and dusty plasmas.
In order to find the periodic wave solution of (4.1), we use the transformations
$u = v^2,\ v(x,t) = V(\eta);\ \eta = kx - \omega t + \eta_0,$, then (4.1) becomes

$$-\omega V V' + k V^2 V' + \delta k^3 [V V''' + 3V' V''] = 0. \tag{4.2}$$

According to the mapping method, we assume that Eq. (4.2) has the following solution:

$$V(\eta) = A_0 + A_1 f(\eta), \tag{4.3}$$

where A_0 and A_1 are constants to be determined and $f(\eta)$ satisfies Eq. (1.6).
Substitution of Eq. (4.3) into Eq. (4.2) and selecting the values of p, q and r, we have the solutions of Eq. (4.1) which was given in [20] as follows:
Case 1. $p = 2,\ q = -4(1+m^2),\ r = 6m^2$. In this case, we have $f(\eta) = \text{sn}^2\eta$. Thus the periodic wave solutions of Eq. (4.1) are

$$\begin{aligned} u_1(x,t) &= 100\delta^2 k^4 \left[1 + m^2 \pm \sqrt{1 - m^2 + m^4} - 3m^2 \text{sn}^2 \eta\right]^2, \\ \eta &= kx \mp 16\delta k^3 \sqrt{1 - m^2 + m^4}\, t + \eta_0. \end{aligned} \tag{4.4}$$

Case 2. $p = \frac{-(1-m^2)^2}{2},\ q = 2(1+m^2),\ r = \frac{-3}{2}$. The solutions of Eq. (1.6) are $f(\eta) = (m\,\text{cn}\eta \pm \text{dn}\eta)^2$. Thus the exact solutions of Eq. (4.1) are

$$\begin{aligned} u_2(x,t) &= \tfrac{25\delta^2 k^4}{4} \left[-2(1+m^2) \pm \sqrt{1 + 14m^2 + m^4} + 3\,(m\,\text{cn}\eta \pm \text{dn}\eta)^2\right]^2, \\ \eta &= kx \mp 4\delta k^3 \sqrt{1 + 14m^2 + m^4}\, t + \eta_0. \end{aligned} \tag{4.5}$$

Case 3. $p = \frac{m^2}{2},\ q = 2(m^2 - 2),\ r = \frac{3m^2}{2}$. The solutions of Eq. (1.6) are
$f(\eta) = \left(\frac{m\,\text{sn}\eta}{1 \pm \text{dn}\,\eta}\right)^2$. So, we obtained the exact solutions of Eq. (4.1)in the form

$$\begin{aligned} u_3(x,t) &= \tfrac{25\delta^2 k^4}{4} \left[2(2 - m^2) \pm \sqrt{16 - 16m^2 + m^4} - 3m^4 \left(\frac{\text{sn}\,\eta}{1 \pm \text{dn}\,\eta}\right)^2\right]^2, \\ \eta &= kx \mp 4\delta k^3 \sqrt{16 - 16m^2 + m^4}\, t + \eta_0. \end{aligned} \tag{4.6}$$

There are several exact solutions for the Eq. (4.1) which are omitted here for simplicity. As $m \to 1$, these solutions reduce to the solitary wave solutions

$$\begin{aligned} u_4(x,t) &= 900\delta^2 k^4 \text{sech}^4(kx - 16\delta k^3 t + \eta_0), \\ u_5(x,t) &= 100\delta^2 k^4 [2 - 3\,\text{sech}^2(kx + 16\delta k^3 t + \eta_0)]^2. \end{aligned} \tag{4.7}$$

$$u_6(x,t) = \frac{225\delta^2 k^4}{4}\left[1-\left(\frac{\tanh(kx-4\delta k^3 t+\eta_0)}{1+\operatorname{sech}(kx-4\delta k^3 t+\eta_0)}\right)^2\right]^2. \tag{4.8}$$

The KdV equation in two dimensions, known as Kadomtsev Petviashivili (KP) equation [32], was derived for ion-acoustic waves in a non magnetized plasma by Kako and Rowlands [33]. Therefore the modified KP equation containing a square root nonlinearity is very attractive model for the study of ion-acoustic waves in plasma and dusty plasma [34- 36]. Extensive work has been devoted to the study of nonlinear waves associated with the dust ion-acoustic waves, particularly the dust ion-acoustic solitary and shock waves in dusty plasmas in which dust particles are stationary and provide only the neutrality [37]. The KP equation is derived [38] for the propagation of nonlinear waves in warm dusty plasmas with variable dust charge, two-temperature ions and nonthermal electrons by using the reductive perturbation theory. Consider the modified KP equation

$$(u_t + \alpha u^{1/2} u_x + \beta u_{xxx})_x + \delta u_{yy} = 0, \tag{4.9}$$

where α and β are constants. The modified KP equation (4.9) for ion-acoustic waves in a multi species plasma consisting of non-isothermal electrons have been derived by Chakraborty and Das [34]. We applied the mapping method with the ansatz solution (4.3) and the solutions of auxiliary equation (1.6) to find the solutions of equation (4.9) (see [39]).

5. The ZK equation and modified ZK equation

The equation

$$u_t + \beta u^2 u_x + u_{xxx} + u_{yyx} = 0, \tag{5.1}$$

is the modified ZK in (2+1) dimensions which is a model for acoustic plasma waves [40, 41]. The ZK equation was first derived for describing weakly nonlinear ion- acoustic waves in a strongly magnetized lossless plasma in two dimension [41]. The ZK equation and modified ZK equation possess traveling wave structures [28, 42]. Peng [42] studied the exact solutions of ZK equation by using extended mapping method. Various types of solutions of Schamel-KdV equation and modified ZK equation arising in plasma and dust plasma are presented in [43].

We apply the F-expansion method to the modified ZK equation. Thus, Eq. (5.1) has a solution in the form

$$u(\xi) = a_0 + a_1 F(\xi), \quad \xi = k(x + ly - \omega t) + \xi_0.$$

Substituting this equation into Eq. (5.1), we obtain the following classes of exact solutions of the modified ZK equation:

$$\begin{aligned} u &= m\sqrt{\frac{6\omega}{(m^2+1)\beta}}\,\operatorname{sn}\left(\sqrt{\frac{-\omega}{(m^2+1)(1+l^2)}}\,(x+ly-\omega t+\xi_0)\right), \\ u &= m\sqrt{\frac{6\omega}{(2-m^2)\beta}}\,\operatorname{dn}\left(\sqrt{\frac{\omega}{(2-m^2)(1+l^2)}}\,(x+ly-\omega t+\xi_0)\right). \end{aligned} \tag{5.2}$$

In the following we apply the mapping method to the ZK equation

$$u_t + \alpha u\, u_x + u_{xxx} + u_{yyx} = 0. \tag{5.3}$$

In this case, we have $n = 1$. Thus Eq. (5.3) has a solution in the form

$$u(\eta) = A_0 + A_1 f(\eta), \quad \eta = kx + ly - \omega t + \eta_0.$$

Substituting this equation into Eq. (5.3) to determine A_0, A_1, k, ω and using the solutions of auxiliary equation (1.6), we obtained the following classes of exact solutions of the ZK equation [39]:

$$u_1(x,y,t) = \frac{\omega}{k\alpha} + \frac{4(1+m^2)(l^2+k^2)}{\alpha} - \frac{12m^2(l^2+k^2)}{\alpha}\,\mathrm{sn}^2(kx+ly-\omega t+\eta_0),$$
$$u_2(x,y,t) = \frac{\omega}{k\alpha} + \frac{4(m^2-2)(l^2+k^2)}{\alpha} + \frac{12(l^2+k^2)}{\alpha}\,\mathrm{dn}^2(kx+ly-\omega t+\eta_0), \qquad (5.4)$$

$$u_3 = \frac{\omega}{k\alpha} - \frac{2(1+m^2)(l^2+k^2)}{\alpha} + \frac{3(l^2+k^2)}{\alpha}[m\,\mathrm{cn}(\eta) \pm \mathrm{dn}(\eta)]^2, \qquad (5.5)$$

$$u_4(x,y,t) = \frac{\omega}{k\alpha} - \frac{2(1+m^2)(l^2+k^2)}{\alpha} - \frac{3(1-m^2)(l^2+k^2)}{\alpha}\left(\frac{\mathrm{cn}(kx+ly-\omega t+\eta_0)}{1\pm\mathrm{sn}(kx+ly-\omega t+\eta_0)}\right)^2. \qquad (5.6)$$

When $m \longrightarrow 1$, some of these solutions degenerate as solitary wave solutions of ZK equation. The solutions (5.3) are coincide with the solutions given in [44].
Recently, some properties of the quantum ion-acoustic waves were also investigated in dense quantum plasmas by studying the quantum hydrodynamical equations in different conditions, which includes the quantum Zakharov Kuznetsov equation, the extended quantum Zakharov Kuznetsov equation, and the quantum Zakharov system [45]. The three-dimensional extended quantum Zakharov Kuznetsov (QZK) equation [46] was investigated in dense quantum plasmas which arises from the dimensionless hydrodynamics equations describing the nonlinear propagation of the quantum ion-acoustic waves. The three-dimensional extended QZK equation was given in [46]

$$\Phi_t + (A\Phi + B\,\Phi^2)\Phi_x + C\,\Phi_{zzz} + D\,(\Phi_{xxz} + \Phi_{yyz}) = 0, \qquad (5.7)$$

where A, B, C and D are constants. This equation has the following JEF solutions (see [45, 46]):

$$\Phi_1 = -\frac{A}{2B} + m\,k\sqrt{\frac{-6E}{B}}\,\mathrm{sn}(k(x+ly+\gamma z-\omega t+\eta_0)), \quad \omega = -\frac{4BEk^2(1+m^2)+A^2}{4B}, \quad BE<0,$$
$$\Phi_2 = -\frac{A}{2B} + mk\sqrt{\frac{6E}{B}}\,\mathrm{cn}(k(x+ly+\gamma z-\omega t+\eta_0)), \quad \omega = -\frac{4BEk^2(1-2m^2)+A^2}{4B}, \quad BE>0, \qquad (5.8)$$

with $E = C\gamma^2 + D(1+l^2)$. Moreover, many types of analytical solutions of the extended QZK equation are constructed in terms of some powerful ansatze, which include doubly periodic wave solutions, solitary wave solutions, kink-shaped wave solutions, rational wave solutions and singular solutions [46].

6. The modified fifth order KdV equation

Higher order KdV equations have many applications in different fields of mathematical physics. For example the fifth-order KdV equations can be derived in fluid dynamics and in magneto-acoustic waves in plasma and its exact solutions was given in [47-51]. The higher-order KdV equation can be derived for magnetized plasmas by using the reductive perturbation technique. Traveling wave solutions of Kawahara equation and modified Kawahara equation have been studied [9, 48, 49]. Moreover, the solitary wave solutions of nonlinear equations with arbitrary odd-order derivatives were studied by many authors [47, 51].
Consider the modified fifth order KdV equation

$$u_t + \beta u^2 u_x + c_3 u_{xxx} + c_5 u_{xxxxx} = 0, \qquad (6.1)$$

where β, c_3 and c_5 are constants. Here, we review the exact traveling wave solutions of equation (6.1) using exact solutions of the auxiliary equation (1.5) and applied the mapping method. Thus, Eq. (6.1) has the solutions in the form

$$u(\eta) = A_0 + A_1 f(\eta), \quad \eta = kx - \omega t + \eta_0, \tag{6.2}$$

Substituting equation (6.2) into (6.1) and equating the coefficients of like powers of f to zero, yields a system of algebraic equations for A_0, A_1, k and ω and then solve it. Therefore, the solutions of the modified fifth order KdV equation (6.1) was given in [39] as follows:

$$\begin{aligned} u_1 &= \pm\frac{(10k^2c_5(1+m^2)+c_3)}{\sqrt{-10\beta c_5}} \mp k^2\sqrt{\frac{-90\,c_5}{4\beta}}\left(m\,\mathrm{cn}\,\eta \pm \mathrm{dn}\,\eta\right)^2, \\ \eta &= k\left[x + \frac{(15c_5^2k^4(m^4+14m^2+1)+c_3^2)}{10c_5}\,t\right] + \eta_0. \end{aligned} \tag{6.3}$$

If we choose $A_0 = 0$, equation (6.3) takes the form

$$\begin{aligned} u_2 &= \pm\frac{3c_3}{2(1+m^2)\sqrt{-10\,\beta\,c_5}}\left(m\,\mathrm{cn}\,\eta \pm \mathrm{dn}\,\eta\right)^2, \\ \eta &= \pm\sqrt{\frac{-c_3}{10c_5(1+m^2)}}\left[x + \frac{(23m^4+82m^2+23)\,c_3^2}{200c_5(1+m^2)^2}\,t\right] + \eta_0. \end{aligned} \tag{6.4}$$

Moreover, we have obtained the exact solutions

$$\begin{aligned} u_3 &= \pm\frac{3m^2c_3}{2(m^2-2)\sqrt{-10\,\beta\,c_5}}\left(\frac{m\,\mathrm{sn}\,\eta}{1\pm\mathrm{dn}\,\eta}\right)^2, \\ \eta &= \pm\sqrt{\frac{c_3}{10c_5(2-m^2)}}\left[x + \frac{(23m^4-128m^2+128)\,c_3^2}{200c_5(m^2-2)^2}\,t\right] + \eta_0, \end{aligned} \tag{6.5}$$

$$\begin{aligned} u_4 &= \pm\frac{3c_3}{2(1-2m^2)\sqrt{-10\,\beta\,c_5}}\left(\frac{\mathrm{sn}\,\eta}{1\pm\mathrm{cn}\,\eta}\right)^2, \\ \eta &= \pm\sqrt{\frac{-c_3}{10c_5(1-2m^2)}}\left[x + \frac{(128m^4-128m^2+23)\,c_3^2}{200c_5(1-2m^2)^2}\,t\right] + \eta_0. \end{aligned} \tag{6.6}$$

There are several other JEFs of Eq. (6.1) which are omitted here for simplicity. When $m \longrightarrow 1$, then (6.4)-(6.6) become the solitary wave solutions

$$u_5 = \pm\frac{3c_3}{\sqrt{-10\beta c_5}}\,\mathrm{sech}^2\Big(\frac{1}{2}\sqrt{\frac{-c_3}{5c_5}}\,(x + \frac{4\,c_3^2}{25c_5}\,t) + \eta_0\Big), \tag{6.7}$$

$$u_6 = \mp\frac{3c_3}{2\sqrt{-10\,\beta\,c_5}}\left(\frac{\tanh\eta}{1\pm\mathrm{sech}\,\eta}\right)^2, \quad \eta = \pm\sqrt{\frac{c_3}{10c_5}}\left[x + \frac{23\,c_3^2}{200c_5}\,t\right] + \eta_0. \tag{6.8}$$

We notice that Eq. (6.7) is the solution given by Example 2 in Ref. [47].
Finally, we can construct various types of exact and explicit solutions of the generalized ZK equation

$$u_t + (\alpha + \beta\,u^p)\,u^p u_x + u_{xxx} + \delta u_{yyx} = 0, \tag{6.9}$$

by using suitable method and using an appropriate transformation. Also, we can study the exact solution of the generalized KdV equation ($\delta = 0$) which studied by many authors [22, 23, 31]. The generalized ZK equation was first derived for describing weakly nonlinear ion-acoustic waves in strongly magnetized lossless plasma in two dimensions and governs the behavior of weakly nonlinear ion-acoustic waves in plasma comprising cold ions and hot isothermal electrons in the presence of a uniform magnetic field. Eq. (6.9) includes considerable interesting equations, such as KdV equation, mKdV equation, ZK equation and mZK equation. Exact traveling wave solutions for the generalized ZK equation with higher-order nonlinear terms have obtained in [52-54]. Moreover, we can use the symbolic computations and apply the mapping method with the ansatz solution (1.5) to find the several classes of traveling wave solutions of the fifth order KdV equation

$$u_t + c_1 u\, u_x + c_2\, u_{xxx} + \delta u_{xxxxx} = 0.$$

This equation appears in the theory of shallow water waves with surface tension and the theory of magneto-acoustic waves in plasmas [9]. Wazwaz [55] studied soliton solutions of fifth-order KdV equation. We can use a suitable method to construct the exact solutions of some special types of nonlinear evolution equations aries in plasma physics such as Liouville, sine-Gordon and sinh-Poisson equations.

7. References

[1] G. Whitham, Linear and Nonlinear Waves, New York, Wiley (1974).

[2] R. Davidson, Methods in Nonlinear Plasma Theory, New York, Academic Press (1972).

[3] M. J. Ablowitz and P. A. Clarkson, Solitons, Nonlinear Evolution Equations and Inverse Scattering Transform, *Cambridge, Cambridge University Press* (1991).

[4] W. Malfliet, Solitary wave solutions of nonlinear wave equations, *Am. J. Phys.* 60 (1992) 650-654;
W. Malfliet, The tanh method: a tool for solving certain classes of nonlinear evolution and wave equations, *J. Comput. Appl. Math.* 164-156 (2004) 529-541.

[5] E.G. Fan, Extended tanh-function method and its applications to nonlinear equations, *Phys. Lett.* A 277 (2000) 212-218;
E.G. Fan and Y.C. Hong, Generalized tanh method to special types of nonlinear equations, *Z. Naturforsch.* A 57 (2002) 692-700.

[6] M. Wang, Exact solutions for a compound KdV - Burgers equation, *Phys. Lett.* A 213 (1996) 279-287.

[7] S. K. Liu, Z. T. Fu, S. D. Liu, and Q. Zhao, Jacobi elliptic function expansion method and periodic wave solutions of nonlinear wave equations, *Phys. Lett. A* 289 (2001) 69-74;

[8] Z.T.Fu, S. K. Liu, S. D. Liu, and Q. Zhao, New Jacobi elliptic function expansion method and new periodic solutions of nonlinear wave equations, *Phys. Lett. A* 290 (2001) 72-76.

[9] E.J. Parkes, B.R. Duffy and P.C. Abbott, The Jacobi elliptic-function method for finding periodic wave solutions to nonlinear evolution equations, *Phys. Lett.* A 295 (2002) 280-286.

[10] H. T. Chen and H. Q. Zhang, Improved Jacobin elliptic method and its applications, *Chaos, Solitons and Fractals* 15 (2003) 585-591.

[11] Y. Peng, Exact periodic wave solutions to a new Hamiltonian amplitude equation, *J. Phys. Soc. Japan* 72 (2003) 1356-1359;
Y. Peng, New exact solutions to a new Hamiltonian amplitude equation II, *J. Phys. Soc. Japan* 73 (2004) 1156-1158.

[12] Y. Peng, Exact periodic wave solutions to the Melnikov equation, *Z. Naturforsch A* 60 (2005) 321-327.

[13] Y. B. Zhao, M. L. Wang and Y. M. Wang, Periodic wave solutions to a coupled KdV equations with variable coefficients, *Phys. Lett. A* 308 (2003) 31-36.

[14] J. Liu and K. Yang, The extended F-expansion method and exact solutions of nonlinear PDEs, *Chaos, Solitons and Fractals* 22 (2004) 111-121.

[15] H. T. Chen and H. Q. Zhang, New double periodic and multiple soliton solutions of the generalized (2+1)-dimensional Boussinesq equation, *Chaos, Solitons and Fractals* 20 (2004) 765-769.
[16] S. A. Elwakil, S. K. El-labany, M. A. Zahran and R. Sabry, Modified extended tanh-function method for solving nonlinear partial differential equations, *Phys. Lett. A.* 299 (2002) 179-188.
[17] M. A. Abdou and S. Zhang, New periodic wave solutions via extended mapping method, *Commun. Nonlinear Sci. Numer. Simul.* 14 (2009) 2-11.
[18] D. Baldwin, Ü. Göktas, W. Hereman et al., Symbolic computions of exact solutions expressible in hyperbolic and elliptic functions for nonlinear PDEs, *J. Symb. Comput.* 37 (2004) 669-705.
[19] Taogetusang and Sirendaoerji, The Jacobi elliptic function-like exact solutions to two kinds of KdV equations with variable coefficients and KdV equation with forcible term, *Chinese Phys.* 15 (2006) 2809-2818.
[20] A. H. Khater, M. M. Hassan, E. V. Krishnan and Y.Z. Peng, Applications of elliptic functions to ion-acoustic plasma waves, *Eur. Phys. J. D* 50 (2008) 177-184.
[21] M. Abramowitz and I. A. Stegun, Handbook of Mathematical Functions, *Dover, New York* (1965).
[22] P. G. Drazin and R. S. Johnson, Solitons,: An Introduction, *Cambridge University press, Cambridge* (1989).
[23] A. Jeffrey and T. Kakutani, Weak nonlinear dispersive waves: A discussion centered around the Korteweg de Vries equation, *SIAM Rev.* 14 (1972) 582-643.
[24] H. Washimi and T. Taniuti, Propagation of ion acoustic solitary waves of small amplitude, J Phys Rev Lett. 17 (1966) 996 -998.
[25] E.K. El-Shewy, H.G. Abdelwahed and H.M. Abd-El-Hamid, Computational solutions for the KortewegŰdeVries equation in warm Plasma, Computat. Methods in Sci. Technology 16 (2010) 13-18.
[26] S. K. El-Labany, Contribution of higher-order nonlinearity to nonlinear ion acoustic waves in a weakly relativistic warm plasma. Part1. isothermal case, J. Plasma Phys. 50 (1993) 495 .
[27] J. F. Zhang, New solitary wave solution of the combined KdV and m KdV equation, *Int. J. Theoret. Phys.* 37 (1998) 1541-1546.
[28] A. H. Khater and M. M. Hassan, Travelling and periodic wave solutions of some nonlinear wave equations, *Z. Naturforsch.* 59 a (2004) 389-396.
[29] H. Schamel, A modified Korteweg de Vries equation for ion acoustic waves due to resonant electrons, *J. Plasma Phys.* 9 (1973) 377-387.
[30] A. H. Khater, M. M. Hassan and R. S. Temsah, Exact solutions with Jacobi elliptic functions of two nonlinear models for ion-acoustic plasma waves, *J. Phys. Soc. Japan* 74 (2005) 1431-1435.
[31] M. M. Hassan, Exact solitary wave solutions for a generalized KdV - Burgers equation, *Chaos, Solitons and Fractals* 19 (2004) 1201-1206.
[32] B. B. Kadomtsev and V. I. Petviashvili, On the stability of solitary in weakly dispersive media, *Soviet Phys. Dokl.* 15 (1970) 539-541.
[33] M. Kako and G. Rowlands, Two-dimensional stability of ion acoustic solitons, *Plasma Phys.* 18 (1976) 165-170.
[34] D. Chakraborty and K. P. Das, Stability of ion acoustic solitons in a multispecies plasma consisting of non-isothermal electrons, *J. Plasma Phys.* 60 (1998) 151-158.
[35] A. H. Khater, A. A. Abdallah, O. H. El-Kalaaway and D. K. Callebaut, Backlund transformations, a simple transformation and exact solutions for dust-acoustic solitary

waves in dusty plasma consisting of cold dust particles and two-temperature isothermal ions, *Phys. Plasmas* 6 (1999) 4542-4547.

[36] A. H. Khater and M. M. Hassan, Exact Jacobi elliptic function solutions for some special types of nonlinear evolution equations, *Il Nuovo Cimento* 121 B (2006) 613- 622.

[37] A. A. Mamun and P.K. Shukla, Cylindrical and spherical dust ion-acoustic solitary waves, *Phys. Plasmas 9* (2002) 1468.

[38] H.R. Pakzad, Soliton energy of the Kadomtsev Petviashvili equation in warm dusty plasma with variable dust charge, two-temperature ions, and nonthermal electrons, *Astrophys Space Sci.* 326 (2010) 69 -75.

[39] A. H. Khater, M. M. Hassan and D. K. Callebaut, Travelling wave solutions to some important equations of mathematical physics, *Reports on Math. Phys.* 66 (2010) 1-19.

[40] K. P. Das and F. Verheest, Ion acoustic solitons in magnetized multi-component plasmas including negative ions, *J. Plasma. Phys.* 41 (1989) 139-155.

[41] V. E. Zakharov and E. A. Kuznetsov, On three-dimensional solitons, *Sov. Phys. JETP* 39 (1974) 285-286.

[42] Y.Z. Peng, Exact travelling wave solutions of the Zakharov Kuznetsov equation, *Appl. Math. Comput.* 199 (2008) 397- 405.

[43] M. M. Hassan, New exact solutions of two nonlinear physical models, *Commun. Theor. Phys.* 53 (2010) 596-604.

[44] M. M. Hassan, Exact solutions for some models of nonlinear evolution equations, *J. Egypt Math. Soc.* 12 (2004) 31- 43.

[45] R. Sabry, W.M. Moslem, F. Haas, S. Ali, P.K. Shukla, Nonlinear structures: Explosive, soliton and shock in a quantum electron- positron-ion magentoplasma, Phys. Plasmas 15 (2008) 122308.

[46] Z. Yan, Periodic, Solitary and rational wave solutions of the 3D extended quantum Zakharov Kuznetsov equation in dense quantum plasmas, *Phys. Let.* A 373 (2009) 2432-2437.

[47] E.J. Parkes, Z. Zhu, B.R. Duffy and H. C. Hang, Sech-polynomail traveling solitary-wave solutions of odd-order generalized KdV equations, *Phys. Lett.* A 248 (1998) 219-224.

[48] Z.J. Yang, Exact solitary wave solutions to a class of generalized odd-order KdV equations, *Int. J. Theor. Phys.* 34 (1995) 641-647.

[49] D. Zhang, Doubly periodic solution of modified Kawahara equation, *Chaos, Solitons and Fractals* 25 (2005) 1155-1160.

[50] A. H. Khater, M. M. Hassan and R. S. Temsah, Cnoidal wave solutions for a class of fifth-order KdV equations, *Math. Comput. Simulation* 70 (2005) 221-226.

[51] J. Sarma, Solitary wave solution of higher-order Korteweg de Vries equation, *Chaos, Solitons and Fractals* 39 (2009) 277 - 281.

[52] L.- H. Zhang, Travelling wave solutions for the generalized Zakharov-Kuznetsov equation with higher-order nonlinear terms, *Appl. Math. Comput.* 208 (2009) 144-155.

[53] C. Deng, New exact solutions to the Zakharov -Kuznetsov equation and its generalized form, *Commun. Nonlinear Sci. Numer. Simulat.* 15 (2010) 857 - 868.

[54] SUN Yu-Huai, MA Zhi-Min and LI Yan, Explicit solutions for generalized (2+1)-dimensional nonlinear Zakharov- Kuznetsov equation, *Commun. Theor. Phys.* 54 (2010) 397-400.

[55] A.-M., Wazwaz, Soliton solutions for the fifth-order KdV equation and the Kawahara equation with time-dependent coefficients, *Phys. Scr.* 82 (2010) 035009.

Part 2

Acoustic Waves as Investigative Tools

6

Acoustic Waves: A Probe for the Elastic Properties of Films

Marco G. Beghi
Politecnico di Milano, Energy Department and NEMAS Center, Milano
Italy

1. Introduction

Films and thin films are exploited by an ever increasing number of technologies. The properties of films can be different from those of the same material in bulk form, and can depend on the preparation process, and on thickness. Specific techniques are needed for their measurement. Whenever films or thin layers have structural functions, as in micro electro-mechanical systems (MEMS), a precise characterization of their stiffness is crucial for the design of devices. The same can be said for the design of devices which exploit thin layers to support surface acoustic waves (surface acoustic wave filters). More generally, knowledge of the elastic properties is interesting because such properties depend on the structural properties.

The most widespread technique for the mechanical characterization of films, instrumented indentation, induces both elastic and inelastic strains. It also characterizes irreversible deformation, but the extraction of the information concerning the elastic behaviour is non trivial. To overcome this difficulty, several measurement methods have been developed, which exploit vibrations as a probe of the material behaviour. These methods intrinsically involve only elastic strains, and are non destructive. This is true at any length scale, and is peculiarly useful at micrometric and sub-micrometric scales, where the exploitation of other types of probes can become critical.

Both propagating waves and standing waves can be exploited, with excitation which can be either monochromatic (e.g. resonance techniques) or impulsive, therefore broadband, requiring an analysis of the response either in the time domain (the so called picoseconds ultrasonics) or in the frequency domain (the so called laser ultrasonics). The propagation velocities of vibrational modes are obtained, from which the stiffness is derived if an independent value of the mass density (the inertia) is available.

Older resonance techniques have been developed to be operated with thin slabs, also exploiting optical measurements of displacement. In the measurement of films and small structures, the advantages of light, a contact-less and inertia-less probe, are substantial, and are increasingly exploited. The laser ultrasonics technique, commercially available since some years ago, measures waves travelling along the film surface. The so called picosecond ultrasonics technique measures waves travelling across the film thickness; it is a relatively sophisticated optical technique, which exploits femtosecond laser pulses in a pump-and-probe measurement scheme. For best performance it needs the deposition of an interaction layer, possibly combined with microlithography techniques to obtain a patterned layer.

Techniques for optical detection of acoustic vibrations include inelastic scattering of light: Brillouin scattering. Brillouin spectroscopy does not excite vibrations, but relies on thermal excitation, which has the broadest band but small amplitude, and measures the spectrum of inelastically scattered laser light. Brillouin Spectroscopy and Surface Brillouin Spectroscopy are relatively simple optical techniques, which do not require a specific specimen preparation. They operate at sub-micrometric acoustic wavelengths, which are selected by the scattering geometry, and have been exploited to characterize the elastic properties of bulk materials and of films.

An overview of this variety of techniques is presented here, underlining analogies and differences. The increasing demand of precise characterization raised the point of precision and accuracy achievable by vibration based techniques, and specifically by the optical techniques. In the overview, attention is drawn to the steps or the parameters which are the limiting factor for the achievable precision, and to the way of characterizing them. The effects of inaccuracies of the mass density are common to all the techniques based on vibrations, while other sources of uncertainty are more specific to each technique.

2. Acoustic modes in elastic solids

In the continuum description the instantaneous configuration of a solid undergoing deformation can be represented by the displacement vector field $\mathbf{u}(\mathbf{r},t)$, where $\mathbf{u}=(u_1,u_2,u_3)$, $\mathbf{r}=(x_1,x_2,x_3)$ and t is time. The local state of the solid being represented by the strain and stress tensors, the linear elastic behaviour is characterized by the tensor of the elastic constants C_{ijmn}, which can be conveniently represented by the matrix of the elastic constants C_{ij}. When other phenomena, like e.g. viscoelasticity or elasticity of higher orders, can be neglected, the tensor of the elastic constants fully characterizes the stiffness. Inertia is characterized by the mass density ρ. Within a homogeneous linear elastic solid, in the absence of body forces, the equations of motion for the displacement vector field are homogeneous, and read (Auld, 1990; Every, 2001; Kundu, 2004)

$$\rho\frac{\partial^2 u_i}{\partial t^2}=\sum_{j,m,n} C_{ijmn}\frac{\partial^2 u_m}{\partial x_j \partial x_n},\quad i=1,2,3 \quad . \tag{1}$$

These equations describe vibrational elastic excitations, which are typically called acoustic also in the ultrasonic frequency range. The basic solutions of Eqs.(1), and the most important ones when boundary effects are irrelevant, are the plane acoustic waves, or modes (Auld, 1990; Kundu, 2004), of the form

$$\mathbf{u}=\mathbf{e}\Re\left\{\tilde{A}\exp\left[i(\mathbf{k}\cdot\mathbf{r}-\omega t)\right]\right\} \quad , \tag{2}$$

where $\mathbf{k}$ is the wavevector, $\omega = 2\pi f$ the circular frequency, f the frequency, $\tilde{A}$ an arbitrary complex amplitude, and $\mathbf{e}$ the polarization vector, which is normalized. The continuum description, underlying Eq. (1), is appropriate until the wavelength $\lambda=2\pi/|\mathbf{k}|$ is much larger than the interatomic distances. The three translational degrees of freedom of each infinitesimal volume element correspond, for each wavevector $\mathbf{k}$, to three independent

modes, having different polarization vectors and frequencies. In general the phase velocity $v = \omega / |\mathbf{k}| = \lambda f$ depends on both the direction of **k** and the polarization vector **e**. In an infinite homogeneous medium, travelling waves of the type given by Eq. (2) exist for any frequency f compatible with the mentioned lower limit for wavelength.
In the simplest case, the isotropic solid, the matrix of the elastic constants is fully determined by only two independent quantities; the only non null matrix elements are $C_{11} = C_{22} = C_{33}$, $C_{44} = C_{55} = C_{66}$, $C_{12} = C_{13} = C_{23} = C_{11} - 2C_{44}$. In this case the shear modulus G coincides with C_{44}, while Young modulus E, Poisson's ratio ν and bulk modulus B are respectively given by (Every, 2001; Kundu, 2004)

$$E = \frac{C_{44}(3C_{12} + 2C_{44})}{C_{12} + C_{44}} = \frac{C_{44}(3C_{11} - 4C_{44})}{C_{11} - C_{44}}, \tag{3}$$

$$\nu = \frac{C_{12}}{C_{11} + C_{12}} = \frac{C_{11} - 2C_{44}}{2(C_{11} - C_{44})} = \frac{E}{2G} - 1, \tag{4}$$

$$B = \frac{C_{11} + 2C_{12}}{3} = C_{11} - \frac{4}{3}C_{44}, \tag{5}$$

In the isotropic case the phase velocities are independent from the direction of **k**, only depending on the relative orientation of **e** with respect to **k**; one of the three modes is longitudinal ($\mathbf{e} \parallel \mathbf{k}$) and has velocity v_l, the other two are transversal ($\mathbf{e} \perp \mathbf{k}$), are independent (the two polarization vectors are orthogonal) and degenerate: they have the same velocity v_t (Auld, 1990; Kundu, 2004). The two velocities are

$$v_l = \sqrt{C_{11} / \rho} \quad \text{and} \quad v_t = \sqrt{C_{44} / \rho}\ . \tag{6}$$

In the non isotropic case more than two independent quantities are needed to determine the matrix of the elastic constants, and the phase velocities, beside depending on the direction of **k**, have a more complex dependence on the C_{ij} values.
In a finite geometry the search for standing waves having the harmonic time dependence of the type $e^{-i\omega t}$ transforms Eq. (1) into an eigenfunction / eigenvalue equation of the Helmholtz type (Auld, 1990); an appropriate set of basis functions allows to transform this equation into a matrix eigenvalue problem (Nakamura et al., 2004). The eigenvalues are proportional to ω^2, the square of the frequencies of the acoustic modes of the structure, or natural frequencies of the structure. In other words, the finiteness of the geometry converts the continuum spectrum of frequencies of the modes of the infinite medium, given by Eq. (2), into the discrete spectrum of the natural frequencies. These frequencies depend on the (C_{ij}/ρ) values and on the geometry.
In a schematic way: also in non isotropic media the acoustic velocities depend on stiffness and inertia as in Eqs. (6): $v^2 = C / \rho$, indicating generically by C the relevant combination of elastic constants and, possibly, direction cosines of **k**. In the simplest case, the one dimensional geometry of length L, the standing waves are identified by the constructive self interference condition $L = n\lambda / 2 = (n/2) v / f$ (n is an integer number), such that $f = (n/2)\sqrt{C/\rho} / L$. Therefore, a measurement of the frequencies of the acoustic modes

allows to derive $C = \rho(fL)^2 / (n/2)^2$. Also in more complex geometries, the dependence is of the same type

$$C = \rho(fL)^2 N \tag{7}$$

where L is now a characteristic length of the structure (for a slender rod, essentially one dimensional, the length), and N is a dimensionless numerical factor which, beside the mode order n, can depend on dimensionless quantities like geometrical aspect ratios or Poisson's ratio. The factor N also depends on the character of the mode whose frequency f is being measured, and therefore on the specific modulus C which is involved.

Structures can be finite in one or two dimensions and practically infinite in others, as it happens e.g. in a slab or a long cylinder. The free surface of an otherwise homogeneous solid is a case of semi-infiniteness along a single dimension. The translational symmetry is broken in the direction perpendicular to the surface, and new phenomena, absent in the infinite medium, are found: the reflection of bulk waves, and the existence of surface acoustic waves. Namely, at a stress free surface Eqs. (1) admit a further solution: the Rayleigh wave, the paradigm of the surface acoustic waves (SAWs). Such waves have peculiar characters (Farnell & Adler, 1972): a displacement field confined in the neighborhood of the surface, with the amplitude which declines with depth, a wavevector parallel to the surface, and a velocity lower than that of any bulk wave, such that the surface wave cannot couple to bulk waves, and does not lose its energy irradiating it towards the bulk. Pseudo surface acoustic waves can also exist, which violate this last condition. The velocity v_R of the Rayleigh wave cannot be given in closed form; in the isotropic case a good approximation is (Farnell & Adler, 1972)

$$v_R \cong v_t \frac{0.862 + 1.14\nu}{1+\nu} \quad . \tag{8}$$

The continuum model of a homogeneous solid does not contain any intrinsic length scale. Accordingly, all the solutions for this model are non dispersive, meaning that the velocities (Eqs. (6) and (8)) are independent from wavelength (or frequency).

More complex modes occur in non homogeneous media. Layered media are a particularly relevant case, in which new types of acoustic modes can occur; namely, modes confined around the interfaces and modes which are essentially guided by one layer or another, like the Sezawa waves. In this case, also in the continuum model the physical system has an intrinsic length scale, identified by the layer thicknesses. For wavelengths much smaller than the thicknesses wave propagation occurs within each layer as if it was infinite, with reflections and refractions at the surfaces. Instead, for wavelengths comparable to, or larger than, the thicknesses, the acoustic modes extend over several layers, and are modes of the whole structure. Such modes are dispersive: their velocities depend on wavelength, or, more precisely, on the wavelength to thickness ratio(s). Also the simplest surface wave, the Rayleigh wave of a bare homogeneous substrate, is modified by a layer deposited on it, and becomes dispersive: the propagation velocity depends on wavelength, therefore on frequency. The velocities of the acoustic modes in layered structures can be numerically computed, as non trivial functions of the properties of the substrate and the layer(s), and of the wavelength to thickness ratio. The dispersion relations $\omega(k)$ or $v(f)$ are thus obtained.

3. Stiffness measurements

3.1 Vibration based methods

It has always been recognized that since the phase velocities of acoustic waves and the natural frequencies of the acoustic modes depend on stiffness and inertia, their measurement gives access, by Eq.(6) or Eq. (7), to the elastic constants C_{ij}, if the mass density ρ, and possibly the geometry, are known. Many experimental methods have been devised, which exploit vibrations to measure the elastic properties of solids. These methods measure the dynamic, or adiabatic, elastic moduli; these moduli do not coincide with the isothermal moduli which are measured in monotonic tests (if strain rate is not too high), but in elastic solids the difference between adiabatic and isothermal moduli seldom exceeds 1% (Every, 2001). Furthermore, when the elastic constants are needed to design a device which operates dynamically, like most microdevices, the dynamic moduli are exactly the ones which are needed in the design process.

Some methods measure the wave propagation velocity by measuring the transit time over a finite, macroscopic distance, other methods measure the frequency of standing modes defined by the sample geometry, or the frequency of propagating waves of well defined wavelength. The excitation can be either monochromatic, at a frequency which typically should be adjustable until resonance conditions are achieved, or broadband. The latter is typically obtained by an impulsive excitation, which can be provided by a mechanical percussion or by a laser pulse. Generally, the response to a broadband excitation is spectrally analyzed. The availability of ultrafast lasers (femtosecond laser pulses) also allows an analysis in the time domain, by pump-and-probe techniques.

In homogeneous specimens each propagation velocity, or the frequency of each standing wave, has a single value, from which the corresponding elastic modulus can be derived. In non homogeneous specimens, typically in supported films, each propagation velocity can depend on wavelength or frequency. Dispersion relations $\omega(k)$ or $v(\lambda)$ can be measured over a finite interval of frequency or wavelength, and the film properties can be obtained fitting the computed dispersion relations to the measured ones.

3.2 Vibration excitation and detection techniques

The various experimental methods operate in different frequency ranges. The range is determined by both the excitation and the detection techniques, and is strictly correlated to the spatial resolution. It is worth remembering that the acoustic velocities in typical elastic solids like metals and ceramics are of the order of a few km/s = mm×MHz, and that a phase velocity v links the frequency f to a characteristic length L, which can be a characteristic dimension of a structure supporting standing waves, or the wavelength of a travelling wave.

Characteristic lengths of centimetres imply frequencies in the tens of kHz range, which are easily excited by a mechanical percussion and measured by a microphone; Nieves et al. (2000) estimate at around 0.1 MHz the upper limit of the frequencies excited by the mechanical percussion, with a steel ball of a very few millimetres. Piezoelectric actuators, and sensors are also available.

Characteristic lengths of several micrometers correspond to frequencies in the tens of MHz range. Structures of this size can be built by micromachining techniques, and their vibration can be excited and detected by capacitive actuators and sensors. Measurement techniques of this type are essentially a miniaturization of the vibrating reed technique (Kubisztal, 2008). Czaplewski et al. (2005) built flexural and torsional resonators of tetrahedral amorphous carbon (also known as amorphous diamond), by standard techniques for the production of micro-electromechanical systems. They exploited piezo-electric actuation, and an interferometric technique to measure the oscillation. They were able to perform measurements at variable temperatures, determining the elastic moduli of this material as function of temperature. They analyzed the uncertainty sources, finding that the leading contribution to the uncertainty comes, for the flexural oscillator, from the value of the mass density of this material, while for the torsional oscillator it comes from the exact dimensions of the thin member undergoing torsion.

In larger structures, waves at frequencies in the tens of MHz range can be excited and detected by piezoelectric elements, possibly operating simultaneously as actuators and sensors. Excitation can also be performed by a laser pulse; if the pulse is short enough, the upper limit of the measurable frequency range can be set by the piezoelectric sensor. Optical detection techniques are also available. Specific devices like interdigitated transducers (IDTs) can be built by lithographic techniques on, or within, an appropriate layer stack, which must include a piezoelectric layer. Such devices emit and receive waves at the wavelength which resonates with the periodicity of the transducer, typically at micrometric scale. This configuration was exploited to measure the material properties (Bi et al. 2002; Kim et al., 2000), but it is seldom adopted, because it requires the production of a dedicated micro device.

Micrometric and sub-micrometric wavelengths correspond to frequencies in the GHz to tens of GHz range. Detection of such frequencies requires optical techniques; excitation of such frequencies can be obtained by laser pulses of short enough duration.

The variety of vibration based methods to measure the stiffness of solids the can be classified according to various criteria. In this chapter methods are reviewed grouping them by the main vibration excitation techniques: mechanical excitation, either periodic or by percussion, laser pulse excitation, and inelastic light scattering (Brillouin spectroscopy). Similarly to Raman spectroscopy, Brillouin spectroscopy does not excite vibrations at all, but relies on the naturally occurring thermal motion. This gives access to the broadest band, but with small vibration amplitudes, which require time consuming measurements.

3.3 Precision and accuracy

In all the techniques based on vibrations the elastic constants themselves are not the direct outcome of the measurement, but are derived from direct measurements of a primary quantity like frequency or velocity, and 'auxiliary' quantities like thickness, or mass density. The uncertainty to be associated to the resulting value of each elastic constant must be evaluated considering the uncertainties associated to each of the raw measurements. For a quantity q which is derived from directly measured quantities a, b and c, the uncertainty σ_q depends on the 'primary' uncertainties σ_a, σ_b and σ_c. For a functional dependence of the type $f = Aa^{\alpha}b^{\beta}c^{\gamma}$, where A is a numerical constant, the usual error propagation formula can be written in terms of the relative uncertainties (σ_a / a), (σ_b / b), (σ_c / c) as

$$\left(\frac{\sigma_q}{q}\right)^2 = \alpha^2\left(\frac{\sigma_a}{a}\right)^2 + \beta^2\left(\frac{\sigma_b}{b}\right)^2 + \gamma^2\left(\frac{\sigma_c}{c}\right)^2 \tag{8}$$

However, the various uncertainties can have different meanings and consequences. The frequency is typically measured either identifying the frequency of a periodic signal which achieves resonance, or by the spectral analysis of the response to a broadband excitation. In both cases each frequency reading is associated to a finite degree of uncertainty, mainly due to random errors. In a set of repeated measurements such errors are uncorrelated, and tend to be cancelled by an averaging process; the error of each measurement affects the dispersion of results around the average, but not the average itself. In other words, these errors affect precision, but not accuracy. The accuracy of results is at most affected by the finite accuracy in the calibration of the frequency meter, of whichever nature it be. When then deriving the elastic moduli, the frequency reading can be exploited as such (see e.g. Eq.(7)), or via the determination of a propagation velocity. In both cases, the obtained moduli also depend on further 'auxiliary' parameters. In a very simple case, from Eq. (6) we have $C_{11} = \rho v_l^2$, and the resulting value of C_{11} is also affected by the finite uncertainty of the best available value of ρ, exploited in the derivation. However, in a set of repeated measurements the uncertainty of ρ does not contribute to the dispersion of results around the average, but it affects the average itself. This means that it affects accuracy, but not precision. The same can be said for the sample geometry (see Eq. (7)).

4. Mechanical excitation

Mechanical excitation can be either impulsive and broadband, as obtained by a percussion, or narrow band, as provided by a periodic excitation. Most methods exploiting mechanical excitation rely on the identification of the natural frequencies, or resonances, of a structure. With a periodic excitation, such frequencies are identified scanning the excitation frequency until resonance conditions (maximum oscillation amplitude for given excitation force) are detected. With a broadband excitation the response (measured amplitude) is frequency analyzed to identify the resonant frequencies.
Among the methods adopting harmonic excitation, acoustic microscopy (Zinin, 2001) exploits a piezoelectric actuator, typically in the form of an acoustic lens, and often operating also as a transducer. The acoustic lens is mechanically coupled to the sample by a liquid drop. Acoustic microscopy can be operated with imaging purposes; in the quantitative acoustic microscopy version (Zinin, 2001) it aims at measuring the acoustic properties of the sample.
Beside acoustic microscopy, two main types of methods have been developed. The first one measures the bulk properties. It adopts macroscopic homogeneous samples, and can exploit either broadband or narrow band excitation. These methods have also been ruled by norms (ASTM, 2008, 2009). A second group of methods, collectively called Resonance Ultrasound Spectroscopy, aims at measuring the properties of thin supported films. It almost invariably exploits a periodic excitation, whose frequency is swept in order to achieve resonance conditions.

4.1 Measurement of bulk properties

Macroscopic homogeneous samples are self supporting. They can be tested as free standing samples, provided the disturbances to free oscillations are minimized. Such a minimization includes sample suspension by thin threads, or specimen support by adequate material (cork, rubber), the supports having contact of minimum size, and positions at the nodes of the fundamental vibrational mode of interest, either flexural or torsional. Also the sensor contact, if oscillation is detected by a contact device, must be devised aiming at the minimization of the disturbance induced by the contact. Non contact detection techniques are available, including all optical techniques, and acoustic techniques: in the proper frequency range, oscillations can be detected through the air, by a microphone, and even excited, with harmonic excitation, by a similar technique, exploiting an audio oscillator and an audio amplifier. The optical techniques, intrinsically contact-less and inertia-less, have the broadest band, only limited by the light detection and analysis apparatus.

Since the full characterization of the elasticity of an isotropic medium requires two independent parameters, it typically requires excitation of at least two modes of different nature. For a specific simple geometry, the slender rod of length L, test methods have been regulated by norms (American Society for Testing and Materials [ASTM], 2008, 2009, and other ASTM norms cited by these two). Mainly flexural and torsional modes are considered, for a slender rod of mass m, and either rectangular section of width b and thickness t, or circular section of diameter D. Mass density being measured as $\rho = m/(tbL)$, Eq.(7) takes in this case, for the rectangular section, the forms (ASTM, 2008)

$$E = \left[m/(tbL)\right]\left(f_f L\right)^2 \times 0.9465 (L/t)^2 T_E(\nu, L/t) \tag{9a}$$

$$G = \left[m/(tbL)\right]\left(f_t L\right)^2 \times 4 T_G(b/t, b/L) \tag{9b}$$

where f_f and f_t are the frequencies of respectively the fundamental flexural and torsional modes, and T_E and T_G are numerical factors, functions of the indicated dimensionless quantities. Similar formulas hold for rods of circular section. The fundamental frequencies can be identified by either sweeping the frequency of a periodic excitation (ASTM, 2008), or by the broadband excitation by a mechanical percussion (ASTM, 2009), followed by the frequency analysis of the response. In both cases the displacement can be sensed by a contact transducer or by a microphone. The estimates for E and G being coupled by the value of Poisson's ratio (Eqs. 4 and 9a), an iterative procedure is indicated, to obtain consistent estimates. The algorithm of Eq.(8) applied to Eq. (9a) gives

$$\left(\frac{\sigma_E}{E}\right)^2 = \left(\frac{\sigma_m}{m}\right)^2 + 3^2\left(\frac{\sigma_t}{t}\right)^2 + \left(\frac{\sigma_b}{b}\right)^2 + 3^2\left(\frac{\sigma_L}{L}\right)^2 + 2^2\left(\frac{\sigma_f}{f}\right) \tag{10}$$

similar expressions being obtained for (σ_G/G) and for rods of circular section; such equations allow to derive the uncertainties to be associated to the obtained moduli, from the intrinsic uncertainties of the primary quantities. It was estimated (ASTM, 2008) that the major sources of uncertainty come from the fundamental frequency f and from the smallest specimen dimension (thickness or diameter). Uncertainties of the moduli in the 1% range are achievable.

Other free standing sample geometries were considered, and analyzed by detailed finite elements computations, to identify the appropriate values of the numerical factor N (see Eq. (7)). Nieves et al. (2000) consider a cylinder with length equal to diameter ($L = D$). They excite vibrations by a longitudinal percussion, and detect the displacement by an optical technique; several modes are typically observed. Alfano & Pagnotta (2006) consider instead a thin square plate, and, analyzing the displacement distributions of the first modes, they identify the best positions for the plate supports and for perpendicular percussion. They measure the response by a microphone. D'Evelyn & Taniguchi (1999) similarly exploited thin disks, exciting different modes by different impact points of a hollow zirconia bead, and measuring the response by a microphone. They exploit computations of the resonant frequencies of thin disks performed by others, and they estimate the accuracies of these computations to 1 % or better.

In all these cases, from the numerical computations the mode frequencies are tabulated or interpolated; Nieves et al. exploit the scaling parameter $\sqrt{G/\rho}/(\pi L)$. The ratios of the mode frequencies depend essentially on Poisson's ratio: the ratios of the observed frequencies allow therefore to identify the modes and to evaluate Poisson's ratio. The frequency values allow then, by the scaling parameter, to derive the shear modulus.

Both Nieves et al. and Alfano & Pagnotta perform detailed analyses of the measurement uncertainties. They both find that, also with their experimental set-up, the frequency measurement has a crucial role in the precision of the obtained moduli. Since Alfano & Pagnotta consider thin plates, they find that the precision of the thickness t is also crucial, simply because, thickness t being much smaller than the plate size a, a small value of the relative uncertainty σ_t/t is more difficultly achievable than for σ_a/a.

For comparison purposes, Nieves et al. (2000), beside considering the axial modes, also excite, by tangential percussion, the torsional modes, whose frequencies can be computed in closed form. Once the torsional modes are discriminated from the bending modes, the results agree to better than 1%, indicating a precision of this order. They also perform measurements by the pulse-echo method, finding instead discrepancies of 2% or more; they suggest that this method, which measures the propagation velocity of travelling waves, might become intrinsically less accurate when performed in a confined geometry of small size.

4.2 Resonance ultrasound spectroscopy

The Resonance Ultrasound Spectroscopy (RUS) methods have been developed (Migliori et al., 1993; Ohno, 1976; Schwarz et al., 2005; So et al., 2003) aiming in particular at the measurement of the properties of supported thin films. A recent implementation (Nakamura et al., 2004, 2010) exploits, as film support, a thin plate which, to be measured, is located on a tripod. One of the three legs is rigid, and contains a thermocouple which monitors the sample temperature; the second leg is a piezoelectric actuator, feeding a harmonic excitation whose frequency is swept, the third one is a piezoelectric sensor, which detects the oscillation amplitude. The resonance spectrum, i.e. the oscillation amplitude as function of the frequency, is measured sweeping the excitation frequency. Several peaks are found, sometimes with partial overlaps; their amplitudes, but not their frequencies, depend on the position of the piezoelectric sensor. The measurement is precise enough to clearly detect the difference between measurements in air and in vacuum, and the reproducibility of the resonance frequencies is at the 0.1 % level (Nakamura et al., 2010). The elastic constants are found fitting the computed frequencies to the measured ones; to this end, mode identification is crucial. Identification is performed keeping the excitation frequency at resonance, and scanning the specimen surface by a laser-Doppler interferometer. The map of the out-of-plane displacement of the vibrating specimen is thus obtained, which allows an unambiguous mode identification.

The elastic constants of the substrate are previously found by performing the same type of measurement on a bare substrate, and the elastic constants of the film are derived from the measured modifications of the resonance spectrum of the substrate. Since the vibration amplitude of the standing waves in a plate is maximum at the surface, the sensitivity of the method is better than the ratio of film thickness to the support thickness. In particular, the sensitivity of each resonance frequency to each elastic constant is assessed evaluating numerically the derivatives $\partial f / \partial C_{ij}$, and from these values the uncertainties ΔC_{ij} are derived from the estimated uncertainties Δf as

$$\Delta C_{ij} = \frac{1}{\partial f / \partial C_{ij}} \Delta f \ . \tag{11}$$

Deposited thin films often have a significant texture, with one crystalline direction preferentially oriented perpendicularly to the substrate surface, and random in-plane orientations, resulting, at a scale larger than that of the single crystallite, in in-plane isotropy, with different out-of-plane properties. This type of symmetry corresponds to the hexagonal symmetry, in which the tensor of the elastic constants is determined by five independent quantities. Among these, the resonance frequencies turn out to be almost insensitive to C_{44}, which therefore remains not determined, while the highest sensitivity is to C_{11} (Nakamura et al., 2010).

5. Laser pulse excitation

A laser pulse which is absorbed induces a sudden local heating which, by thermal expansion, produces an impulsive mechanical loading. Such a mechanical impulse excites waves in a broad frequency range, which are then detected. The accessible frequency band can be limited by either the excitation bandwidth or the detection device. According to the nature of the material being investigated, the deposition of an interaction layer can be needed. In particular, a short absorption depth is required, in order to excite a pulse which has small spatial and temporal duration, and therefore a broad band. The power density threshold for ablation must also be considered: measurements are typically conducted with high repetition rate pulses, and if ablation occurs the specimen undergoes a continuous modification during the measurement.

Two main configurations have been developed up to maturity; they differ for the propagation geometry and for the technique to detect vibrations. They are respectively called laser ultrasonics and picoseconds ultrasonics.

5.1 Laser ultrasonics

The so called laser ultrasonics technique is mainly exploited to characterize thin supported films. Oscillations are excited by a focused laser pulse, and propagation along the surface is detected, measuring the surface displacement at a distance from excitation pulse ranging from millimetres to centimetres. It is mainly the Rayleigh wave, modified by the presence of the film, which is excited and detected. The laser pulse, typically of nanosecond duration, is focused by a cylindrical lens on a line. The sudden expansion of this line source launches surface waves of limited divergence, propagating along the surface, perpendicularly to the focusing line. The component of the surface displacement normal to the surface itself can be measured, at various distances from the line source, by optical interferometry (Neubrand & Hess, 1992; Withfield et al., 2000), or, in a simpler and more robust way, by a piezoelectric sensor (Lehmann et al., 2002; Schneider et al., 1997, 1998, 2000).

The recorded displacement is frequency analyzed, yielding the dispersion relation $v(f)$ for a frequency interval that can extend over a full frequency decade (e.g. 20 to 200 MHz).

Fitting the computed dispersion relation to the measured one allows to derive the film properties. The width of the measured frequency interval can allow to to derive the Young modulus and also the film thickness (Schneider et al., 1997, 2000). The uncertainties of the results is evaluated numerically by the fitting procedure. Since the observed propagation distance is of the order of millimetres, the obtained properties are representative of an average over the propagation distance. The performance of the method could be pushed to the measurement of the properties of diamond-like carbon films having thickness down to 5 nm (Schneider et al., 2000). By stretching the observed propagation path to 20 mm it was possible to reduce the uncertainty of the measured propagation velocity of the Rayleigh wave to below 0.25 m/s. It was thus possible to detect the small variation of the Rayleigh velocity (5081 m/s for the bare (001) silicon substrate) induced by the presence of the film.
In a different configuration, the laser pulse is focused on a point instead of a line, with consequent excitation of waves which expand in all the radial directions, and a common path interferometer is adopted, whose light collection point scans the specimen surface (Sugawara et al., 2002). It is thus possible to visualize the wavefronts, circular for isotropic samples and non circular for anisotropic ones, with time resolution of the order of picoseconds and spatial resolution of a few micrometres.

5.2 Picosecond ultrasonics

The so called picosecond ultrasonics technique owes its name to the picosecond laser pulses which were available at the time it was introduced (Thomsen et al., 1984, 1986). It is nowadays implemented by femtosecond laser pulses, and is intrinsically suited to characterize thin supported films and multilayers. It follows the optical pump-and-probe scheme (Belliard et al., 2009; Bienville et al., 2006; Bryner et al. 2006; Vollmann et al. 2002). The pump beam, a femtosecond laser pulse, is focused, by a spherical lens, at the specimen surface and, at least partially, absorbed. The focusing spot, a few to tens of micrometers wide, is orders of magnitude larger than the characteristic lengths of the involved phenomena: mainly the optical absorption length, and also the thermal diffusion length and the acoustic wave propagation length, both for a femtosecond time scale. Bryner et al. (2006) estimate that with an aluminium surface, a near infrared laser (800 nm), and a pulse width of 70 fs, the absorption depth and therefore the dominant acoustic wavelength are of the order of 10 nm. Quite consistently, for a similar laser pulse but for Pt and Fe ultra thin films, Ogi et al. (2007) estimate at several THz the upper bound of the frequencies excited by the laser pulse.
The thermal and mechanical fields are thus (almost perfectly) laterally uniform, and essentially one dimensional: the absorbed pulse has a depth of the order of nanometres, and propagates like an acoustic wave with a plane wavefront which travels perpendicular to the surface, towards the specimen depth. At each interface this wave is partly transmitted and partly reflected, according to the acoustical impedances of the layers, and gives rise to echoes, which return to the surface, where they again are reflected back.
The surface is then probed by the probe beam, much weaker than the pump beam, which reaches the surface with a variable delay, controlled by a delay line. The probe beam, similarly to the acoustic wave, is partly transmitted and partly reflected at each interface, leading, in each layer, to a forward and a backward electromagnetic field, whose complex amplitudes can be calculated using a transfer-matrix formalism. The external reflectivity of the surface is the ratio of the backward to the forward complex amplitudes in the outer

space. This reflectivity is modified by the acoustic strain, by two mechanisms. Firstly, each interface is displaced by the acoustic wave, and, secondly, the refractive index in each layer is modified by the elastic strain, by the acousto-optic (also called photoelastic) coupling. In particular, it must be remembered that the traveling acoustic pulse extends over a depth of a few nanometers, inducing a localized perturbation of the refractive index, which partially scatters the optical beam. Interference can occur between the beam reflected at the outer surface and that reflected by the traveling acoustic pulse.

Interferometric techniques allow to measure the variation of both amplitude and phase of the reflected beam, thus measuring the variation of the complex reflection coefficient. With probe pulses in the femtosecond range, and varying the probe pulse delay, the time evolution of the surface reflectivity can be monitored with high temporal resolution. This time evolution typically shows several features. The diffusion, towards the sample depth, of the heat deposited by the laser pulse gives a slowly varying reflectivity background. The echoes of the acoustic pulse which, after partial reflection at the film/substrate interface, are again reflected at the outer surface are generally visible. The so called Brillouin oscillations, due to the interference between the beam reflected at the outer surface and that reflected by the traveling acoustic pulse, can then be found.

The analysis of the various features allows to characterize the waves which cross the layers travelling perpendicularly to the surface. In the derivation of the film properties, the knowledge of film thickness, typically obtained by X-ray reflectivity, has a crucial role; the uncertainty about thickness is one of the leading terms in the uncertainty of the final results. The achievable resolution depends on the excited wavelength. In copper the absorption depth is larger than the value cited above for aluminium. Since the smallness of the absorption depth determines the localization of the acoustic pulse and the achievable resolution, the deposition of an aluminium interaction layer, which guarantees a very small absorption depth, is a common practice. The interaction layer, typically a few tens of nanometres thick, then participates to the vibrational behaviour of the structure being investigated. Accurate measurements of stiffness therefore require consideration of the effects of the interaction layer, e.g. by measurements with layers of different thicknesses, followed by an extrapolation to null thickness (Mante et al., 2008). Obviously this deconvolution of the effects of the interaction layer contributes to the uncertainty of the final results.

Near infrared lasers are a common choice, because at shorter wavelength more complex phenomena can occur, which were attributed to electronic interband transitions (Devos & Cote, 2004); obviously, if one is interested in elastic properties, electronic transitions are a spurious effect to be avoided.

By picoseconds ultrasonics it was possible to characterize a layer stack, including a buried layer of about 20 nm thickness (Bryner et al., 2006). The uncertainty for the elastic constants of this layer is estimated at 20-25%, which is however remarkable for a layer of this type. The lowest limit for layer detection is also estimated at about 10 nm thickness. In a different configuration, namely a single Pt or Fe layer on a silicon substrate or a borosilicate glass substrate, Ogi et al. (2007) were able to characterize metallic films of thickness down to 5 nm. They found, at nanometric thicknesses, a dependence of the elastic moduli on thickness. This was explained by the impossibility of plastic flow at such low thicknesses: the elastic strains can thus reach levels which are non reachable in thicker samples, such that higher order elastic constants are no longer negligible.

Periodic Mo/Si multilayers (superlattices) were investigated by Belliard et al. (2009), exploring various periodicities in the nanometric range. They detect bulk waves crossing

back and forth the multilayer; such waves show pulses of the order of 10 ps, which correspond to propagation lengths much larger than the superlattice period. These waves therefore see the whole multilayer as an effective medium. The theoretical prediction for the properties of the effective medium and the reflection coefficient at the superlattice / Si substrate interface are confirmed by the experimental findings. They also detect higher frequency oscillations, which correspond to localized waves. The periodic multilayer acts as a Bragg reflector, and opens forbidden gaps in the spectrum. It can confine a mode in the neighbourhood of the outer surface (acoustic-phonon surface modes), but only if the outer layer is the lower acoustic impedance layer (in this case Si), which therefore acts as a perfect reflector. The properties of such modes could be correctly predicted only taking into account the nanometric top silicon oxide layer, which spontaneously forms at he silicon surface. The presence of this additional layer is also consistent with the X-ray reflectivity measurements. The behaviour of a Mo cavity sandwiched between Mo/Si mirrors was also analyzed.
The picoseconds ultrasonics technique was also exploited to investigate non laterally homogeneous specimens (Bienville et al., 2006; Mante et al., 2008). One limitation of this technique in the measurement of the elastic constants is that it involves only plane waves travelling perpendicular to the surface, thus allowing only the out-of-plane elastic characterization of the film. To overcome this limitation, Mante et al. and Robillard et al. (2008) proposed a technique by which the film to be characterized, and the aluminium interaction layer deposited on it, are cut by lithographic techniques to obtain a periodic square lattice of square (200 nm × 200 nm) pillars. As confirmed by the measurements performed on square lattices of different lattice constants, the pump pulse also excites, in this nanostructured film, acoustic collective modes of the pillars, which propagate along the surface in various directions. Various branches are measured, from which also the in-plane properties of the film can be measured, achieving a complete elastic characterization.

6. Brillouin spectroscopy

Aggregates of atoms, from molecules to clusters, to nanoparticles and nanocrystals, up to mesoscopic and macroscopic aggregates, can interact with electromagnetic waves either elastically or inelastically. Inelastic interactions include emission/absorption phenomena, and inelastic scattering. We consider here inelastic scattering by vibrational excitations. At the molecular scale the atomic structure of matter has a crucial role, and quantum phenomena are relevant. At this level, vibrational excitations are the vibrations of molecules, or, in a crystal, the vibrations of the internal degrees of freedom of each unit cell, which form the so called optical branches of the dispersion relation, or optical phonons. Broadly speaking, inelastic scattering by these excitations is called Raman scattering.
Aggregates above the nanometric scale also support collective vibrational excitations which begin to resemble to acoustic waves, and can be described by a continuum model. In a crystal, the vibrations of the degrees of freedom of the centre of mass of each unit cell form the so called acoustic branches of the dispersion relation, or acoustic phonons. In the long wavelength limit they are the are acoustic waves, accurately described by the continuum model (Eq. (1)). Broadly speaking, inelastic scattering by these excitations is called Brillouin scattering.
Visible light has sub micrometric wavelength. In media, either crystalline or amorphous, which are homogeneous, and therefore translationally invariant, over at least a few micrometres, vibrational excitations of sub-micrometric wavelength have a well defined

wavevector (see Eq. (2)). Due to translational invariance, the kinematics of scattering selects the excitations whose wavevector is close to the Brillouin zone center (**k** = 0). The acoustic and optical phonon branches have very different behaviours close to the zone center (**k** → 0). The acoustic branch frequency ω_a goes to zero, with phase velocity ω_a/k and group velocity $\partial\omega_a / \partial k$ which tend to coincide with the sound velocity (which depends on polarization and, in an anisotropic crystal, on the wavevector direction), while the optical branch frequency ω_o typically goes to a maximum, with group velocity $\partial\omega_o / \partial k$ which goes to zero.

Correspondingly, with visible light and with typical properties of solids, the two types of branches produce inelastic scattering with frequency shifts ranging from a fraction of cm^{-1} to a few cm^{-1} (i.e from a few GHz to tens of GHz) for Brillouin scattering, and from hundreds to thousands of cm^{-1} (i.e from THz to tens of THz) for Raman scattering. The spectral analysis of so widely different frequency ranges requires different types of spectrometer. However, for both types of scattering the experiments are performed without exciting the vibrations, but relying on the naturally occurring thermal motion.

Brillouin spectrometry thus offers a fully optical, and therefore contact-less, method to measure the dispersion relations of bulk and surface acoustic waves, whose wavelength is determined by the scattering geometry and the optical wavelength, and is typically sub-micrometric. The frequency results from the medium properties, and typically falls in the GHz to tens of GHz range. Measurements are performed illuminating the sample by a focused laser beam, and analyzing the spectrum of scattered light, which is dominated by the elastically scattered light, but also contains weak Stokes/anti-Stokes doublets due to inelastic scattering by thermally excited vibrations (Beghi et al., 2004; Comins, 2001; Every, 2002; Grimsditch, 2001; Sandercock, 1982).

In sufficiently transparent materials scattering can occur in the bulk, by bulk acoustic waves. The coupling mechanism is the elasto-optic (or acousto-optic) effect: the periodic modulation of the refractive index by the periodic strain of the acoustic wave. In both transparent and opaque materials scattering can also occur by surface acoustic waves, by the ripple mechanism: the periodic corrugation of the surface due to the surface wave.

In more detail: the incident beam, of angular frequency Ω_i and wavelength λ_0, impinges on a sufficiently transparent sample with wavevector $\mathbf{q}_i$ and is refracted into the wavevector $\mathbf{q}'_i$. Scattered light, of wavevector $\mathbf{q}'_s$, emerges with wavevector $\mathbf{q}_s$. The probed wavevector, $\mathbf{k} = \pm(\mathbf{q}'_s - \mathbf{q}'_i)$, is determined by λ_0, the directions of $\mathbf{q}_i$ and $\mathbf{q}_s$, and the refractive index n. Light inelastically scattered by a vibrational excitation of angular frequency $\omega(\mathbf{k})$ gives a Stokes/anti-Stokes doublet at frequencies $\Omega_s = \Omega_i \pm \omega$. Detection, in the spectrum of scattered light, of such a doublet allows to measure $\omega = |\Omega_s - \Omega_i|$ and to derive the excitation velocity $v = \omega / |\mathbf{k}|$. In both transparent and opaque samples scattering occurring by surface waves only depends on the components of wavevectors parallel to the surface: the probed wavevector is $\mathbf{k}_{||} = \pm(\mathbf{q}'_s - \mathbf{q}'_i)_{||}$, and the surface wave velocity is $v = \omega / |\mathbf{k}_{||}|$. In other words, the spontaneous thermal motion can be viewed as spatially Fourier transformed into an incoherent superposition of harmonic waves having all the possible wavevectors; the scattering geometry (the directions of wavevectors $\mathbf{q}_i$ and $\mathbf{q}_s$) selects a specific wavevector $\mathbf{k}$ or $\mathbf{k}_{||}$ which is probed by the inelastic light scattering event.

Although also other geometries have been exploited (Beghi et al. 2011), in Brillouin spectroscopy the most frequently adopted scattering geometry is backscattering: $\mathbf{q}_s = -\mathbf{q}_i$. For bulk scattering it corresponds to $\mathbf{k} = \pm 2\mathbf{q}'_i$, such that

$$|\mathbf{k}| = 2\frac{2\pi}{\lambda_0}n \ , \tag{12}$$

which depends on the refractive index, but depends on geometry only when the sample is anisotropic, while for surface scattering it means $\mathbf{k}_{\parallel} = \pm 2\mathbf{q}_{i\parallel}$, i.e.

$$|\mathbf{k}_{\parallel}| = 2\frac{2\pi}{\lambda_0}\sin\theta \ , \tag{13}$$

where θ is the incidence angle (the angle between the incident beam and the surface normal). In this case the probed wavevector depends the incidence angle, but not on the refractive index, because Snell's law implies that upon refraction the optical parallel components $\mathbf{q}_{\parallel}$ remain unchanged.

The data analysis for Brillouin spectroscopy results is common to all the methods, like laser ultrasonics, which measure the velocity of travelling waves. In the simplest cases the velocity is a function of the elastic constants which can be given in closed form. For instance, if scattering by the longitudinal bulk wave, of velocity $v_l = \sqrt{C_{11}/\rho}$ (Eq. (6)) is detected, C_{11} is directly obtained as $C_{11} = \rho v_l^2 = \rho\omega^2/k^2$, and its uncertainty $\sigma_{C_{11}}$ is evaluated by Eq. (8). In other cases, and namely in the case of supported films, the mode velocity can be computed as function of the elastic constants only numerically. In that case it must be remembered that the stiffness of an elastic solid is determined by as many independent parameters as are needed to completely identify the tensor of the elastic constants. In other words the stiffness is identified by a point in a multidimensional space, the dimensionality being 2 in the simplest case of the isotropic medium, and being higher for lower symmetry media.

Focusing here on the isotropic case, the stiffness can be represented, among other possible choices, by the (E,ν) couple (see Eqs. (3) and (4)). In Brillouin spectroscopy, for each mode the velocity is measured for various wavevectors k (in laser ultrasonics it is measured for various frequencies) as $v_m(k)$ with uncertainty $\sigma_{v(k)}$; since it can also be computed as $v_c(E,\nu,k)$, the stiffness, represented by the (E,ν) couple, can be determined by a standard least squares minimization procedure. The sum of squares is computed as

$$\chi^2(E,\nu) = \sum_k \left(\frac{v_c(E,\nu,k) - v_m(k)}{\sigma_{v(k)}} \right)^2 \ , \tag{14}$$

where, for each wavevector k, the sum is further extended to all the detected acoustic modes. Following standard estimation theory, the minimum of $\chi^2(E,\nu)$ identifies the most probable value $(\overline{E,\nu})$ of the (E,ν) couple, and the isolevel curves of the normalized estimator $\left(\chi^2(E,\nu) - \chi^2(\overline{E,\nu})\right)/\chi^2(\overline{E,\nu})$ identify the confidence region at any predetermined confidence level (Beghi et al., 2001, 2004, 2011; Lefeuvre et al., 1999). In some cases a well defined minimum of $\chi^2(E,\nu)$ is found, allowing a good identification of the parameters (Beghi et al., 2001; Comins et al., 2000; Zhang et al., 2001a), while in other cases a broad, valley-shaped minimum is found. In such cases a good identification of the parameters is not possible (Beghi et al., 2002; Zhang et al., 1998), although sometimes some combination of the parameters can be identified with better precision than individual

parameters (Comins et al., 2000; Zhang et al., 1998, 2001a). When several acoustic modes are measured, the larger amount of available information allows a precise and complete elastic characterization, as obtained e.g. for SiC films of micrometric thickness (Djemia et al, 2004).
In Eq. (14) each value $v_m(k)$, being obtained as ω/k, has an uncertainty $\sigma_{v(k)}$ which depends in turn on the uncertainty of the frequency of each spectral doublet, and on the precision of the incidence angle (see Eq. (13)) or of the refractive index (see Eq. (12)). It can be noted that the uncertainties of frequency and angle are of the random type, which affects precision but not accuracy, while the uncertainty of the refractive index affects accuracy but not precision (see Section 3.3). As with other techniques, also the uncertainties concerning the mass density and the layer thickness(es) affect accuracy but not precision. These uncertainties were the object of detailed investigations (Beghi et al., 2011; Stoddart et al., 1998). The effects of the uncertainties of the quantities which are directly measured ('primary uncertainties') on the values of the elastic constants which are finally obtained were evaluated. It was found that with appropriate sets of measurement uncertainties at the 1% level are reachable (Beghi et al., 2011).
As already noted, Brillouin spectroscopy measures the acoustic modes at frequencies of the order of GHz to tens of GHz, therefore at wavelengths much shorter than those corresponding to frequencies of tens to hundreds of MHz, typically observed with piezoelectric excitation and/or detection. This gives Brillouin spectroscopy an intrinsically higher sensitivity to the properties of films, or to the perturbation induced by the presence of very thin films. Brillouin spectroscopy was exploited to characterize tetrahedral amorphous carbon films of thicknesses of hundreds of nanometres (Chirita et al., 1999), tens of nanometres (Ferrari et al., 1999), down to a few nanometres (Beghi et al., 2002). It was also shown that inelastic light scattering can be sensitive to nanometric thickness differences (Lou et al., 2010). By Brillouin spectroscopy it was also possible to characterize buried layers in silicon-on-insulator structures (Ghislotti & Bottani, 1994).
On the other hand, the techniques which excite vibrations operate with oscillation amplitudes significantly larger than those measured by Brillouin spectroscopy; this allows more precise measurements of frequencies, which at least partially compensates for the lower intrinsic sensitivity due to the larger wavelengths. Combinations of techniques were also exploited: thicker tetrahedral amorphous carbon films films (3 micron) were characterized combining Brillouin spectroscopy and laser ultrasonics. A wide range of frequencies was thus accessible, allowing a detailed characterization of the elastic properties of the film (Berezina et al., 2004). A combination of Brillouin spectroscopy and picosecond ultrasonics was instead exploited to characterize superlattices formed by periodic multilayers of permalloy/alumina, with various periodicities at the nanometric scale (Rossignol et al., 2004). Picosecond ultrasonics characterizes the out-of-plane properties by waves travelling normal to the surface, while Brillouin spectroscopy characterizes the in-plane properties by waves travelling along the surface. The combination of techniques elucidated the effects of the interfaces.
Another algorithm for data analysis, different from that outlined above, has also been recently proposed (Every et al., 2010). Both algorithms refer to the types of waves most frequently measured in Brillouin spectroscopy of films or layered structures: surface acoustic wave, or pseudo surface acoustic waves, which essentially travel parallel to the surface, or however have a significant wavevector component parallel to the surface. It can also be mentioned that it was also possible, by Brillouin spectroscopy, to detect standing

acoustic waves trapped within a film, which are reflected back and forth, crossing the film perpendicularly to its surface (Zhang, 2001b).

Brillouin spectroscopy lends itself to the characterization of structures other than films or layers. In particular, single-walled carbon nanotubes were characterized, measuring Brillouin scattering by a free-standing film of pure, partially aligned, single-walled nanotubes, and analyzing the results in terms of continuum models (Bottani et al., 2003). The dependence of the measured spectra on the angle between the exchanged wavevector and the preferential direction of the tubes shows that the tube-tube interactions are weak: the tubes are vibrationally almost independent. The tubes are modelled as continuous membranes, at two different levels: at the first one the membrane is infinitely flexible, only able to transmit in-plane forces, while at the second level of approximation the tube wall is treated as also able to transmit shear forces and torques not belonging to the shell surface. In both cases scattering was essentially due to longitudinal waves travelling along the tubes. Taking into account that AFM images suggest that the tube segments contributing to scattering are not in the infinite tube length approximation, it was possible to derive the 2D Young modulus for the tube wall, achieving the first dynamic estimation of the stiffness of the tube wall. Scattering from carbon nanotubes was observed also in a different geometry, with an ordered array of tubes, clamped at one end (Polomska et al., 2007).

Due to its intrinsic contact-less nature, Brillouin spectroscopy is the natural choice for the measurement of elastic properties in conditions, like high temperature and/or high pressure, in which physical contact with the specimen is difficult, if possible at all. Brillouin spectroscopy only requires optical access, which can be obtained by an appropriate window, and even in the extreme conditions achievable in a diamond anvil cell, optical access is guaranteed by the transparency of the same diamond anvils.

Measurements were performed at high temperatures (Pang, 1997; Stoddart, 1995; Zhang et al., 2001a), as well as at low temperatures, which were crucial to single out a particular mechanism of hypersound propagation in alkali-borate glasses (Carini et al., 2008). After pioneering experiments at high pressures (Crowhurst et al., 1999; Whitfield et al., 1976,), in recent years dedicated Brillouin spectrometers were built at synchrotron facilities, allowing simultaneous performance of high resolution x-ray diffraction and Brillouin spectroscopy (and possibly other optical investigations like Raman spectroscopy, fluorescence, absorption) on specimens subjected to extreme pressures in a diamond anvil cell, with possible heating (Murakami et al., 2009; Prakapenka et al., 2010; Sinogeikin et al., 2006). This set-up, of particular interest for geophysicists since it allows to characterize the behaviour of minerals in the conditions which are found in the Earth's interior, was exploited to perform measurements on SiO_2 glass (Murakami & Bass, 2010) and other minerals, but also on polymers (Stevens et al., 2007) and liquid methane (Li et al., 2010).

7. Conclusion

The stiffness of films, characterized by the elastic constants, depends on the film microstructure, and its precise characterization is crucial when thin layers have structural functions. The interest in the measurement of the elastic constants is witnessed by the number of new techniques, or of improvements of existing techniques, being proposed.

The techniques which exploit either propagating acoustic waves or standing oscillations involve exclusively elastic strains: they therefore offer the most direct and clean access to the elastic properties, and potentially the most accurate measurements. Among the methods

based on vibrations, those which exploit, for excitation and/or detection, the contact-less and inertia-less nature of light, have an important role.
An overview of the variety of existing methods was presented here, trying to present a unified picture, and underlining the peculiarities of each of them, in particular for what concerns the experimental uncertainties. It turns out that, under appropriate conditions and experimental procedures, several techniques can achieve significant precision and accuracy.

8. References

Alfano, M. & Pagnotta, L., (2006). Measurement of the dynamic elastic properties of a thin coating, *Review of Scientific Instruments*, Vol.77, Paper No. 056107

ASTM E1875-08 (2008). *Standard Test Method for Dynamic Young's Modulus,Shear Modulus, and Poisson's Ratio by Sonic Resonance*, ASTM International, West Conshohocken, PA.

ASTM E1876-09 (2009). *Standard Test Method for Dynamic Young's Modulus, Shear Modulus, and Poisson's Ratio by Impulse Excitation of Vibration*, ASTM International, West Conshohocken, PA.

Auld, B. A. (1990). *Acoustic fields and Waves in Solids*, Robert E. Krieger Publishing Company, Malabar, Florida

Beghi, M. G., Bottani, C. E. & Pastorelli, R. (2001). High accuracy measurement of elastic constants of thin films by surface Brillouin scattering. In *Mechanical properties of structural films*, C. Muhlstein, C. & Brown, S.B. (Eds.), pp. 109-126. ASTM STP 1413, American Society for Testing and Materials, Conshohoken, PA

Beghi, M.G., Ferrari, A. C., Teo, K.B.K., Robertson, J., Bottani, C.E., Libassi, A. & Tanner, B.K. (2002). Bonding and mechanical properties of ultrathin diamond-like carbon films, *Applied Physics Letters*, Vol.81, pp. 3804-3806

Beghi, M.G., Every, A.G. & Zinin, P.V. (2004). Brillouin scattering measurement of SAW velocities for determining near-surface elastic properties, in *Ultrasonic nondestructive evaluation*, T. Kundu (Ed.), pp. 581-651, CRC Press, Boca Raton, FL. Revised edition in press (2011)

Beghi, M. G., Di Fonzo, F., Pietralunga, S., Ubaldi, C. & Bottani, C. E. (2011). Precision and accuracy in film stiffness measurement by Brillouin spectroscopy, *Review of Scientific Instruments*, Vol.102, Paper No. 053107

Belliard, L., Huynh, A., Perrin, B., Michel, A., Abadias, G. & Jaouen, C. (2009). Elastic properties and phonon generation in Mo/Si superlattices, *Physical Review B* Vol.80, Paper No. 155424

Berezina, S., Zinin, P. V., Schneider, D., Fei, D. & Rebinsky, D. A. (2004). Combining Brillouin spectroscopy and laser-SAW technique for elastic property characterization of thick DLC films, *Ultrasonics*, Vol.43, pp. 87 - 93

Bi, B., Huang, W.-S., Asmussen, J. & Bolding, B. (2002). Surface acoustic waves on nanocrystalline diamond, *Diamond and Related Materials*, Vol.11, pp. 677--680.

Bienville, T., Robillard, J.F., Belliard, L., Roch_Jeune, I, Devos, A. & Perrin, B. (2006). Individual and collective vibrational modes of nanostructures studied by picosecond ultrasonics, *Ultrasonics*, Vol.44, pp. e1289-e1294

Bottani, C.E., Li Bassi, A., Beghi, M.G., Podesta, A., Milani, P., Zakhidov, A., Baughman, R., Walters, D.A. & Smalley, R.E. (2003). Dynamic light scattering from acoustic modes in single-walled carbon nanotubes, *Physical. Review. B* , Vol.67, paper n. 155407

Bryner, J., Profunser, D.M., Vollmann, J., Mueller, E. & Dual, J. (2006). Characterization of Ta and TaN diffusion barriers beneath Cu layers using picosecond ultrasonics, *Ultrasonics,* Vol.44, pp. e1269-e1275

Carini, G., Tripodo, G. & Borjesson, L. (2008). Thermally activated relaxations and vibrational anharmonicity in alkali-borate glasses: Brillouin scattering study. *Physical Review B,* Vol.78, paper n. 024104

Chirita, M., Sooryakumar, R., Xia, H., Monteiro, O. R. & Brown, I. G. (1999). Observation of guided longitudinal acoustic modes in hard supported layers. *Physical Review B,* Vol.60, pp. 5153-5156.

Comins, J. D., Every, A. G., Stoddart, P. R., Zhang, X., Crowhurst, J. C. & Hearne, G. R. (2000). Surface Brillouin scattering of opaque solids and thin supported films. *Ultrasonics,* Vol.38, pp. 450-458.

Comins J.D. (2001). Surface Brillouin scattering, in *Handbook of Elastic Properties of Solids, Liquids, and Gases,* M. Levy, H. E. Bass and R. R. Stern & V. Keppens (Eds.), Volume I: *Dynamic Methods for Measuring the Elastic Properties of Solids,* pp. 349-378, Academic Press, New York

Crowhurst, J.C., Hearne, G.R., Comins, J.D., Every, A.G. & Stoddart, P.R.. (1999). Surface Brillouin scattering at high pressure: Application to a thin supported gold film. *Physical Review B,* Vol.60, pp. R14990-R14993

Czaplewski, D.A., Sullivan, J.P., Friedmann, T.A. & Wendt, J.R. (2005). Temperature dependence of the mechanical properties of tetrahedrally coordinated amorphous carbon thin films, *Applied Physics Letters,* Vol.87, paper n. 161915

D'Evelyn M.P. & Taniguchi, T. (1999). Elastic properties of translucent polycrystalline cubic boron nitride as characterized by the dynamic resonance method, *Diamond and Related Materials,* Vol.8, pp. 1522-1526

Devos, A. & Côte, R. (2004). Strong oscillations detected by picoseconds ultrasonics in silicon: evidence for an electronic structure effect, *Physical Review B,* Vol.70, paper n. 125208

Djemia, P., Roussigné, Y., Dirras, G.,F. & Jackson, K.M. (2004). Elastic properties of SiC films by Brillouin light scattering, *Journal of Applied Physics,* Vol. 95, pp. 2324-2330

Every, A.G. (2001). The Elastic Properties of Solids: Static and Dynamic Principles, In: *Handbook of Elastic Properties of Solids, Liquids, and Gases,* M. Levy, H. Bass, R. Stern & V. Keppens (Eds.), Volume I: *Dynamic Methods for Measuring the Elastic Properties of Solids,* pp. 3-36, Academic Press, New York

Every, A.G. (2002). Measurement of the near surface elastic properties of solids and thin supported films, *Measurement Science and Technology,* Vol.13, pp. R21-R39

Every, A.G., Kotane, L.M. & Comins, J. D. (2010). Characteristic wave speeds in the surface Brillouin scattering measurement of elastic constants of crystals. *Physical Review B,* Vol.81, paper n. 224303

Farnell, G.W. & Adler, E.L. (1972). Elastic wave propagation in thin layers. In: *Physical Acoustics,* W.P. Mason & R.N. Thurston (Eds.), Vol. 9, pp. 35-127, Academic, New York

Ferrari, A. C., Robertson, J., Beghi, M. G., Bottani, C. E., Ferulano, R. & Pastorelli, R. (1999). Elastic constants of tetrahedral amorphous carbon films by surface Brillouin scattering, *Applied Physics Letters,* Vol.75, pp. 1893-1895.

Ghislotti, G. & Bottani, C. E. (1994). Brillouin scattering from shear horizontal surface phonons in silicon-on-insulator structures - Theory and experiment, *Physical Review B*, Vol.50, pp. 12131-12137.

Grimsditch, M., (2001). Brillouin scattering, in *Handbook of Elastic Properties of Solids, Liquids, and Gases,* M. Levy, H. E. Bass and R. R. Stern & V. Keppens (Eds.), Volume I: *Dynamic Methods for Measuring the Elastic Properties of Solids,* pp. 331-347, Academic Press, New York

Kim, J. Y., Chung, H. J., Kim, H. J., Cho, H. M., Yang, H. K. & Park, J. C. (2000). Surface acoustic wave propagation properties of nitrogenated diamond-like carbon films, *Journal of Vacuum Science and Technology A*, Vol.18, pp. 1993—1997

Kubisztal, M., Kubisztal, J., Chrobak, A., Haneczok, B., Budniok, A. & Rasek, J. (2008). Elastic properties of Ni and Ni + Mo coatings electrodeposited on stainless steel substrate, *Surface and Coatings Technology*, Vol.202, pp. 2292–2296

Kundu, T. (2004). Mechanics of elastic waves and ultrasonics non-destructive evaluation, in *Ultrasonic nondestructive evaluation,* T. Kundu (Ed.), pp. 1-142, CRC Press, Boca Raton, FL. Revised edition in press (2011)

Lefeuvre, O., Pang, W., Zinin, P., Comins, J.D., Every, A.G., Briggs, G.A.D., Zeller, B.D. & Thompson, G.E. (1999). Determination of the elastic properties of a barrier film on aluminium by Brillouin spectroscopy. *Thin Solid Films*, Vol.350, pp. 53-58.

Lehmann, G., Hess, P., Weissmantel, S., Reisse, G., Scheible, P. & Lunk, A. (2002). Young's modulus and density of nanocrystalline cubic boron nitride films determined by dispersion of surface acoustic waves, *Applied Physics A*, Vol.74, pp. 41—45

Li, M., Li, F.F., Gao, W., Ma, C.L., Huang, L.Y., Zhou, Q.A. & Cui, Q.L. (2010). Brillouin scattering study of liquid methane under high pressures and high temperatures. *Journal of Chemical Physics*, Vol.133, paper n. 044503

Lou, N., Groenen, J., Benassayag, G. & Zwick, A. (2010). Acoustics at nanoscale: Raman-Brillouin scattering from thin silicon-on-insulator layers, *Applied Physics Letters*, Vol.97, article n. 141908

Mante, P.A., Robillard, J.F. & Devos, A. (2008). Complete thin film mechanical characterization using picoseconds ultrasonics and nanostructured transducers: experimental demonstration on SiO_2, *Applied Physics Letters*, Vol.93, paper n. 071909

Migliori, A., Sarrao, J.L., Visschera, W.M., Bella, T.M., Leia, M., Fisk, Z. & Leisure, R.G. (1993). Resonant ultrasound spectroscopic techniques for measurement of the elastic moduli of solids, *Physica B*, Vol.183, pp. 1–24.

Murakami, M., Asahara, Y., Ohishi, Y., Hirao N. & Hirose, K. (2009). Development of in situ Brillouin spectroscopy at high pressure and high temperature with synchrotron radiation and infrared laser heating system: Application to the Earth's deep interior. *Physics of the Earth and Planetary Interiors*, Vol.174, pp. 282-291

Murakami, M. & Bass, J.D. (2010). Spectroscopic Evidence for Ultrahigh-Pressure Polymorphism in SiO_2 Glass. *Physical Review Letters*, Vol.104, pp. 025504.

Nakamura, N., Ogi, H. & Hirao, M. (2004). Resonance ultrasound spectroscopy with laser-Doppler interferometry for studying elastic properties of thin films, *Ultrasonics*, Vol.42, pp. 491–494

Nakamura, N., Nakashima, T., Oura, S., Ogi, H. & Hirao, M. (2010). Resonant-ultrasound spectroscopy for studying annealing effect on elastic constant of thin film, *Ultrasonics*, Vol.50, pp. 150–154

Neubrand, A. & Hess, P. (1992). Laser generation and detection of surface acoustic waves: Elastic properties of surface layers, *Journal of applied physics,* Vol.71, pp. 227-238

Nieves, F.J., Gascòn, F. & Bayòn, A. (2000). Precise and direct determination of the elastic constants of a cylinder with a length equal to its diameter, *Review of Scientific Instruments,* Vol.71, pp. 2433-2439

Ogi, H., Fujii, M., Nakamura, N., Shagawa, T. & M. Hirao, M. (2007). Resonance acoustic-phonon spectroscopy for studying elasticity of ultrathin films, *Applied Physics Letters,* Vol. 90, paper n. 191906.

Ohno, I. (1976). Free vibration of a rectangular parallelepiped crystal and its application to determination of elastic constants of orthorhombic crystal, *Journal of Physics of the Earth,* Vol.24, pp. 355–379.

Pang, W., Stoddart, P.R., Comins, J.D., Every, A.G., Pietersen, D. & Marais P.J. (1997). Elastic properties of TiN hard films at room and high temperatures using Brillouin scattering, *International Journal of Refractory Metals and Hard Materials,* Vol.15, pp. 179-185.

Polomska, A.M., Young, C.K., Andrews, G.T., Clouter, M.J., Yin, A. & Xu, J. M. (2007). Inelastic laser light scattering study of an ordered array of carbon nanotubes. *Applied Physics Letters,* Vol.90, paper n. 201918

Prakapenka, V. (2010). On-line Brillouin Spectroscopy at GSECARS: Basic Principles and Application for High Pressure Research, *Synchrotron Radiation News* Vol.23, pp. 14-15.

Robillard, J.-F., Devos, A., Roch-Jeune, I. & Mante, P. A. (2008). Collective acoustic modes in various two-dimensional crystals by ultrafast acoustics: Theory and experiment, *Physical Review B,* Vol.78, paper n. 064302

Rossignol, C., Perrin, B., Bonello, B., Djemia, P., Moch, P., & Hurdequint, H. (2004). Elastic properties of ultrathin permalloy/alumina multilayer films using picosecond ultrasonics and Brillouin light scattering, *Physical Review.B,* Vol.70, paper n. 094102

Sandercock J.R. (1982). Trends in Brillouin scattering - Studies of opaque materials, supported films, and central modes, in *Light Scattering in solids III,* M. Cardona and G. Güntherodt (Eds.), pp. 173-206, Springer, Berlin

Schneider, D., Schwarz, T., Scheibe, H.-J. & Panzner, M. (1997) Non destructive evaluation of diamond-like carbon films by laser induced surface acoustic waves, *Thin Solid Films, Vol.*295, pp. 107—116

Schneider, D., Schultrich, B., Scheibe, H.-J., Ziegele, H. & Griepentrog, M. (1998). A laser acoustic method for testing and classifying hard surface layers, *Thin Solid Films,* Vol. 332, pp. 157—163.

Schneider, D., Witke, T.H., Schwarz, T.H., Schoneich, B. & Schultrich, B. (2000). Testing ultra-thin films by laser-acoustics, Surface and Coatings Technology, Vol.126, pp.136-141

Schwarz, R.B., Hooks, D.E., Dick, J.J. & Archuleta, J.I. (2005). Resonant ultrasound spectroscopy measurement of the elastic constants of cyclotrimethylene trinitramine, *Journal of Applied Physics,* Vol.98, paper N. 056106.

Sinogeikin, S., Bass, J., Prakapenka, V., Lakshtanov, D., Shen, G.Y., Sanchez-Valle, C & Rivers, M. (2006). Brillouin spectrometer interfaced with synchrotron radiation for simultaneous X-ray density and acoustic velocity measurements. *Review of Scientific Instruments,* Vol.77, article n. 103905

So, J.H., Gladden, J.R., Hu, Y.F., Maynard, J.D. & Qi Li (2003). Measurements of elastic constants in thin films of colossal magnetoresistance material, *Physical Review Letters*, Vol.90, paper N. 036103.

Stevens, L.L., Orler, E.B., Dattelbaum, D.M., Ahart, M. & Hemley, R. J. (2007). Brillouin scattering determination of the acoustic properties and their pressure dependence for three polymeric elastomers. *Journal of Chemical Physics*, Vol.127, article n. 104905

Stoddart, P.R., Comins, J.D. & Every, A.G. (1995). Brillouin-scattering measurements of surface-acoustic-wave velocities in silicon at high-temperatures. *Physical Review B*, Vol.51, pp. 17574-17578

Stoddart, P.R., Crowhurst, J.C., Every, A.G. & Comins, J. D. (1998). Measurement precision in surface Brillouin scattering, *Journal of the Optical Society of America B*, Vol.15, pp. 2481-2489.

Sugawara, Y., Wright, O.B., Matsuda, O. & Gusev, V.E. (2002). Spatiotemporal mapping of surface acoustic waves in isotropic and anisotropic materials, *Ultrasonics*, Vol.40, pp. 55-59

Thomsen, C., Strait, J., Vardeny, Z., Maris, H.J., Tauc, J. & Hauser, J.J. (1984). Coherent phonon generation and detection by picoseconds light pulses, *Physical Review Letters*, Vol. 53, pp. 989-992

Thomsen, C., Grahn, H.T., Maris, H.J. & Tauc, J. (1986). Surface generation and detection of phonons by picoseconds light pulses, *Physical Review B*, Vol. 34, pp. 4129-4138

Vollmann, J., Profunser, D.M. & Dual, J., (2002). Sensitivity improvement of a pump-probe set-up for thin film and microstructure metrology, *Ultrasonics*, Vol.40, pp. 757-763

Whitfield, C.H., Brody, E.M. & Bassett, W.A. (1976). Elastic moduli of NaCl by Brillouin scattering at high pressure in a diamond anvil cell. *Review of Scientific Instruments*, Vol.47, pp. 942-947

Whitfield, M. D., Audic, B., Flannery, C. M., Kehoe, L. P., Crean, G. M. & Jackman, R. B. (2000). Charactrization of acoustic Lamb wave propagation in polycrystalline diamond film by laser ultrasonics, *Journal of Applied Physics*, Vol.88, pp. 2984—2993.

Zhang, X., Comins, J.D., Every, A.G. & Stoddart, P.R. (1998). Surface Brillouin scattering studies on vanadium carbide, *International Journal of Refractory Metals and Hard Materials*, Vol.16, pp. 303-308

Zhang, X., Stoddart, P.R., Comins, J.D. & Every, A. G. (2001a). High-temperature elastic properties of a nickel-based superalloy studied by surface Brillouin scattering, *Journal of Physics: Condensed. Matter*, Vol.13, pp. 2281-2294

Zhang, X., Sooryakumar, R., Every, A.G. & Manghnani, M.H. (2001b). Observation of organpipe acoustic excitations in supported thin films. Physical Review B, Vol.64, paper n. 081402.

Zinin, P.V (2001). Quantitative acoustic microscopy of solids, in *Handbook of Elastic Properties of Solids, Liquids, and Gases,* M. Levy, H. E. Bass and R. R. Stern & V. Keppens (Eds.), Volume I: *Dynamic Methods for Measuring the Elastic Properties of Solids,* pp. 187-226, Academic Press, New York

7

Machinery Faults Detection Using Acoustic Emission Signal

Dong Sik Gu and Byeong Keun Choi
Gyeongsang National University
Republic of Korea

1. Introduction

Application of the high-frequency acoustic emission (AE) technique in condition monitoring of rotating machinery has been growing over recent years. This is particularly true for bearing defect diagnosis and seal rubbing (Mba et al., 1999, 2003, 2005; Kim et al., 2007; Siores & Negro, 1997). The main drawback with the application of the AE technique is the attenuation of the signal and as such the AE sensor has to be close to its source. However, it is often practical to place the AE sensor on the non-rotating member of the machine, such as the bearing or gear casing. Therefore, the AE signal originating from the defective component will suffer severe attenuation before reaching the sensor. Typical frequencies associated with AE activity range from 20 kHz to 1 MHz.

While vibration analysis on gear fault diagnosis is well established, the application of AE to this field is still in its infancy. In addition, there are limited publications on application of AE to gear fault diagnosis. Siores explored several AE analysis techniques in an attempt to correlate all possible failure modes of a gearbox during its useful life. Failures such as excessive backlash, shaft misalignment, tooth breakage, scuffing, and a worn tooth were seeded during tests. Siores correlated the various seeded failure modes of the gearbox with the AE amplitude, root mean square, standard deviation and duration. It was concluded that the AE results could be correlated to various defect conditions (Siores et al., 1997). Sentoku correlated tooth surface damage such as pitting to AE activity. An AE sensor was mounted on the gear wheel and the AE signature was transmitted from the sensor to data acquisition card across a mercury slip ring. It was concluded that AE amplitude and energy increased with increased pitting (Sentoku, 1998). In a separated study, Singh studied the feasibility of AE for gear fault diagnosis. In one test, a simulated pit was introduced on the pitch line of a gear tooth using an electrical discharge machining (EDM) process. An AE sensor and an accelerometer for comparative purposes were employed in both test cases. It was important to note that both the accelerometer and AE sensor were placed on the gearbox casing, it was observed that the AE amplitude increased with increased rotational speed and increased AE activity was observed with increased pitting. In a second test, periodically occurring peaks were observed when natural pitting started to appear after half an hour of operation. These AE activities increased as the pitting spread over more teeth. Singh concluded that AE could provide earlier detection over vibration monitoring for pitting of gears, but noted it could not be applicable to extremely high speeds or for

unloaded gear conditions. (Singh et al., 1996) Tan offered that AE RMS (Root Mean Square) levels from the pinion were linearly correlated to pitting rates; AE showed better sensitivity than vibration at higher toque level (220Nm) due to fatigue gear testing using spur gears. He made sure that the linear relationship between AE, gearbox running time and pit progression implied that the AE technique offers good potential in prognostic capabilities for monitoring the health of rotating machines. (Tan et al., 2005, 2007)
On the other hand, the signal processing method for AE signal was studied using bearing and gearbox. In the results of the research (Sheen, 2008, 2010; Yang et al., 2007), the envelope analysis was found to be useful to detect fault in rolling element bearing. The fault detection frequency of bearing can be presented in the power spectrum. Wavelet transform was used for the signal processing method for the gearboxes (Wu et al., 2006, 2009), but wavelet transforms can give the different results with the envelope analysis. It can be shown the defect frequency, but the efficiency is lower than that of envelope analysis. Thus, the signal processing method for AE signal has not been completed until now, and it must be developed in the future.
Therefore, in this paper, a signal processing method for AE signal by envelope analysis with discrete wavelet transforms is proposed. For the detection of faults generated by gear systems and a cracked rotor using the suggested signal processing, these were installed in each test rig system. In gearbox, misalignment was created by a twisted case caused by arc-welding to fix the base and bearing inner race fault was generated by severe misalignment. Through the 15 days test using AE sensor, misalignment was observed and bearing faults were also detected in the early fault stage. To identify the sensing ability of the AE, vibration signal was acquired through an accelerometer and compared with the AE signal. Also, to find the advantage of the proposed signal processing method, it was compared to traditional envelope analysis. The detection results of the test were shown by the power spectrum and comparison of the harmonics level of the rotating speed. Modal test and zooming by a microscope were performed to prove the reason of the other faults. And the crack was seeded by wire cutting with 0.5 mm depth. The shaft was coupled with motor and non-drive-end was left 6.5 mm by lifting tool. During rotating the shaft, AE signals were acquired by AE sensor with 5MHz sampling frequency and 0.5 seconds storing time. The AE signals were transformed by FFT to create the power spectrums, and in the spectrums several peaks were occurred by the crack growth. Along the growth of the crack, the characteristic of the power spectrum was changed and displayed different frequencies.

2. Signal processing method

Envelope analysis typically refers to the following sequence of procedure: (1) band-pass filtering (BPF), (2) wave rectification, (3) Hilbert transform or low-pass filtering (LPF) and (4) power spectrum. The purpose of the band-pass filtering is to reject the low-frequency high-amplitude signals associated with the *i*th mechanical vibration components and to eliminate random noise outside the pass-band. Theoretically, in HFRT (High Frequency Resonance Technique) analysis, the best band-pass range includes the resonance of the bearing components. This frequency can be found through impact tests or theoretical calculations involving the dimensions and material properties of the bearing. However, it is very difficult to predict or specify which resonant modes of neighboring structures will be excited. It will be costly and unrealistic in practice to find the resonant modes through experiments on rotating machinery that may also alter under the different operational conditions. In addition, it is also difficult to estimate how these resonant modes are affected in the assembly of a complete bearing and mounting in a specific housing, even if the resonant frequencies of individual bearing elements can be tested or calculated theoretically (Misiti et al., 2009). Therefore, most researchers decide on the band-pass range as on option. To recover the disadvantage of this option, wavelet analysis is included in the process of traditional envelope analysis in this paper.

Wavelet theory (Burrus et al., 1997) is introduced that is a tool for the analysis of transient, non-stationary, or time-varying phenomena. Wavelet analysis is also called wavelet transform. There are two kinds of wavelet transform: continuous wavelet transform (CWT) and discrete wavelet transform (DWT). CWT is defined as the sum over all time of the signal multiplied by scaled, shifted versions of the wavelet function. To use CWT, one signal can be decomposed into a series of "small" waves belonging to a wavelet family. The wavelet family is composed of scaling functions, $\phi(t)$ deduced by father wavelet and wavelet functions, $\psi(t)$ deduced by mother wavelet. The scaling function can be represented by the following mathematical expression:

$$\phi_{j,k}(k) = 2^{j/2}\phi\left(2^j t - k\right)$$

where j represents the scale coefficient and k represents shift coefficient. Scaling a wavelet simply means stretching (or compressing) it. Shifting a wavelet simply means delaying (or hastening) its onset. Mathematically, delaying a function $f(t)$ by k is represented by $f(t + k)$. Similarly, the associated wavelet function can be generated using the same coefficients as the scaling function.

$$\psi_{j,k}(t) = 2^{j/2}\psi\left(2^j t - k\right)$$

The scaling functions are orthogonal to each other as well as the wavelet functions as shown in the following equations:

$$\int_{-\infty}^{\infty}\phi(2t-k)\cdot\phi(2t-l)dt = 0 \text{ for all } k \neq l$$
$$\int_{-\infty}^{\infty}\psi(t)\cdot\phi(t)dt = 0$$

Using an iterative method, the scaling function and associated wavelet function can be computed if the coefficients j and k are known.

For many signals, the low-frequency content is the most important part. It is what gives the signal its identity. The high-frequency content, on the other hand, imparts flavour or nuance that is often useful for singular signal detection. In wavelet analysis, we often speak of approximations and details. The approximations are the high-scale, low frequency components of the signal. The details are the low-scale, high-frequency components. A signal can be decomposed into approximate coefficients $a_{j,k}$, through the inner product of the original signal at scale j and the scaling function.

$$\alpha_{j,k} = \int_{-\infty}^{\infty} f_j(t)\cdot\phi_{j,k}(t)dt$$

Similarly the detail coefficients $d_{j,k}$ can be obtained through the inner product of the signal and the complex conjugate of the wavelet function.

$$d_{j,k} = \int_{-\infty}^{\infty} f_j(t) \cdot \psi_{j,k}(t)\, dt$$

However, CWT takes a long time due to calculating the wavelet coefficient at all scales and it produces a lot of data. To overcome such a disadvantage, we can choose scales and positions based on powers of two - the so-called dyadic scales and positions - then wavelet analysis will be much more efficient and just as accurate. Such an analysis is obtained from the discrete wavelet transform (DWT). The approximate coefficients and detail coefficients decomposed from a discredited signal can be expressed as

$$\alpha_{(j+1),k} = \sum_{k=0}^{N} \alpha_{j,k} \int \phi_{j,k}(t) \cdot \phi_{(j+1),k}(t)\, dt = \sum_{k} \alpha_{j,k} \cdot g[k]$$
$$\alpha_{(j+1),k} = \sum_{k=0}^{N} \alpha_{j,k} \int \phi_{j,k}(t) \cdot \psi_{(j+1),k}(t)\, dt = \sum_{k} \alpha_{j,k} \cdot h[k]$$

The decomposition coefficients can therefore be determined through convolution and implemented by using a filter. The filter, $g[k]$, is a low-pass filter and $h[k]$ is a high-pass filter. The decomposition process can be iterated, with successive approximations being decomposed in turn, so that one signal is broken down into many lower resolution components. This is called the wavelet decomposition tree as shown in Fig. 1.

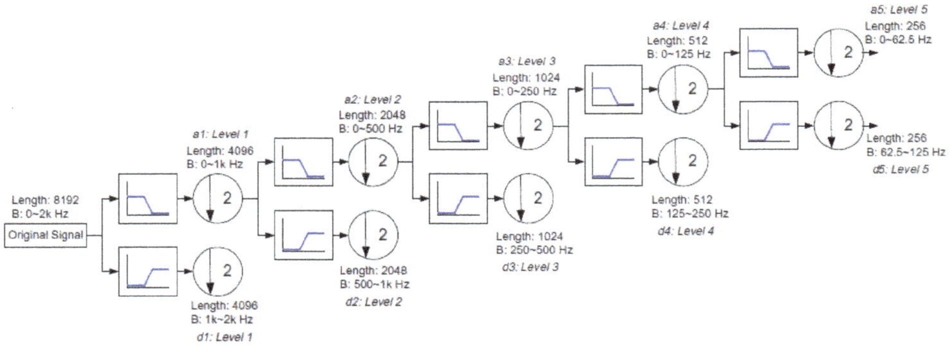

Fig. 1. Wavelet decomposition tree

DWT has a de-noise function and a filter effect focused on impact signal. To make up the weak point of BPF of the envelope analysis, DWT was intercalated on typical envelope analysis, between BPF and wave rectification exactly. The signal by DWT will be separated to different band widths by decomposition level and adapted to the signal with impact.

For more complicated signals which are expressible as a sum of many sinusoids, a filter can be constructed which shifts each sinusoidal component by a quarter cycle. This is called a Hilbert transform filter. Let $H_t\{x\}$ denotes the output at time t of the Hilbert-transform filter applied to the signal x. Ideally, this filter has magnitude 1 at all frequencies and introduces a phase shift of $-\pi/2$ at each positive frequency and $+\pi/2$ at each negative frequency. When a real signal $x(t)$ and its Hilbert transform $y(t) = H_t\{t\}$ are used to form a new complex signal $z(t)= x(t)+jy(t)$, the signal $z(t)$ is the (complex) analytic signal corresponding to the real signal $x(t)$. In other words, for any real signal $x(t)$, the corresponding analytic signal $z(t) = x(t)+jH_t\{x\}$ has the property that all 'negative frequencies' of $x(t)$ have been 'filtered out' (Douglas & Pillay, 2005). Hence, the coefficients of complex term in the corresponding analytic signal were used for FFT.

Fig. 2 shows an analytic signal of the Hilbert transform for envelope analysis. The solid line is a time signal and the dash is its envelope curve. A high frequency signal modified by wavelet transform is modulated to a low frequency signal with no loss of the fault information due to envelope effect. According to that, the fault signals in the low frequency region can be detected using the analytic signal. That is an important fact for the proposed signal processing method. Therefore, the proposed signal processing method in this paper is an envelope analysis with DWT and using the coefficients of the complex term in Hilbert transform.

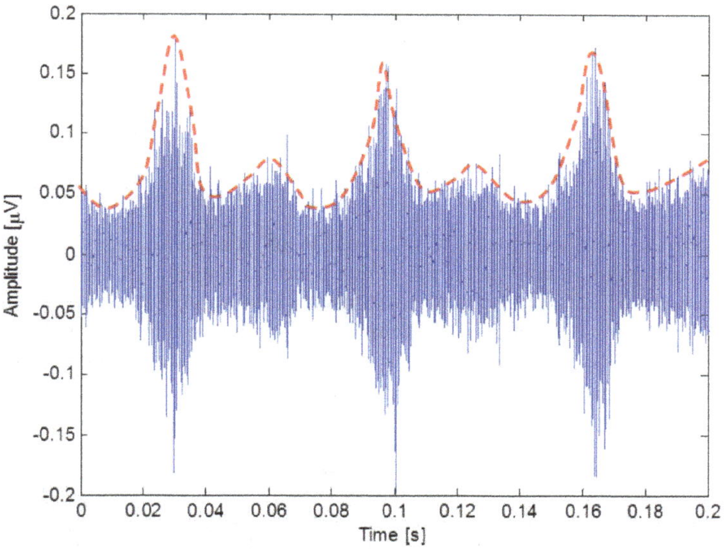

Fig. 2. Analytic signal (dash) of the envelope effect

Furthermore, to reduce the noise level in the power spectrum, the spectrum values were presented as the mean value of each day. Fig. 3 shows the power spectrums of the two different signal processing method. Fig. 3(a) is from envelope analysis, and Fig. 3(b) shows the envelope analysis intercalated DWT using Daubechies mother function between BPF and wave rectification. In Fig. 3, the DWT has an effect the amplifying sidebands peaks, especially about gear mesh frequencies, so the peaks of the harmonics of the rotating speed (f_r) and gear mesh frequencies (f_m) are bigger than another, and we can check up them easily. Therefore, in the following result, the power spectrum through envelope analysis with DWT will be shown.

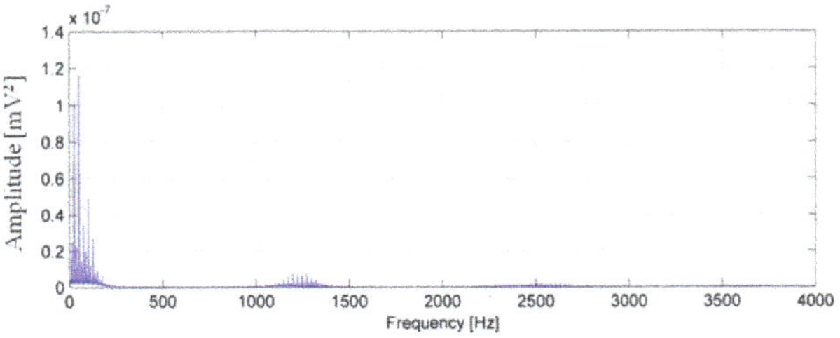

(a) Envelope analysis

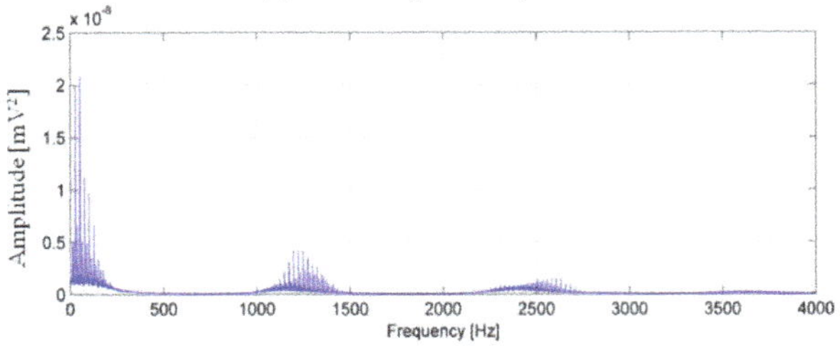

(b) Envelope analysis with DWT

Fig. 3. The comparison of Power spectrums in Envelope analysis with/without DWT

3. Gearbox

3.1 Test-rig

The test-rig employed for this investigation consists of one identical oil-bath lubricated gearbox, 3 HP-motor, rigid coupling, tapper-roller bearing, pinion, gear, control panel and break system, as seen Fig. 4. The pinion was made from steel with heat treatment, the number of teeth is 70, and diameter is 140 mm. The gear was made from steel, but it was produced without any heat treatment process during manufacturing. The number of gear teeth is 50 and diameter is 100 mm, and module is 2 mm for the gear and pinion, respectively.

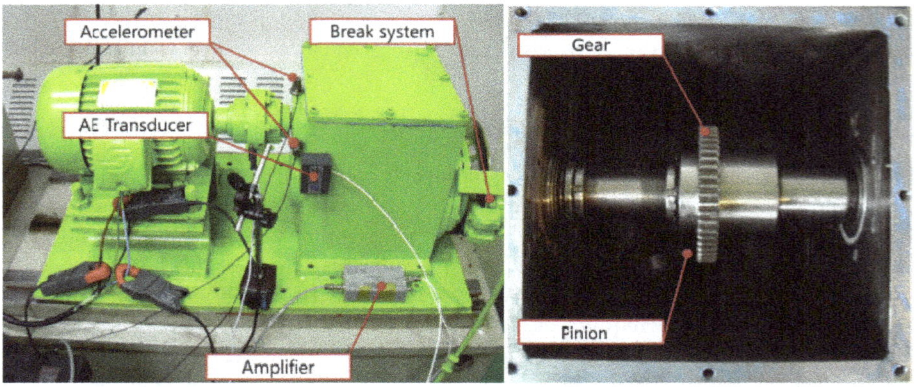

Fig. 4. Test-rig

A simple mechanism that permitted a break of disk-pad type to be rotated relative to each other was employed to apply torque to the gear. Contact ratio (Pinion/Gear) of the gears was 1.4. The motor used to drive the gearbox was a 3-phase induction motor with a maximum running speed of 1800 rpm respectively and was operated for 15 days with 1500rpm. The torque on the output shaft was 1.2 kN·m while the motor was in operation, and other specifications of the gearbox are given as in Table 1.

<table>
<tr><td></td><td colspan="2">Gear</td><td colspan="2">Pinion</td></tr>
<tr><td>No. of teeth</td><td colspan="2">50</td><td colspan="2">70</td></tr>
<tr><td>Speed of shaft</td><td colspan="4">25.01 rev/s</td></tr>
<tr><td>Meshing frequency</td><td colspan="2">1250 Hz</td><td colspan="2">1750 Hz</td></tr>
<tr><td colspan="5">Bearing (NSK HR 32206J)</td></tr>
<tr><td>No. of rolling element</td><td>17</td><td>Type</td><td>Defect Freq. (f_d)</td><td>Fault Freq. (f_d X f_r)</td></tr>
<tr><td>Diameter of outer race</td><td>62 mm</td><td rowspan="2">BPFO
BPFI
FTF
BSF</td><td rowspan="2">8.76 Hz
11.24 Hz
3.84 Hz
0.44 Hz</td><td rowspan="2">219.3 Hz
281.38 Hz
96.13 Hz
11.01 Hz</td></tr>
<tr><td>Diameter of inner race</td><td>30 mm</td></tr>
<tr><td colspan="5">BPFO : ball pass frequency of outer race
BPFI : ball pass frequency of inner race
FTF : fundamental train frequency
BSF : ball spin frequency</td></tr>
</table>

Table 1. Specification of gearbox and bearing

3.2 Acquisition system and test procedures

AE sensors used in this paper are a broadband type with a relative flat response in the range frequency from 10 kHz to 1 MHz. They are placed on the right side of the gearbox cases near the coupling in the horizontal direction at the same height with the shaft center (Fig. 4).

AE signals are pre-amplified by 60 dB and the output from the amplifier is collected by a commercial data acquisition card with 10 MHz sampling rate during the test. Prior to the analog-to-digital converter (ADC), anti-aliasing filter is employed that can be controlled DAQ software. And Table 2 is shown the detail specifications of the data acquisition system. Before the test, attenuation test on the gearbox components was taken in order to understand the characteristics of the test-rig. The gearbox was run for 30 minutes prior to acquiring AE data for the unload condition. Based on the sampling rate of 10 MHz, the available recording acquisition time was 2 sec.

2 Channel AE system on PCI-Board	18-bit A/D conversion 10M samples/s rate (on one channel, 5M samples/s on 2 AE channels)
AE Sensor (Wideband type)	Peak sensitivity [V/μbar] : -62 dB Operating frequency range : 100-1,000 kHz Resonant freq. [V/ μbar] : 650 kHz Directionality : ±1.5 dB
Preamplifier Gain	Wide dynamic range < 90 dB Single power/signal BNC or optional separate power/signal BNC 20/40/60 dB selectable gain

Table 2. Specifications of data acquisition system

3.3 Experiment result and discussion

In general, the misaligned gear which almost always excites higher order f_m harmonics is shown as in Fig. 5. Often, only small amplitudes will be at the fundamental f_m, but much higher levels will be at 2 f_m and/or 3 f_m. The sideband spacing about f_m might be 2 f_r or even 3 f_r when gear misalignment problems are involved. When significant tooth wear occurs, not only will sidebands appear about f_m, but also about the gear natural frequencies. In the case of those around f_m, the amplitude of the sidebands themselves is a better indicator for wear than the amplitude of f_m.

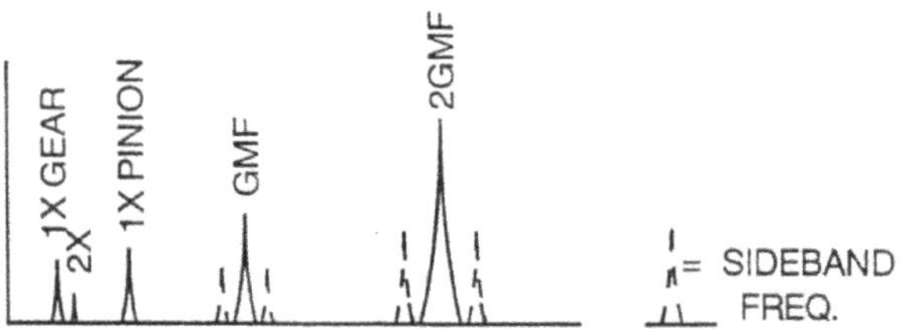

Fig. 5. Spectrum indicating misalignment of gear

As for significant gear eccentricity and/or backlash, these problems display the following characteristics:

- Both eccentricity and backlash excite the gear natural frequencies as well as f_m. They also may generate a number of sidebands about both the natural and gear mesh frequencies.
- If a gear is eccentric; it will modulate the natural frequency and gear mesh frequencies, both of which will be sidebanded around the f_r of the eccentric gear. An eccentric gear can generate significant forces, stresses and vibration if it is forced to bottom out with the meshing gears. (James & Bery, 1994)

In the results of the envelop analysis with DWT, the high harmonics of f_m occurred by strong wearing phenomena caused by misaligned teeth. In the power spectrum (Fig. 6), 25Hz (f_r) and its harmonics are generated and 11.32Hz was the ball pass frequency of inner race (BPFI [f_d]). In Fig. 6(c) and (d), the center dash line is shown for f_m and 2 f_m, and their side lines are the sidebands with difference 25Hz (f_r).

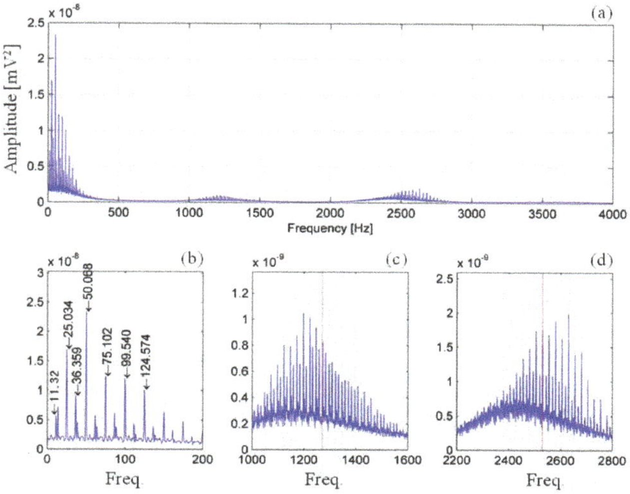

Fig. 6. Power spectrum of the second day

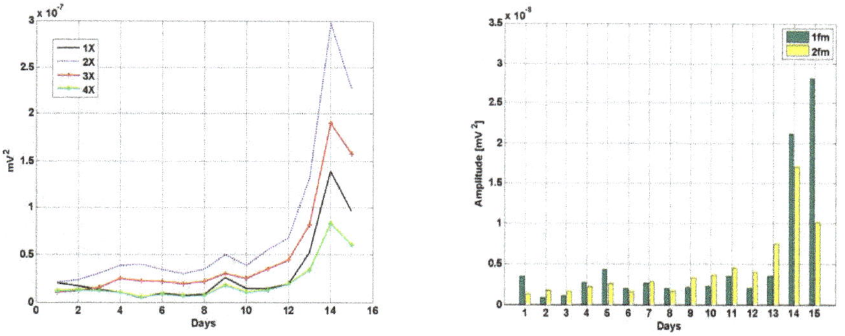

(a) Harmonics of f_r

(b) Harmonics of f_m

Fig. 7. Peak level trend among days

Fig. 8. Gear tooth weaned by misalignment

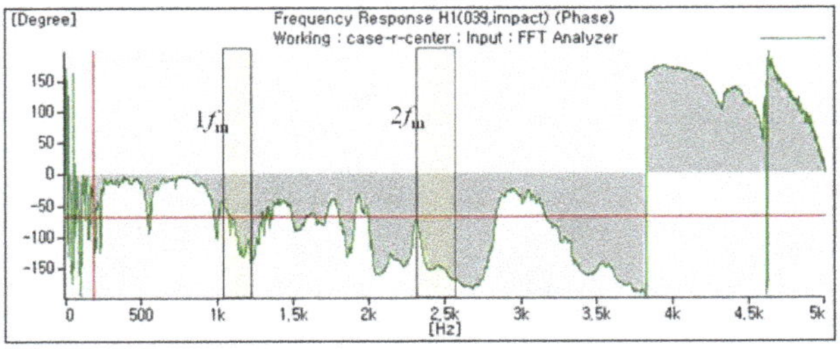

(a) Phase

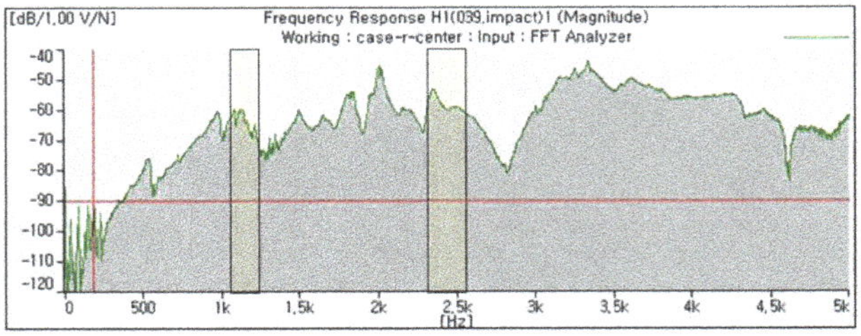

(b) Frequency response function

Fig. 9. Modal test result

In condition monitoring for general rotating machinery, the harmonics ($2f_r$, $3f_r$, $4f_r$ ...) of f_r are occurred higher than f_r when the misalignment was happened. According to the phenomena of misalignment as shown in Fig.5, high level harmonics of f_r were generated such as in Fig. 6(b), and $2f_r$ was always bigger than f_r as shown in Fig. 7(a). The level of $2f_m$ from second to thirteenth day was higher than or similar to f_r as shown in Fig. 7(b). Thus, it is easily catching up to the misalignment that occurred in this test rig. However, it might be that faults of this system are not only misalignment but also resonance trouble, looseness, bearing fault, etc.

Wearing effect by misalignment pollutes the lubrication oil. In Fig. 8, it could be found by the worn teeth and the spots near the pieces of gear teeth. The dripped pieces from the unloading surface raised the wearing effect on the loading surface, and then the gap between gear and pinion was increased. In addition, we could know that the impact marks on the unloading surface (Fig. 8(b)) were generated by misalignment; the impacting force was strong in the initial condition. In this way, the gear teeth were seriously damaged as in Fi g. 8. In Fig. 6(c) and (d), the sidebands are created on wide-spread frequency range near f_m and $2f_m$. That is similar to a state excited by impact force. To confirm the natural frequencies of the test-rig, modal test was fulfilled. The result of the modal test as in Fig. 9 show that f_m and $2f_m$ exist on the exiting frequency range. On the other hand, partial frequency bands close to f_m and $2f_m$ were excited by the impact force, but it is not an exact natural frequency because the phase did not shift enough. Therefore, the peaks near f_m and $2f_m$ were amplified and have many sidebands of f_r and 11.32Hz (BPFI [f_d]). Therefore, it is considered that excessive backlash occurred. Moreover, Fig. 10(a) shows the zooming power spectrum of Fig. 6(a) focused on f_r harmonics. We could clearly know that if the sidebands were caused by BPFI [f_d], then the inner race had some kind of fault. To find out the fault, the surface of the bearing inner race was carried out and viewed by a microscope with 100X zoom as shown in Fig. 11. Small spots were found on the surface, and small cracks were found out on the spots. However, this trouble was not seeded and existed from the initial condition. Thus, it is as assumed that the problem happened in assembly and/or was caused by misalignment.

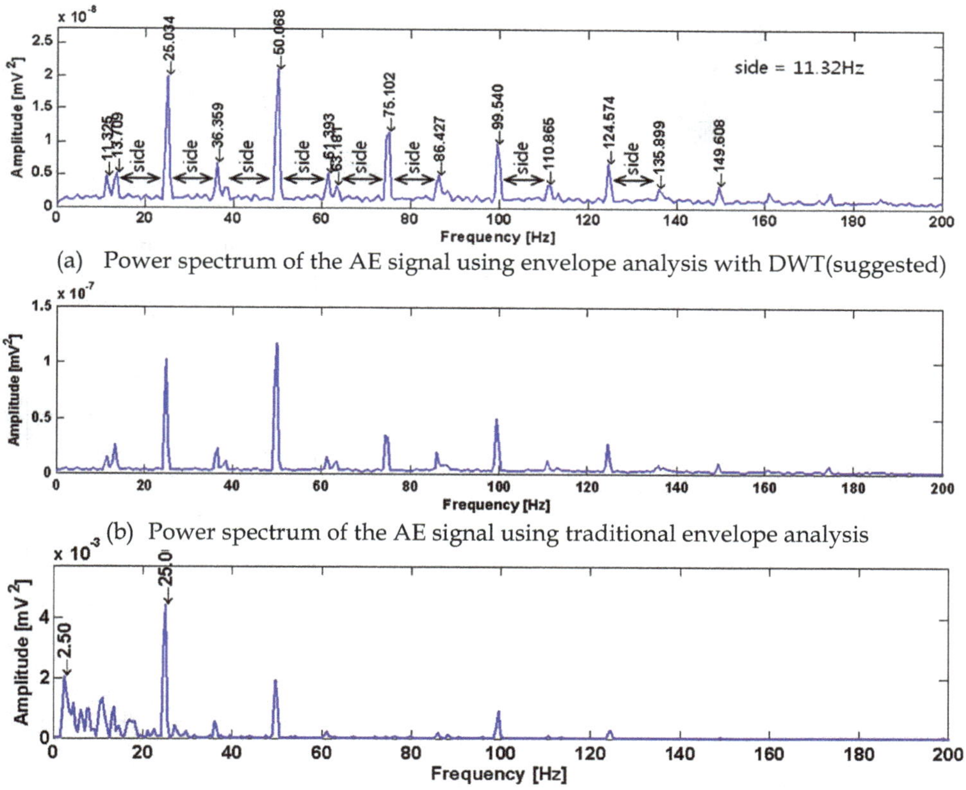

(a) Power spectrum of the AE signal using envelope analysis with DWT(suggested)

(b) Power spectrum of the AE signal using traditional envelope analysis

(c) Power spectrum of the signal from accelerometer

Fig. 10. Sidebands BPFI in the first day power spectrum

To identify the sensing ability of the AE, vibration signal was acquired through accelerometer and compared with the AE signal. Also, to find the advantage of the proposed signal processing method, it was compared to traditional envelope analysis.

The power spectrum of the AE signal using traditional envelope analysis is shown in Fig. 10(b), and the power spectrum using the vibration signal by accelerometer is displayed on Fig. 10(c). The vibration signal was treated by the same method with AE signal. The harmonics of f_r are generated, and 2 f_r for detecting the misalignment is created and can be found in all spectrums (Fig. 10) but the power spectrum of the AE signal, Fig. 10(a) and (b), can explicitly display the defect frequencies as compared to the accelerometer signal (Fig. 10(c)). For example, in Fig. 10(c), the sidebands of BPFI are not easily found because of the higher level of noise in the low frequency range below f_r than in the AE signal with or without DWT.

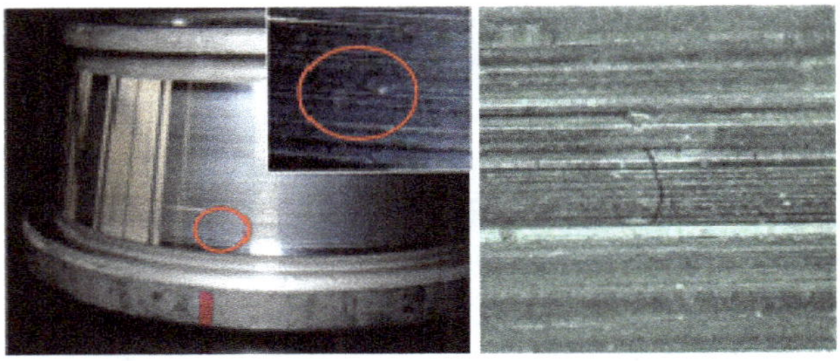

Fig. 11. Zooming of the inner race surface of the fault bearing

Frequency [Hz]		Traditional Method	Proposed method
13.709	1X-BPFI	0.2146	0.2459
25.034	1X	0.8727	0.9525
36.360	1X+BPFI	0.1936	0.3196
38.750	2X- BPFI	0.0952	0.1435
50.070	2X	1.0000	1.0000
61.393	2X+ BPFI	0.1277	0.2470
63.181	3X- BPFI	0.0970	0.1560
75.102	3X	0.3044	0.5465
86.427	3X+ BPFI	0.1699	0.2352
99.540	4X	0.4269	0.4616
110.866	4X+ BPFI	0.1062	0.1675
124.574	5X	0.2312	0.3253
135.899	5X+ BPFI	0.0714	0.1443

Table 3. Ratio of peaks versus the maximum peak in respective spectrum

According to the above results, we can understand that the AE signal can detect the fault more easily than accelerometers and can be used in the condition monitoring system for early detection fault. Moreover, as shown in Table 3 which is the ratio of peaks versus the maximum peak in the respective spectrum, the peak levels of the harmonics of f_r and sidebands caused by BPFI are highly generated in the proposed signal processing method (Fig. 10(a)) than the traditional method. This can lead good feature values to evaluate the condition of the machinery. Therefore, the power spectrum of the proposed envelope analysis using AE signal can be shown the clean result with harmonics and sidebands and is a better technique for condition monitoring system.

4. Cracked rotor

4.1 Experiment system

Test rig consisted of a motor, a flexible coupling, rolling element bearings (NSK6200), three steel bearing housings, a lifting tool and a cracked shaft. The transverse crack was seeded by

wire-cutting with 0.5 mm depth on the shaft made from SM45C. As shown in Fig. 12, the crack was positioned at 5mm near to the second drive-end bearing, and the non-driven end of the shaft was left 6.5 mm with bearing housing by the lifting tool.
AE signal was acquired by an AE sensor and transferred to amplifier, analog-filter, DAQ board and HDD of a desktop. AE sensor is a wideband type with a relative flat response in the range frequency range from 100 kHz to 1 MHz. AE signals were pre-amplified with 60 dB and the output from the amplifier was collected by a commercial data acquisition card with 5 MHz sampling rate during the test. The signals were stored 0.5sec by every 30sec until the shaft was fractured, and the rotating speed of the motor was 600rpm (10Hz).

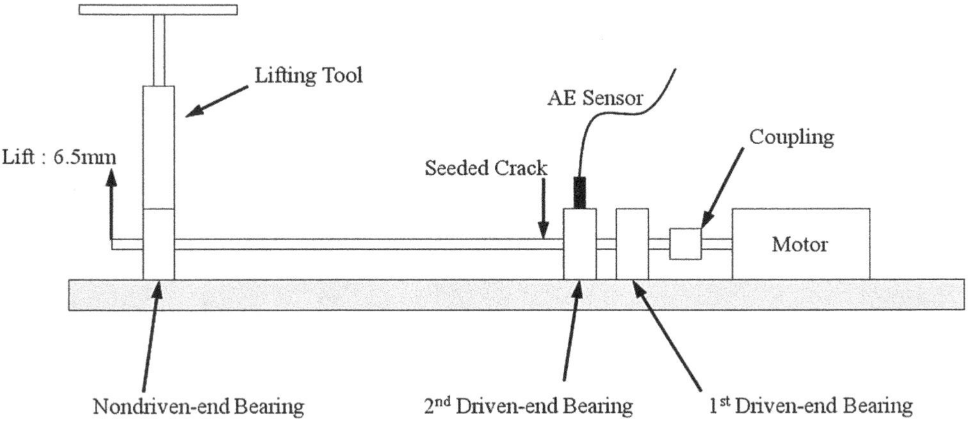

Fig. 12. Experiment system

4.2 Test result and discussion

The operating speed was 600rpm, and the initial radial load for 160N was employed. The radial load was a variable parameter because it was applied by keeping the lifting distance with 6.5mm of the non-drive end of the shaft, and the test terminated on a fracture of the shaft. Fig. 13 shows the observations of continuous feature values as mean value, RMS, peak value and entropy estimation. In information theory, uncertainty can be measured by entropy. The entropy of a distribution is the amount of a randomness of that distribution. Entropy estimation is two stage processes; first a histogram is estimated and thereafter the entropy is calculated. Here, we estimate the entropy of AE signal with using unbiased estimated approach. Fig. 13(a), relatively high level of AE activity was noted from 18 minutes, and it was increased until 60 minutes. But in RMS, Peak and Entropy estimation, the levels were kept to 18minutes beside a peak in around 9 minutes, since that these were continuously decreased to 70 minutes and were increased with hunting with several minute intervals until the fracture.
Interestingly observations of the AE waveform, sampled at 5 MHz showed changing characteristics as a function of time. Fig. 14 shows a contour map of the peaks level of each frequency with time. Rotating speed (9.5Hz) and 3rd harmonic of rotating speed (28.6Hz, 3X) dominated while the test as shown in Fig. 14(b). It is normally known that 3X is caused by misalignment of the bearings created by the loading system for this test (Hatch & Bently,

2002). However, the harmonic component (3X) was kept the level in the wavelet level 6(Fig. 14(b)), but it was increased from 30 minutes in the wavelet level 4(Fig. 14(a)). Additionally, 1X started increasing earlier than 3X as shown in Fig. 14(a).

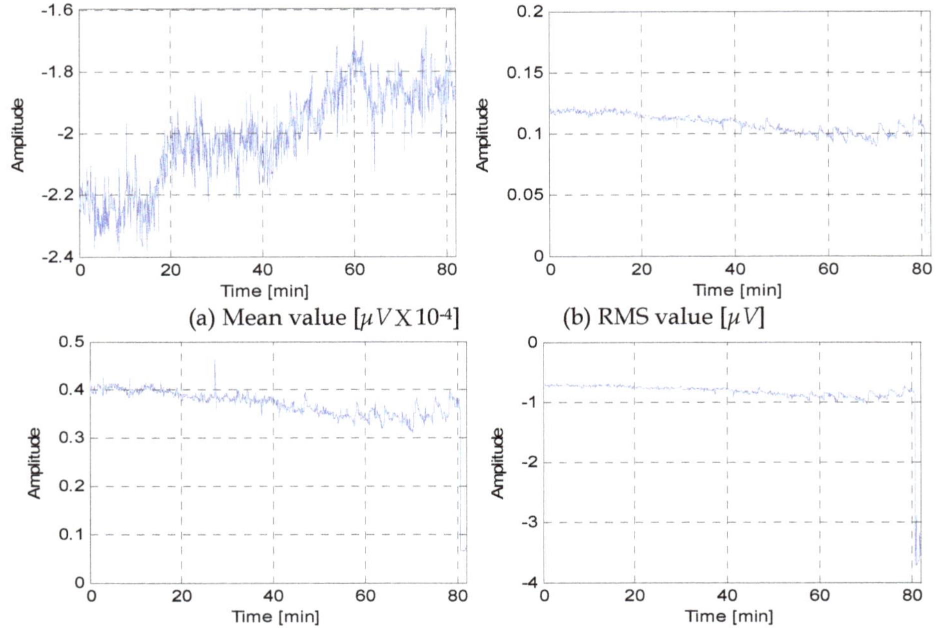

(a) Mean value [$\mu V X 10^{-4}$]
(b) RMS value [μV]
(c) Peak value [μV]
(d) Entropy estimation value

Fig. 13. Shaft test results; run-to-failure

31Hz and 62Hz that were the harmonics of the fundamental train frequency (FTF or cage noise) of the bearing were continuously occurred. Cage noise can be generated in any type of bearing and the magnitude of it is usually not very high. Characteristics of this noise include: (1) it occurs with pressed steel cages, machined cages and plastic cages. (2) It occurs with grease and oil lubrication. (3) It tends to occur if a moment load is applied to the outer ring of a bearing. (4) It tends to occur more often with greater radial clearance. In Fig. 14, 62Hz was continuously detected; 31Hz was detected with 3X. In a general bearing system, the amplitude of the bearing fault frequency is depended on the load grade and is increased along the growing load grade. However, 31Hz of the case noise of this test was not followed the load scale because loading force for this test was decreased with the crack growth. So, it was shown that 31Hz was related with the crack growth.
According to this result, we could know that the reason of 3X (28.6Hz) was the moment load by the loading system; 31Hz was connected with the crack growth. Therefore, the peak levels around 3X and 31Hz was excited by the two frequencies and was increased with the crack growth.
To clear more the characteristic of the crack growth, in addition, PAC energy value was observed. In acoustic emission technology, PAC-Energy is a 2-byte parameter derived from the integral of the rectified voltage signal over the duration of the AE hit (or waveform),

hence the voltage-time units ($\mu V^2 \cdot$ sec). So, PAC energy value was determined by an integral of the square sum total of the transferred time signal in each wavelet level. Fig. 15 shown the energy level of every wavelet levels along time, and its value was transferred to logarithmic value because of too low resolution in linear scale.

Fig. 15 shows the energy trend of wavelet level 1 to 8. Many peaks were created while the test in wavelet level 1(Fig. 15(a)), a high peak was created around 9 minutes existed in wavelet level 1 to 4. Wavelet level 2, 4, 5 and 6 shows a similar trend after 10minutes. The energy level was slowly increased with time until about 30 minutes, and then it was increased fast until 35 minutes (additionally, it was considered that this increasing was related with the growth of the 3X and 31Hz in Fig. 14(a)), after that it was decreased a little for 10 minutes. And it was hunted with every several minutes about 4 minutes until close fracture. In this trend, we had considered of two phenomena, the high peak and the period hunting.

We could mind that the high peak was related with initial crack growth. Because it was shown as follows,

- The increasing ratio of the energy was changed after it was happened.
- The mean value was began fast change (Fig. 13(a)).
- Generated in high frequency range (Fig. 15(a) ~ (d)).

The period hunting was clearly occurred and displayed in wavelet level 2, 4 ~ 8. It was considered that it could indicate a state of the final stage of the fracture in the rotating shaft because of follows,

- It had a period about 4 minutes.
- It was displayed in lower frequency range than the high peak (Fig. 15(e) ~ (h)).
- Its level was increased along near to the fracture.

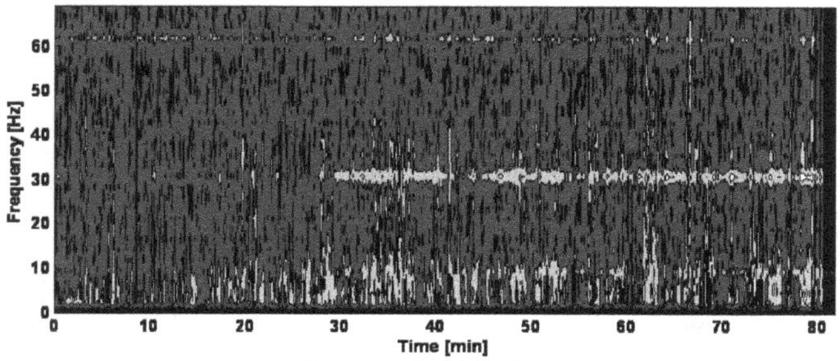

(a) Wavelet level 4

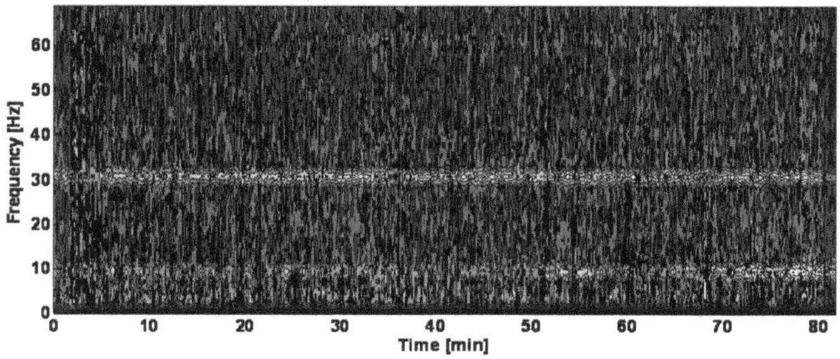

(b) Wavelet level 6

Fig. 14. Peak level trend by frequency

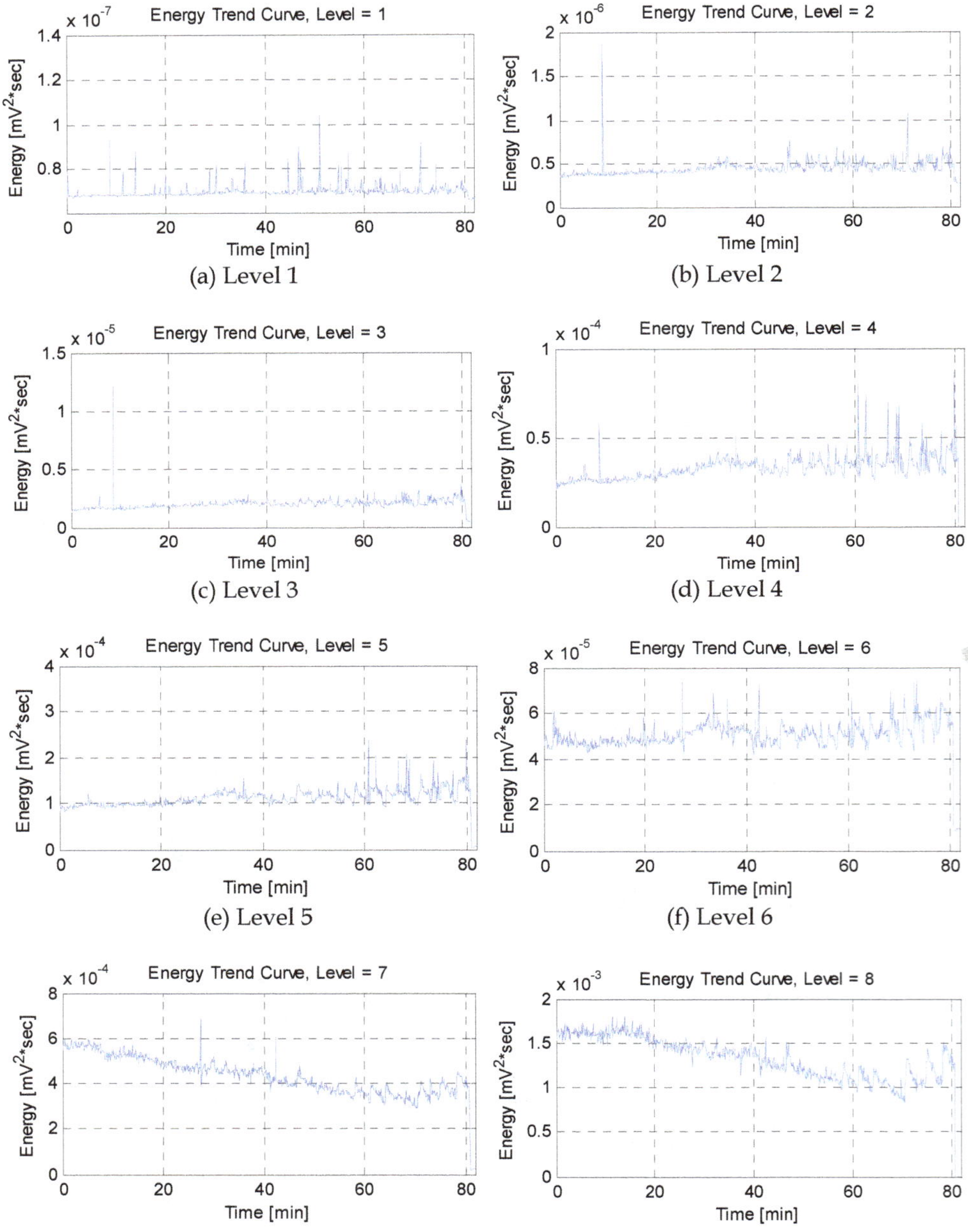

(a) Level 1 (b) Level 2

(c) Level 3 (d) Level 4

(e) Level 5 (f) Level 6

(g) Level 7 (h) Level 8

Fig. 15. PAC energy trend of each wavelet level

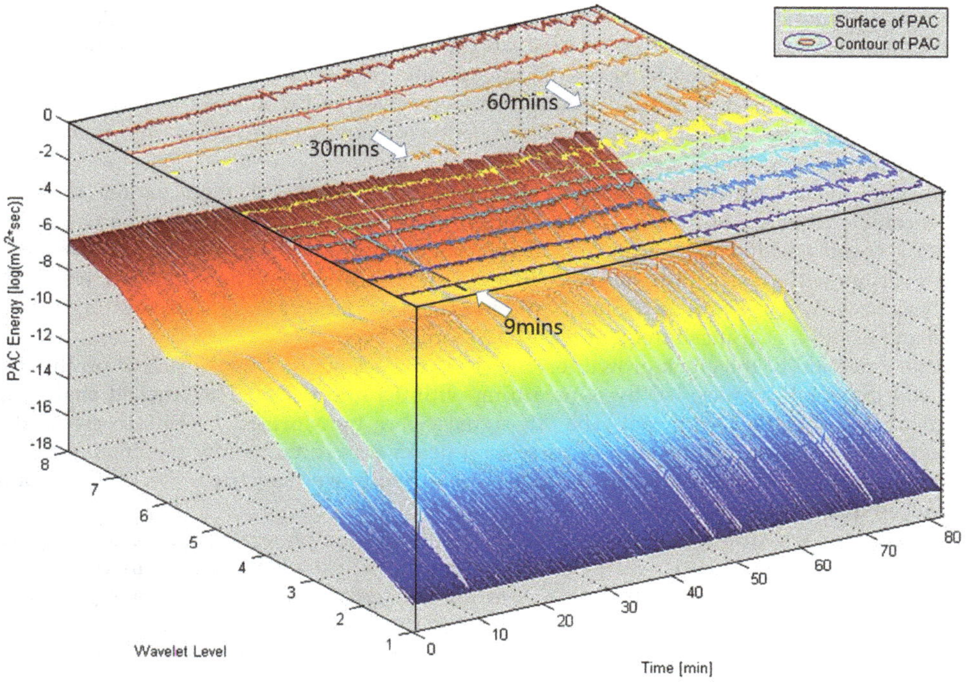

Fig. 16. PAC-Energy level of total wavelet level

To compare absolutely the energy of each wavelet level, all of PAC energy was displayed on a 3D graph with time and wavelet levels as shown in Fig. 16. In wavelet level 5, after approximately 30 minutes, a large transient rise in PAC energy level was observed and this AE activity gradually observed frequently after 60 minutes until the shaft was fractured. In addition, a peak created at 9 minutes was indicated. In here, we could consider that the frequency range of wavelet level 5 could be shown a good relationship between the PAC energy and the crack growth of the middle and final stage. Even so, the trend of the wavelet level 7 and 8 was not clearly connected with the others because the frequency range of wavelet level 7 and 8 was lower than the useful frequency range (100kHz to 1MHz) of AE sensor for this research.

Therefore, the AE signal caused by the crack growth was generated on the whole ultra-sound frequency range; the initial crack could be detected using the PAC energy on wavelet level 1 to 4. In addition, it could be presented on wavelet level 5 until the fracture of the shaft. In the frequency domain, it was shown that the harmonic components of the rotating speed and bearing cage frequency were excited by the crack growth, especially on the 3X (28.6Hz) and 31Hz.

5. Conclusion

In this paper, a signal processing method for AE signal by envelope analysis with discrete wavelet transforms is proposed. For the detection of faults generated from a gear system using the suggested signal processing, a gearbox was installed in the test rig system. Misalignment was created by twisted case caused by arc-welding to fix the base and bearing inner race fault is generated by severe misalignment. To identify the sensing ability of the AE, vibration signal was acquired through accelerometer and compared to the AE signal. Also, to find the advantage of the proposed signal processing method, it was compared with traditional envelope analysis.

According to the experiment result, AE sensor can detect the fault earlier than an accelerometer because of high sensitivity and in the power spectrum, the harmonics of the rotating speed and the gear mesh frequency clearly occurred. Misalignment was observed and bearing faults were also detected in the early fault stage. The proposed envelope analysis is worked to evaluate the faults and indicated the faults frequencies, rotating speed, sideband of BPFI, gear mesh frequency and harmonics, explicitly.

For the detection of the crack growth on the shaft, a cracked shaft was installed on the test rig, and the crack was seeded by wire-cutting with 0.5 mm depth. The cracked shaft was lifted 6.5 mm by the lifting tool. The AE signals were transformed by FFT to create the power spectrums, and in the spectrums several peaks were occurred by the crack growth. Along the growth of the crack, the characteristic of the power spectrum was changed and displayed different frequencies.

In the power spectrum, it was shown that the harmonic components of the rotating speed and bearing cage frequency were excited by the crack growth as shown in the Fig. 6, especially on the 3X (28.6Hz) and 31Hz. And the AE signal caused by the crack growth is generated on the whole ultrasonic frequency range; the initial crack could be detected using the PAC-Energy on wavelet level 1 to 4, and after that, it could be presented on wavelet level 5 until the fracture of the shaft. Therefore, in this paper, it could be shown that the crack growth in rotating machinery is able to be considered and to be detected; in addition, PAE-Energy can be used to detect the early detection of the crack.

Therefore, the proposed signal processing method that is the envelope analysis intercalated DWT using Daubechies mother function between BPF and wave rectification can be shown to provide better result than traditional envelope analysis.

6. Acknowledgment

This work has been supported by Basic Science Research Program through the National Research Foundation of Korea (NRF) funded by the Ministry of Education, Science and Technology (2011-0013652) and the 2nd Phase of Brain Korea 21.

7. References

Burrus, C.S., Gopinath, R.A. & Guo, H. (1997). *Introduction to wavelet and wavelet transforms: A Primer*, Prentice-Hall, ISBN 0134896006, Upper Saddle River, NJ.

Douglas, H. & Pillay, P. (2005). The impact of wavelet selection on transient motor current signature analysis, *Proceedings of 2005 IEEE International Conference on Electric Machines and Drives*, ISBN 0780389875; 978-078038987-8, San Antonio, TX, May 2005.

Hatch, C.T. & Bently, D.E. (2002). *Fundamentals of Rotating Machinery Diagnostics*, Bently Pressurized Bearing Press, ISBN 0-9714081-0-6, Minden, NV.

James, E. & Bery, P.E. (1994). *IRD Advancement Training Analysis II: Concentrated Vibration Signature Analysis and Related Condition Monitoring Techniques*, IRD Mechanalysis Inc., Columbus, Ohio.

Kim, Y.H., Tan, Andy. C.C., Mathew, J., Kosse, V. & Yang, B.S. (2007). A comparative study on the application of acoustic emission technique and acceleration measurements for low speed condition monitoring, *12th Asia-Pacific Vibration Conference*, Hokkaido Univ. Japan, August 2007.

Kim, Y.H., Tan, Andy. C.C., Mathew, J. & Yang, B.S. (2005). Experimental study on incipient fault detection of low speed rolling element bearings: time domain statistical parameters, *12th Asia-Pacific Vibration Conference*, Hokkaido Univ. Japan, August 2007.

Li, H., Zhang, Y. & Zheng, H. (2009). Gear fault detection and diagnosis under speed-up condition based on order cepstrum and radial basis function neural network, *Journal of Mechanical Science and Technology*, Vol. 23, No. 10, (October 2009), pp.2780-2789, ISSN 1738494X.

Mba, D. & Bannister, R. H. (1999). Condition monitoring of low-speed rotating machinery using stress waves: Part 1 and Part 2, *Journal of Process Mechanical Engineering*. Vol. 213, No. 3, (1999), pp.153-185, ISSN 0954-4089(Print), 2041-3009(Online)

Mba, D., Cooke, A., Roby, D. & Hewitt, G. (2003). Opportunities offered by acoustic emission for shaft-seal rubbing in power generation turbines; a case study. *Conference sponsored by the British Institute of NDT. International Conference on Condition Monitoring*, ISBN 1901892174, Oxford, UK, July 2003.

Mba, D., Cooke, A., Roby, D. & Hewitt, G. (2004). The detection of shaft-seal rubbing in large-scale power generation turbines with acoustic emission; Case study, *Proceedings of the Institution of Mechanical Engineers, Part A: Journal of Power and Energy*, Vol. 218, No. 2, (March 2004), pp.71-81, ISSN 0957-6509.

Misiti, M., Misiti, Y., Oppenheim, G. & Poggi, J.M. (2009) *Wavelet Toolbox TM 4 user's guide*, The MathWorks. Inc., Retrieved from <http://www.mathworks.com >

Ronnie, K.M. & V. K. Eric, (2005). *Nondestructive testing handbook Volume 6; Acoustic Emission Testing*, American Society for Nondestructive Testing, ISBN 978-1-57117-137-5, Columbus, Ohio.

Sato, I. (1990). Rotating machinery diagnosis with acoustic emission techniques, *Electrical Engineering of Japanese*, Vol. 110, No. 2, (1990), pp.115-127, ISSN 04247760.

Sentoku, H. (1998). AE in tooth surface failure process of spur gear, *Journal of Acoustic Emission*, Vol. 16 No. 1-4, (August 1998) pp.S19-S24, ISSN 0730-0050.

Sheen, Y.T. (2008). An envelope detection method based on the first-vibration-mode of bearing vibration, *Measurement: Journal of the International Measurement Confederation*, Vol. 41, No. 7, (August 2008), pp.797-809, ISSN 0263-2231.

Sheen, Y.T. (2010). An envelope analysis based on the resonance modes of the mechanical system for the bearing defect diagnosis, *Measurement: Journal of the International Measurement Confederation*, Vol. 43, No. 7, (August 2010), pp.912-934, ISSN 0263-2241.

Shiroishi, J., Li, Y., Lian, S., Danyluk, S. & Kurfess, T. (1999). Vibration analysis for bearing outer race condition diagnostics, *Journal of Brazilian Society of Mechanical Science*, Vol. 21, No. 3, (September 1999), pp.484-492, ISSN 0100-7386.

Singh, A., Houser, D. R. & Vijayakar, S. (1996). Early detection of gear pitting, *American Society of Mechanical Engineers*, Vol. 88, (1996) pp.673-678, ISSN 15214613.

Siores, E. & Negro, A.A. (1997). Condition monitoring of a gear box using acoustic emission testing, *Material Evaluation*, Vol. 55, No. 2, (February 1997), pp.183-187, ISSN 00255327

Tan, C.K. & Mba, D. (2005). Limitation of acoustic emission for identifying seeded defects in gearboxes, *Journal of Non-Destructive Evaluation*, Vol. 24, No. 1, (March 2005), pp.11-28, ISSN 0195-9298.

Tan, C.K., Irving, P. & Mba, D. (2007). A comparative experimental study on the diagnostic and prognostic capabilities of acoustics emission, vibration and spectrometric oil analysis for spur gears, *Mechanical Systems and Signal Processing*, Vol. 21, No. 1, (January 2007), pp.208-233, ISSN 0888-3270

Wu, J.D., Hsu, C.C. & Wu, G.Z. (2009). Fault gear identification and classification using discrete wavelet transform and adaptive neuro-fuzzy inference, *Expert Systems with Applications*, Vol. 36, No. 3, (April 2009), pp.6244-6255, ISSN 0957-4174.

Wu, J.D. & Chen, J.C. (2006). Continuous wavelet transform technique for fault signal diagnosis of internal combustions engines, *NDT&E International*, Vol. 39, No. 4, (June 2006), pp.304-311, ISSN 0963-8695.

Yang, Y., Yu, D. & Cheng, J. (2007). A fault diagnosis approach for roller bearing based on IMF envelope spectrum and SVM, *Measurement: Journal of the International Measurement Confederation*, Vol. 40, No. 9-10, (November 2007), pp.943-950, ISSN 0263-2231.

8

Compensation of Ultrasound Attenuation in Photoacoustic Imaging

P. Burgholzer[1,2], H. Roitner[1,2], J. Bauer-Marschallinger[1,2], H. Grün[2], T. Berer[1,2] and G. Paltauf[3]
[1]*Christian Doppler Laboratory for Photoacoustic Imaging and Laser Ultrasonics,*
[2]*Research Center for Non Destructive Testing (RECENDT),*
[3]*Institute of Physics, Karl-Franzens-University Graz*
Austria

1. Introduction

Photoacoustic imaging is a non-destructive method to obtain information about the distribution of optically absorbing structures inside a semitransparent medium. It is based on thermoelastic generation of ultrasonic waves by the absorption of a short laser pulse inside the sample. From the ultrasonic waves measured outside the object, the interior distribution of absorbed energy is reconstructed. The ultrasonic waves, which transport information from the interior to the surface of the sample, are scattered or absorbed to a certain extent by dissipative processes. The scope of this work is to quantify the information loss which is equal to the entropy production during these dissipative processes and thereby to give a principle limit for the spatial resolution which can be gained in photoacoustic imaging. This theoretical limit is compared to experimental data. In this book chapter state-of-the-art methods for modeling ultrasonic wave propagation in the case of attenuating media are described. From these models strategies for compensating ultrasound attenuation are derived which may be combined with well-known reconstruction algorithms from the non-attenuating case for photoacoustic imaging.

Section 2 gives a short description of photoacoustic imaging, especially photoacoustic tomography, and the available image reconstruction algorithms to reconstruct the interior structure from the detected ultrasound signal at the sample surface. Beside small point-like detectors also large detectors, so called integrating detectors are used for photoacoustic tomography. The latter ones require different image reconstruction algorithms. Spatial resolution is an essential issue for any imaging method. Therefore we describe the influencing factors of the resolution in photoacoustic tomography.

Section 3 is dedicated to acoustic attenuation. The spatial resolution in photoacoustic imaging is limited by the acoustic bandwidth. To resolve small objects shorter wavelengths with higher frequencies are necessary. For such high frequencies, however, the acoustic attenuation increases. This effect is usually ignored in photoacoustic image reconstruction but as small objects or structures generate high frequency components it limits the minimum detectable size, hence the resolution. Several models for acoustic attenuation, especially used for ultrasound propagation in biological tissue, are compared with experimental data.

Section 4 describes two different attempts to compensate this acoustic attenuation: either to include the compensation directly in the image reconstruction, e. g. in a modified time reversal method, or to calculate first the acoustic signal without attenuation from the measured attenuated signal and then perform the conventional photoacoustic image reconstruction. As any compensation of acoustic attenuation is mathematically an ill-posed problem both methods need regularization to prevent small measurement fluctuations from growing infinitely high in the reconstructed image. The possible degree of compensation depends on the size of these fluctuations. On the other hand acoustic attenuation is a dissipative process that causes entropy production equal to a loss of information, which cannot be compensated by any compensation algorithm. Therefore one can use the entropy production caused by acoustic attenuation to determine the minimal fluctuations in the measurement data, which turn out to be equal to thermal fluctuations. In statistical physics this fact is well known as the fluctuation-dissipation theorem, but the information theoretical background as a starting point to derive this theorem was not mentioned before in the literature.

Section 5 uses stochastic processes to understand theoretically how information can be lost and its connection to entropy production. Therefore, the measured pressure signal is treated as a random variable with a certain mean value as a function of time and certain fluctuations around that value. First for the simple model of a damped harmonic oscillator, it is shown how information is lost during a dissipative process and to what extent we can reconstruct the original information after some time. Then attenuated acoustic waves can be treated in a similar way: the spatial Fourier transform of the pressure wave can be described by a similar stochastic process as the damped harmonic oscillator - only in a higher dimension. Each wave vector is represented by a damped oscillator of different frequency.

Thinking about acoustic attenuation as a stochastic process helps to understand how entropy production and loss of information "work" on a microscopic scale. Beside a better theoretical insight the stochastic view on the acoustic wave answers a very important question: which is the best compensation method and the corresponding practicable spatial resolution in photoacoustic imaging? This question can be answered without taking fluctuations on a microscopic scale into account: the entropy production, which can be calculated from macroscopic mean values, is set equal to the information loss.

2. Photoacoustic imaging

In 1880, Alexander Graham Bell discovered that pulsed light striking a solid substrate can produce a sound wave, a phenomenon called the photoacoustic effect (Bell, 1880). Practical imaging methods based on this effect have been developed and reported the last decade (Xu & Wang, 2006). Today, photoacoustic imaging, which is also referred to as optoacoustic imaging or, when using microwaves instead of light for excitation, as thermoacoustic imaging, is attracting intense interest for cross-sectional or three-dimensional imaging in biomedicine.

In photoacoustic imaging, short laser pulses are fired at a sample and the absorbed energy causes local heating (Fig. 1). This heating causes thermoelastic expansion and thereby generation of broadband elastic pressure waves (ultrasound) which can be detected outside the sample, for example by a piezoelectric device or by an optical detector. Two methods are used for photoacoustic imaging: photoacoustic microscopy uses focused ultrasonic detectors and the sample is imaged by scanning the focus through the sample. In photoacoustic tomography (PAT) an unfocused detector is used which detects the pressure from the ultrasound wave arriving from all different locations of the source. A map or "image" of the photo-generated pressure distribution in the sample can be made by collecting the ultrasound at many different locations and processing it using a suitable algorithm e.g. by a filtered backprojection algorithm or by a time reversal algorithm.

Only if the pulse is short enough, thermal expansion causes a pressure rise proportional to the locally absorbed energy density. Short enough means that the pressure wave does not "run out" of the smallest structure which should be resolved in the photoacoustic image during the pulse time. This so called "stress confinement" is therefore fulfilled if the sound velocity multiplied by the pulse time of the laser is small compared to the spatial resolution one wants to achieve in imaging. Another constraint is the "thermal confinement" which is fulfilled if the heat induced in a structure by the absorbed laser pulse does not diffuse out of this structure during the time of the laser pulse. As heat diffusion is usually slower than the propagation of sound the thermal confinement is fulfilled if stress confinement is fulfilled.

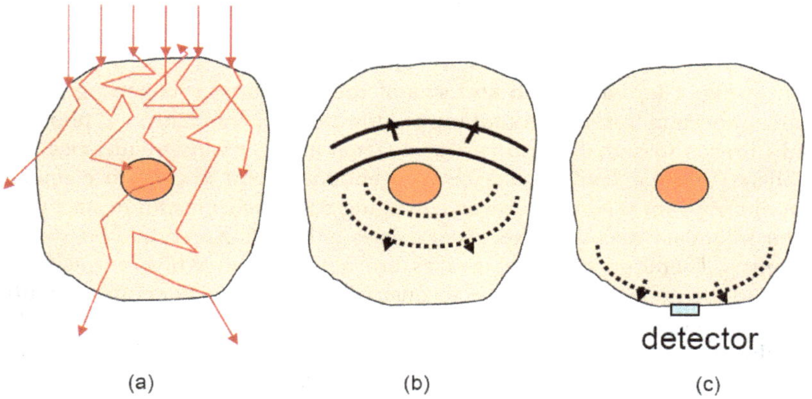

Fig. 1. Photoacoustic Imaging - the spatial resolution is determined by excitation, propagation, and detection of the acoustic wave. (a) Thermoelastic generation of acoustic wave by laser light (arrows indicate scattered photons): excited pressure is proportional to the absorbed optical energy density, if laser pulse is short enough to satisfy thermal and stress confinement. (b) Propagation of ultrasonic wave to sample surface: frequency dependent acoustic attenuation causes entropy production and therefore a loss of information. (c) Detection of ultrasonic wave: bandwidth and size of detector limits spatial resolution

Any photons, either unscattered or scattered (see arrows in Fig. 1), contribute to the absorbed energy as long as the photon excitation is relaxed thermally. Therefore PAT visualizes the product of the optical absorption distribution and the local light fluence.

Using a Nd:YAG laser and an optical parametric oscillator (OPO) light pulses from the infrared to the visible regime can be selected with a repetition rate from 10 Hz up to 100 Hz. Some high speed PAT systems can even go up to 1000 Hz. The pulse duration in the nanosecond range enables a theoretical resolution of several microns in tissue (sound velocity similar to water at approx. 1500 m/s). For biomedical applications the light energy should not exceed 20 mJ/cm^2 in the visible spectral range and 100 mJ/cm^2 in the near infrared light.

If acoustic attenuation and shear waves (in liquid and soft tissue) are neglected, the acoustic pressure p as a function of time and space obeys the equation

$$\Delta p(\mathbf{r},t)-\frac{1}{c^2}\frac{\partial^2}{\partial t^2}p(\mathbf{r},t)=-\frac{\beta}{C_p}\frac{\partial}{\partial t}\mathrm{H}(\mathbf{r},t) \tag{1}$$

where Δ is the three-dimensional Laplace operator, c the sound velocity, β the thermal expansion coefficient, C_p the specific heat capacity and $H(\mathbf{r},t)$ the deposited energy per time and volume ("heating function") caused by the absorption of the electromagnetic radiation in the sample.

For short electromagnetic pulses $H(\mathbf{r},t)=A(\mathbf{r})\cdot\delta(t)$, where $A(\mathbf{r})$ is the energy density of the absorbed electromagnetic radiation and $\delta(t)$ the Dirac delta function. Then the acoustic pressure $p(\mathbf{r},t)$ solves the homogeneous scalar wave equation with the initial conditions $p(\mathbf{r},0)=p_0(\mathbf{r})=\beta c^2/C_p\cdot A(\mathbf{r})\equiv\Gamma\cdot A(\mathbf{r})$ and $\partial/\partial t\, p(\mathbf{r},0)=0$. The initial pressure p_0 at time $t=0$ is therefore directly proportional to the absorbed energy density A with the dimensionless constant Γ, the Grüneisen coefficient.

As shown in Fig. 1 (c) bandwidth and size of the detectors for collecting the ultrasound signals are important for the resolution of this imaging modality. A photoacoustically generated ultrasound signal is a broadband signal and contains frequencies in the range from kilohertz up to a few megahertz. Conventional point like piezo elements such as known from arrays for medical ultrasonic imaging have their maximum sensitivity close to their center frequency and can detect frequencies only within a certain bandwidth around this frequency. Therefore high frequencies are not detected which correspond to small structures and are necessary for image reconstruction with a high spatial resolution. Other approaches are necessary for high resolution photoacoustic imaging. A hydrophone could be one solution (Wang, 2008), or the utilization of optical point like detection as demonstrated e.g. by (Zhang et al., 2009) or (Berer et al., 2010).

Point like detectors show a limit in achievable resolution by their size. The smaller the point detector the better is the spatial resolution. Unfortunately thermal and other fluctuations increase for a smaller detector, which results in a reduction of resolution. A totally different approach is the use of so called integrating detectors which are at least in one dimension larger than the object. This way the drawback of finite dimensions of point like detectors can be overcome. Such an integrating detector for photoacoustic imaging was introduced by (Haltmeier et al, 2004). They showed the mathematical proof of integrating area and line detectors and introduced new reconstruction methods which are necessary when using such a detector. First measurements using an integrating detector were shown by (Burgholzer et al., 2005). The first integrating detector was an area detector which was bigger than the object in two dimensions. For sufficient data for 3D image reconstruction the area detector had to be scanned around the object tangential to the surface of a sphere. This detector movement is difficult to realize. Hence the idea of the integrating line detector was developed. A fragmentation of the area detector into an array of line detectors results in an easier setup with only one rotation axis for the object and a linear motion of the integrating line detector (Fig. 2).

An integrating line detector is a line which has at least a length $\sqrt{8}*D$, where D is the diameter of a circle enclosing the sample and tangentially touching the line detector (Haltmeier et al., 2004). The line detector integrates the pressure along the line on a cylindrical surface with the radius $c \cdot t$ where c is the speed of sound in the medium and t the time. Thus, integrating line detectors arranged in an array around the sample, e.g. in a circle, measure projection images of the object in a first measurement and reconstruction step. By rotating the sample and measuring such projection images from different angles it is possible to reconstruct a 3D image (Fig. 2). Three dimensional image reconstruction from a set of projection images requires only the application of the inverse Radon transform. Therefore 3D imaging using integrating line detectors is not computationally intensive compared with other algorithms which reconstruct a 3D image from a set of signals acquired from point like detectors.

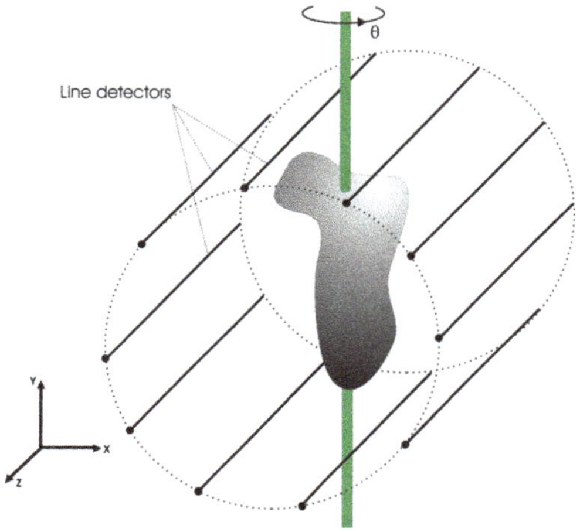

Fig. 2. Line detectors around a sample rotated on one axis. Either one line detector is scanning around the object or a detector array is used

Since the first measurement results from (Burgholzer et al, 2005) the integrating line detector was further improved in sensitivity and spatial resolution. Several types of such line detectors were implemented. One premature approach was a line made of PVDF (piezo foil) which provides high sensitivity but with the drawback of directivity. The next consequential step was an optical line detector realized by an interferometer. A laser beam that is part of an interferometer measures variations of the refractive index induced by the acoustic pressure (elasto-optic effect) (Paltauf et al., 2006). Optical detectors offer a broadband characteristic and due to the circular shape of a laser beam or light guiding fiber there is no such directivity like when using a film of piezo foil. Two main approaches can be distinguished: free-beam implementations and the use of fiber-based interferometers. Independent of the realization different types of the interferometer can be used, e.g. a Mach-Zehnder or a Fabry-Perot interferometer which is in general more sensitive than the first one.

(Paltauf et al., 2006) presented measurements using a free-beam Mach-Zehnder interferometer. They used a focused laser beam as line detector. When placing the object next to the beam waist of the focused laser beam the best spatial resolution due to the smallest beam diameter could be achieved (Paltauf et al., 2008).

(Grün et al, 2010) implemented fiber-based line detectors. The advantages of fiber-based line detectors are the easy handling and the small and constant beam diameter in the fiber. A small diameter of the laser beam is necessary for a good spatial resolution. The smaller the diameter of the detecting part the better is the spatial resolution. A typical single mode fiber for near infrared has a core diameter of 9 microns; single mode fibers for the visible range of detection wavelength have typically about 6 microns. Due to the constant diameter along the whole line detector this type of integrating line detector is dedicated for the imaging of big samples. After first implementations of a Mach-Zehnder and a Fabry-Perot interferometer in glass fibers now polymer fibers are used. Due to the much better impedance matching of polymer fibers to the surrounding water, their sensitivity is higher than for glass fibers, where approximately 2/3 of the incoming signal is reflected before reaching the core (Grün et al., 2010). Furthermore the Young's modulus is much lower in polymer fibers than in glass fibers for which reason the deformation of a polymer fiber is bigger than of a glass fiber applying the same pressure wave. As the strain optic coefficients are in the same order, this results in an enhanced change of refractive index and thus to higher signal amplitudes in the polymer fiber (Kiesel et al., 2007).

(Nuster et al, 2009) did a comparison of the different implementations of an integrating line detector. At this stage of development the free-beam Mach-Zehnder interferometer was the most sensitive integrating line detector. But these measurements showed some new approaches how the fiber-based line detector could be made more sensitive, e.g. by building up a Fabry-Perot interferometer in a single mode polymer fiber.

The next step after developing a sensitive line detector, no matter of which approach is the most sensitive one, is the creation of an array of many integrating line detectors, e.g. 200 detectors arranged in a curve around the sample. This way one could acquire all data for a projection image within one excitation pulse of the laser.

3. Acoustic attenuation

The imaging resolution in photoacoustic imaging is limited by the acoustic bandwidth and therefore by the laser pulse duration as mentioned above, but also by the attenuation of the acoustic wave on its way to the sample surface, and finally by the bandwidth and size of the ultrasonic transducer. The acoustic attenuation can be substantial for high frequencies. This effect is usually ignored in reconstruction algorithms but can have a strong impact on the resolution of small objects or structures within objects. Stokes could already show in 1845 that for liquids with low viscosity, such as water, the acoustical absorption increases by the square of the frequency (Stokes, 1845). If we do not take acoustic attenuation for photoacoustic image reconstruction into account, especially small structures (corresponding to shorter wavelengths and therefore higher frequencies) appear blurred (La Riviere et al., 2006). To what extent this blurring can be compensated by regularization methods (as performed by (La Riviere et al., 2006)) and how much information is lost due the irreversibility of attenuation is investigated in this chapter.

For thin layers (1D), small cylinders (2D), and small spherical inclusions (3D) the effect of attenuation is simulated and experimental results for several types of tissue are given. For photoacoustic tomography a new description of attenuation seems to be useful: like for a standing wave in a resonator the wave number is real but the frequency is complex. The complex part of the frequency is the damping in time. The resulting pressure wave as the solution of the wave equation is of course the same as by decomposing into plane waves with complex wave number. But with the complex frequency description acoustic attenuation can be included in all "k-space" methods well known in photoacoustic tomography just by introducing a factor describing the exponential decay in time (Roitner & Burgholzer, 2011).

Acoustic attenuation is an irreversible process and therefore the wave equation is not invariant to time reversal. Several important reasons for acoustic attenuation have been reported: viscosity, heat conduction, relaxation processes and chemical reactions. Stokes derived the scalar wave equation

$$\Delta p - \frac{1}{c^2}\frac{\partial^2 p}{\partial t^2} + \tau\,\Delta\frac{\partial p}{\partial t} = 0 \tag{2}$$

under the assumption of adiabatic conditions (thus neglecting the loss due to heat conduction) (Stokes, 1845). This equation can also be found in (Shutilov, 1988) and can generally be derived from a relaxation behavior of pressure and density (Royer & Dieulesaint, 2000), where the density change follows the pressure change after a relaxation time τ. If τ is further expressed in terms of viscosity and specific heat, this equation is also known as the thermoviscous equation. Eq. (2) describes acoustic attenuation which is approximately proportional to the square of the frequency. Other wave equations can describe a more general power law frequency dependence of the attenuation of the form $\alpha(\omega) = \alpha_0 \cdot |\omega|^y$ ($0 < y < 3$). (Szabo, 1994) has suggested adding the loss term $L(t) * p(\mathbf{r},t)$ to the wave equation in order to account for such attenuation behavior:

$$\Delta p(\mathbf{r},t) - \frac{1}{c_0^2}\frac{\partial^2}{\partial t^2} p(\mathbf{r},t) + L(t) * p(\mathbf{r},t) = 0 \tag{3}$$

Other models for acoustic wave propagation in acoustic media have been proposed also by (Nachman et al., 1990) and (Treeby & Cox, 2010).

Measurement and simulation of broadband acoustic attenuation:

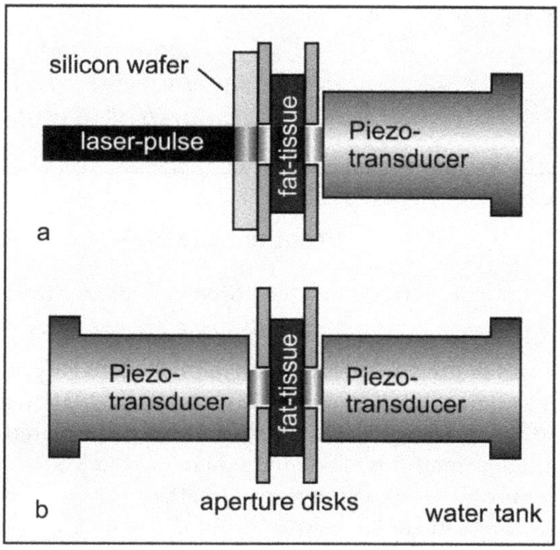

Fig. 3. Experimental set-up to measure broadband high frequency acoustic attenuation in tissue with ultrasound generated by short laser pulses (a) and by piezoelectric transducers (b). In both cases a piezoelectric transducer receives the ultrasonic signals

Fig. 3 shows the set up to determine frequency dependent acoustic attenuation in tissue by a single transmission experiment. High frequency ultrasonic pulses are either generated by a pulse laser (pulse duration 6 ns) heating up the surface of a silicon wafer (Fig. 3a) or by a piezoelectric transducer (Fig. 3b). The resulting planar-like ultrasound waves propagate through a fat tissue with varying thickness and through distilled water as a coupling medium to a piezoelectric transducer. The ultrasonic attenuation α is determined by comparing transmission measurements of the investigated samples and distilled water. In Fig. 4 Fourier transformed attenuation results for subcutaneous fat of pig, human blood and olive oil are shown. A power law $\alpha(f) = \alpha_0 \cdot f^y$ can be applied to those substances, where f denotes the frequency (Bauer-Marschallinger et al., 2011).

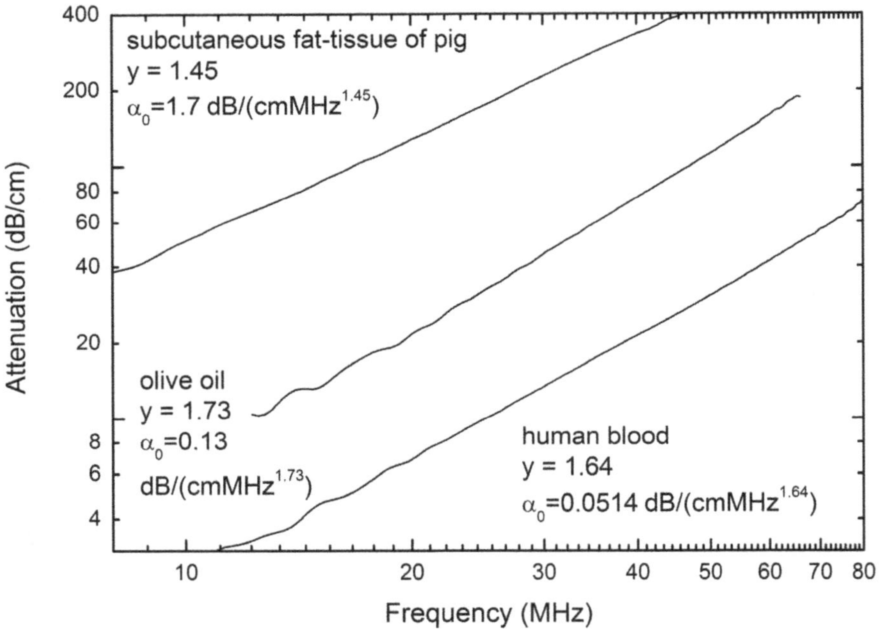

Fig. 4. Attenuation as a function of frequency for three biological substances (double logarithmic scale)

The attenuated planar ultrasound waveform can be calculated by a number of available simulation methods. In a comprehensive study (Roitner et al., 2011) we compared measured waveforms for two fat thicknesses (3.2 mm and 6.2 mm) to simulated waveforms obtained with three current simulation methods. The methods using the Matlab toolbox by (Treeby & Cox, 2010b) or the relation (4) by (LaRiviere et al., 2006) rely on a frequency power law absorption and are described in detail below. A third method by (Nachman et al., 1990) assumes multiple molecular relaxation processes as the cause of attenuation. These processes are characterized by their contributions $\kappa_n (n = 1 \ldots N)$ to the isothermal compressibility of the tissue and the relaxation times $\tau_n (n = 1 \ldots N)$ of their vibrational energies. With these parameters a complex wave number is obtained involving a generalized compressibility $\kappa(t) = \kappa_\infty \delta(t) + \sum_{n=1}^{N} \frac{\kappa_n}{\tau_n} \exp(-t / \tau_n) H(t)$ where $\kappa_\infty = \chi - \sum_{n=1}^{N} \kappa_n$ and $H(t)$ is the Heaviside step function. χ denotes the usual compressibility calculated with the formula $\chi = (c^2 \rho)^{-1}$ which holds in liquids as well as in soft biological tissues of density ρ. Stokes' equation (2) is seen to be the special case of a single relaxation mechanism $N = 1$, $\kappa_1 = \chi$.

All three simulation methods produce waveforms in good agreement to the measured waveform. Of course, the approximation will become less accurate with longer propagation distance in the absorbing medium.

Simulation results may also be validated if we recall that the frequency domain transfer function of the pressure signal propagating through a layer with complex wavenumber $K(\omega)$ and thickness d equals $\exp(iK(\omega)d)$. So the simulated 'water-fat-water' waveform may be obtained from the measured 'water-only' waveform by inverting water attenuation over the fat thickness and then applying fat attenuation over the same fat thickness.

4. Compensation of acoustic attenuation

When taking acoustic attenuation into account, the wave equation, e.g. (2) or (3), is not invariant to time reversal any more. First order terms in time t or higher order odd terms change sign if time is reversed. The equations then do not behave "well" any more, noise is amplified exponentially and regularization methods have to be used for solving these time reversed equations. Using such regularized methods the spatial resolution, that is limited by the frequency-dependent damping, is improved (Burgholzer et al., 2007). To prevent high-frequency noise from growing exponentially, Fourier spectral methods (Trefethen, 2000) in space have been used. They utilize the spatial Fourier transform to calculate the Laplacian and therefore allow for incorporating a damping of higher frequencies when calculating the time reversal. A similar algorithm to compensate acoustic attenuation step by step was proposed by (Treeby et al., 2010c).

In the second strategy, attenuation is compensated in one step where an approximation to the 'un-attenuated' measured signal is calculated from the attenuated measured signal by solving an appropriate integral equation (La Riviere et al., 2006); then an ordinary reconstruction algorithm for photoacoustic tomography is applied as in the absence of attenuation.

The Matlab Toolbox by Treeby et al. implements a state-of-the-art algorithm to simulate ultrasound wave propagation ('the direct problem') and image reconstruction ('the inverse problem') for PAI. For the direct and inverse problems arbitrary source distributions and detector geometries, variable medium densities and sound speeds and an arbitrary constant medium absorption are supported in 1D-3D. The algorithm is first-order finite difference in time and pseudospectral in space (working in spatial Fourier space = 'k-space'). The linearized Euler equations in a fluid $\frac{\partial \mathbf{u}}{\partial t} = -\frac{1}{\rho^*} \nabla p$, $\frac{\partial \rho}{\partial t} = -\rho^* \nabla \cdot \mathbf{u}$ together with an adiabatic equation of state $p = c_0^2 \rho$ are solved up to time T for the sound velocity vector $\mathbf{u}$, pressure p and density ρ. In the case of absorbing media the equation of state is extended to

$p(\mathbf{r},t) = c_0^2 \rho(\mathbf{r},t) + \mathrm{F}^{-1}(\tau k^{y-2} \frac{\partial \hat{\rho}(\mathbf{k},t)}{\partial t} + \eta k^{y-1} \hat{\rho}(\mathbf{k},t))$, where τ is related to attenuation and η to dispersion (Treeby & Cox, 2010) and the inverse Fourier transform (IFT) in d spatial dimensions is defined by $f(\mathbf{r}) = \mathrm{F}^{-1}(\hat{f}(\mathbf{k})) = (2\pi)^{-d} \int_{R^d} \hat{f}(\mathbf{k}) \exp(i\mathbf{k} \cdot \mathbf{r}) d\mathbf{k}$. With this extension of the equation of state frequency-domain ultrasound absorption power laws of the form $\alpha(\omega) = \alpha_0 \cdot |\omega|^y$ can be supported.

For the inverse problem, the same algorithm as for the direct problem is applied, but with zero initial conditions $p(\mathbf{r},0) = 0$, $\mathbf{u}(\mathbf{r},0) = \mathbf{0}$ and time-varying Dirichlet boundary conditions on detector surface points $\mathbf{r}_S$ in the form of the time-reversed sensor data $p(\mathbf{r}_S,t) = p_{meas}(\mathbf{r}_S, T-t)$. Absorption is compensated by inverting the sign of τ but leaving η unchanged. Additionally, a regularization filter is applied in k-space that suppresses the higher frequencies.

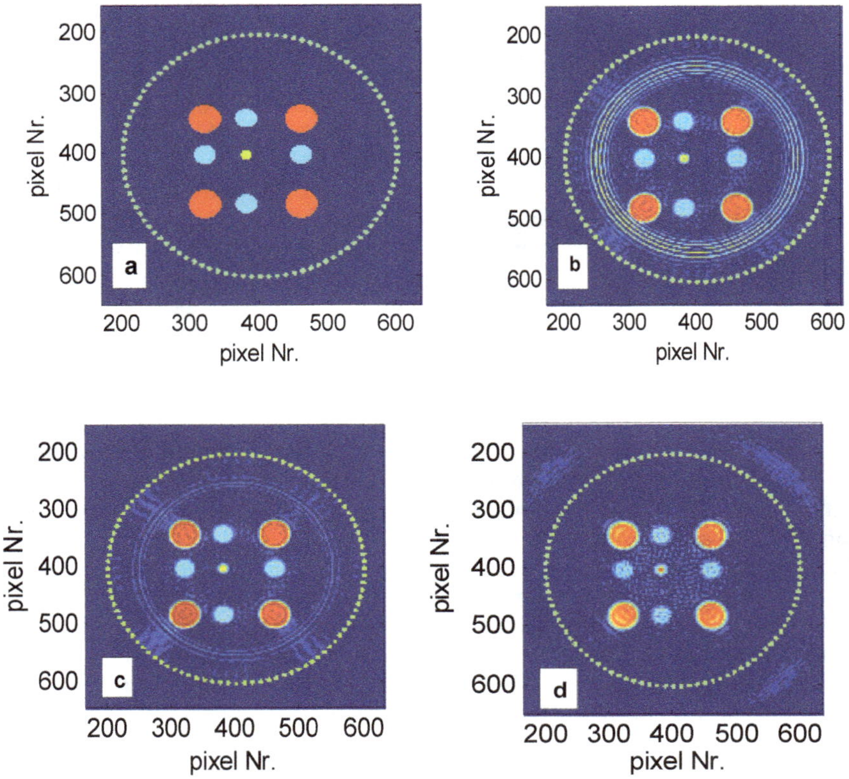

Fig. 5. Reconstruction of nine-circle phantom: (a) initial pressure distribution, (b) reconstruction Matlab toolbox with taper ratio r = 0.5, (c) reconstruction Matlab toolbox with taper ratio r = 0.75, (d) reconstruction one-step compensation plus fast series algorithm

A Tukey (tapered cosine) window is chosen for this filter. This window is parameterized by its cutoff-frequency in k-space (here $2\pi \cdot 10^4 rad/m$, corresponding to $2\pi \cdot 15MHz$ in ω-space) and its taper ratio r. The ratio controls the filter roll-off: raising r suppresses more high-frequency content in the measurement noise. In Fig. 5a we display a complex phantom consisting of nine circles of different sizes and source strengths, surrounded by a circular array of detectors (which could represent line detectors in 3D). This phantom was reconstructed first with taper ratio $r = 0.5$ (Fig. 5b) and then with taper ratio $r = 0.75$ (Fig. 5c). The absorption parameters were an exponent of $y = 1.5$ and $\alpha_0 = 3dB/(cm \cdot MHz^{1.5})$, and the 'measurement noise' consisted only of numerical noise.
Note that in Fig. 5b ring-like artifacts are produced during reconstruction, which are much reduced in Fig. 5c because there the regularization filter is more restrictive. With taper ratio $r = 1$ they disappear altogether. If a simpler phantom is used, e.g. just a single circle, taper ratio $r = 0.5$ does not produce any artifacts. To sum up, complex, fine-structured objects exhibit more signal content and hence more measurement noise at higher frequencies. Then the regularization filter must be parameterized more restrictively than for compact objects.
The formula used for calculating and compensating attenuation in one step was first presented by (LaRiviere et al., 2006) and reads

$$\tilde{p}_{att}(\mathbf{r}_S, \omega) = \frac{\omega}{c_0 \cdot K(\omega)} \int_{-\infty}^{+\infty} p_{ideal}(\mathbf{r}_S, t) \exp(ic_0 K(\omega) \cdot t)\, dt \tag{4}$$

This relation yields the temporal Fourier transform of the attenuated pressure at a given point $\mathbf{r} = \mathbf{r}_S$ (and, via an IFT, the attenuated pressure itself) if the ideal (i.e. un-attenuated) pressure is known over time at that point. This can be exploited for the direct problem where the evaluation involves only two nested numerical integrations. For the inverse problem, the left-hand side of (4) is known and the relation represents an integral equation to be solved for $p_{ideal}(\mathbf{r}_S, t)$. Since this solution is very sensitive to measurement noise, i.e. noise in the spectrum of $p_{att}(\mathbf{r}_S, t)$, regularization is necessary, for instance in the form of truncated singular value decomposition (SVD, see e.g. La Riviere et al., 2006).
If the detector geometry is regular, fast reconstruction algorithms using series expansions in eigenfunctions of the Laplacian on the regular domain are available for the lossless case making the whole reconstruction with compensation much faster than reconstruction with the mentioned toolbox algorithm. In Fig. 5d we display a reconstruction of the nine-circle phantom using this procedure. To apply the SVD, a certain noise level in temporal Fourier space must be assumed. It is desirable to base this assumption on physical reasons - this will be discussed below. Here we assume a white Gaussian noise with a standard deviation of $0{,}1\%$ of the largest amplitude in the spectrum.
Fig. 5d shows that the reconstruction with the one-step compensation is almost of the same quality as the reconstruction with the Matlab toolbox algorithm. However, computation time is reduced by a factor of 10 in comparison to the toolbox algorithm.
We already observed that the quality of reconstructions depends crucially on a good analysis of the amplitude and spectral distribution of the measurement noise. On the source side, advance knowledge of the type of object to be imaged is helpful (more bulky and compact, or more fine-structured). On the detector side, noise is created by fundamental physical processes as well as by technological limitations.

First, for any material volume V at temperature T pressure fluctuations of variance $Var(p) = k_B T / \chi V$ (Landau&Lifshitz, 1980) will occur where χ is the adiabatic compressibility and k_B is the Boltzmann constant. So if detectors are e.g. thin foils with small volume, this may play a role since noise amplitudes grow like $V^{-1/2}$.
Second, the fluctuation-dissipation theorem states that sound absorption processes in media create pressure fluctuations. If the absorption is described by Stokes' equation (2), a straightforward calculation yields a power spectral density of these pressure fluctuations $S_P(\omega) = \frac{2k_B T}{\pi} \cdot \chi^{-2} \frac{\tau}{(1+\omega^2\tau^2)^2}$.
Third, noise is created by the analog-to-digital conversion process during electronic acquisition of the pressure signal. If this quantization noise is assumed to be white, its power spectral density at sampling frequency f_s will be constant and equal $q^2 / 12 f_s$ (Widrow & Kollar, 2008) where q is the quantization interval.
We have given only an outline of the considerations that have to be taken into account for a good estimation of the measurement noise. In a concrete application the relative importance of the mentioned contributions will depend on sizes and shapes of objects and detectors, the physics of the ultrasound propagation medium (compressibility, temperature) and the properties of the measurement devices.

5. Stochastic processes for modeling acoustic attenuation

As for any other dissipative process the energy of the attenuated acoustic wave is not lost but is transferred to heat, which can be described in thermodynamics by an entropy increase. This increase in entropy is equal to a loss of information, as defined by Shannon, and no compensation algorithm can compensate this loss of information. This is a limit given by the second law of thermodynamics. But can this limit be found in the algorithms for compensation of acoustic attenuation given in the previous section? Fluctuations of the measured pressure are "amplified" exponentially during the compensation and therefore we suspect a relation between the entropy production caused by acoustic attenuation and the fluctuations. Indeed such a relation is known in statistical physics as the fluctuation-dissipation theorem and is due to (Callen & Welton, 1951), (Callen & Green, 1952), and (Greene & Callen, 1952). It represents in fact a generalization of the famous (Johnson, 1928) (Nyquist, 1928) formula in the theory of electric noise.
In this section we use the theory of non-equilibrium thermodynamics presented by S.R. de Groot and P. Mazur (De Groot & Mazur). More about random variables and stochastic processes can be found e.g. in the book about Statistical Physics from (Honerkamp, 1998). An introduction to stochastic processes on an elementary level has been published by. (Lemons, 2002), also containing "On the Theory of Brownian Motion" by (Langevin, 1908). An introduction to Markov Processes is given by (Gillespie, 1992).
In this section the measured pressure signal is treated as a time-dependent random variable with a mean value and a variance as a function of time. To be able to use the results of some "model" stochastic processes given in literature (Ornstein-Uhlenbeck process or damped harmonic oscillator) for a model of photoacoustic reconstruction we have changed the initial conditions: instead of a defined initial value (with zero variance) we have taken the stochastic process at equilibrium before time zero and at a time zero a certain perturbation has been applied to the process (e. g. a rapid change in momentum for the damped

harmonic oscillator - called kicked damped oscillator in this chapter). Reconstruction of the size of this perturbation at time t=0 from the measurement after a time t shows how the information about the size of this kick at t=0 gets lost with increasing time if dissipative processes occur. This change of the initial conditions has a significant advantage: it turns out that the variance stays constant in time, while the mean value is a function of time. This facilitates the calculations of the entropy production and of the information loss due to the stochastic process.

Gauss-Markov processes and entropy

In this section we shall assume that the time varying stochastic processes will have Gauss-Markov character. In doing so we do not wish to assume that all dissipative macroscopic processes considered belong to this specific class of Gauss-Markov processes. It may, however, be surmised that a number of real phenomena may, with a certain approximation, be adequately described by such processes (De Groot & Mazur). The advantage of specifying more precisely the nature of the processes considered is that it enables us to discuss, on the level of the theory of random processes, the behavior of entropy production and of information loss.
Following the theory of random fluctuations given e.g. by (De Groot & Mazur), we take as a starting point equations analogous to the Langevin equation used to describe the velocity of a Brownian particle:

$$\frac{d\boldsymbol{\alpha}}{dt} = -\mathbf{M}\cdot\boldsymbol{\alpha} + \boldsymbol{\varepsilon}(t) \tag{5}$$

The components of the vector $\boldsymbol{\alpha}$ are the random variables $a_i (i = 1,2,...,n)$, having zero mean value at equilibrium. The matrix $\mathbf{M}$ of real phenomenological coefficients is independent of time. The vector $\boldsymbol{\varepsilon}(t)$ represents white noise, which is uncorrelated at different times. The distribution density of $\boldsymbol{\alpha}$ turns out to be an n-dimensional Gaussian distribution:

$$f(\boldsymbol{\alpha},t) = \frac{1}{\sqrt{(2\pi)^n |\Sigma|}} e^{-\frac{1}{2}(\boldsymbol{\alpha}-\bar{\boldsymbol{\alpha}}(t))^T \Sigma^{-1} (\boldsymbol{\alpha}-\bar{\boldsymbol{\alpha}}(t))} \tag{6}$$

with the mean value $\bar{\boldsymbol{\alpha}}(t)$ and the covariance matrix Σ, which is usually also a function of time but for the initial conditions used later on the covariance matrix will be constant in time. $|\Sigma|$ is the determinant of the covariance matrix. If the initial value of $\boldsymbol{\alpha}$ is given by a given $\bar{\boldsymbol{\alpha}}_0$, the mean value at a later time t will be:

$$\bar{\boldsymbol{\alpha}}(t) = e^{-\mathbf{M}t}\cdot\bar{\boldsymbol{\alpha}}_0 \tag{7}$$

By an adequate coordinate transformation of $\boldsymbol{\alpha}$ the matrix $\mathbf{M}$ can be diagonalized: the eigenvalues of $\mathbf{M}$ are the elements of the diagonal matrix.
According to the second law of thermodynamics the entropy of an adiabatically insulated system must increase monotonously until thermodynamic equilibrium is established within the system. Then the entropy will be set to zero and the entropy at a time t is:

$$S(t) = -k_B \int f(\boldsymbol{\alpha},t) \ln \frac{f(\boldsymbol{\alpha},t)}{f(\boldsymbol{\alpha},t\to\infty)} d\boldsymbol{\alpha} \tag{8}$$

with the Boltzmann constant k_B. For a constant covariance matrix the above integration results in:

$$S(t) = -\frac{1}{2} k_B \bar{\boldsymbol{\alpha}}(t)^T \Sigma^{-1} \bar{\boldsymbol{\alpha}}(t) \tag{9}$$

Before modeling the attenuated acoustic wave as a Gauss-Markov process we give two simple examples: the Orstein-Uhlenbeck process with only one component of $\boldsymbol{\alpha}$ as a model for the velocity of a Brownian particle and the damped harmonic oscillator with two components of $\boldsymbol{\alpha}$.

Example: kicked Ornstein-Uhlenbeck process

If the random vector $\boldsymbol{\alpha}$ in eq. (5) has only one component we get the Langevin equation

$$\frac{dv(t)}{dt} = -\gamma \cdot v(t) + \sigma\, \eta(t) \tag{10}$$

which was used to describe Brownian motion of a particle. The random variable v is the particle velocity, $-\gamma \cdot v$ is the viscous drag, and σ is the amplitude of the random fluctuations. The Langevin equation governs an Ornstein-Uhlenbeck process, after (Uhlenbeck & Ornstein, 1930), who formalized the properties of this continuous Markov process. Now we assume that initially we have thermal equilibrium with zero mean velocity. At time zero the particle is kicked which causes an immediate change in velocity of v_0. Following eq. (7) the mean value $\bar{v}(t)$ shows an exponential decay:

$$\bar{v}(t) = e^{-\gamma t} \cdot \bar{v}_0 \tag{11}$$

The variance of the velocity $Var(v)$ is $\sigma^2 / 2\gamma$ and is constant in time. In Fig. 6 time and velocity are scaled to be dimensionless and the standard deviation (square root of the variance) of the velocity is normalized.
From eq. (9) the information loss equal to the entropy production till the time t after the kick is:

$$\Delta S(t) = k_B \frac{\gamma}{\sigma^2} \bar{v}(t)^2 \tag{12}$$

On the other hand the entropy production known from thermodynamics is the dissipated energy ΔQ, which is the kinetic energy of the Brownian particle of mass m, divided by the temperature T:

$$\Delta S(t) = \frac{\Delta Q}{T} = \frac{m\bar{v}(t)^2}{2T} \tag{13}$$

The thermodynamic entropy production in eq. (12) has to be equal to the loss of information in eq. (13), and therefore we get for the variance of the velocity:

$$\frac{\sigma^2}{2\gamma} = \frac{kT}{m} \tag{14}$$

Eq. (14) has been derived previously by the equipartition theorem: the equilibrium energy associated with fluctuations in each degree of freedom is $k_B T/2$. We have used the equity of entropy production and information loss. Eq. (14) states a connection between the strength of the fluctuations, given by σ^2, and the strength of the dissipation γ. This is the fluctuation-dissipation theorem in its simplest form for uncorrelated white noise.

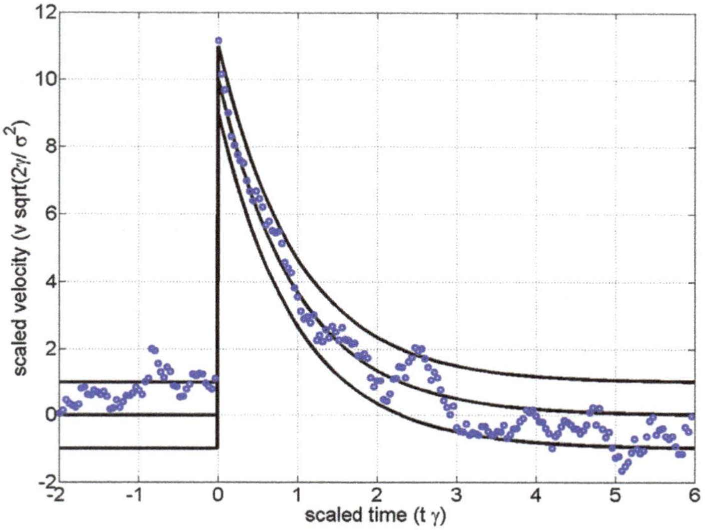

Fig. 6. Points on a sample path of the normalized kicked Ornstein-Uhlenbeck process defined by the Langevin eq. (10). The solid lines represent the mean, and mean ± standard deviation of the scaled velocity coordinate. At the time t=0 a value of v_0=10 has been added to the scaled velocity. After some time the information of the amplitude gets more and more lost due to the fluctuations

It is instructive to determine the least square estimator (Honerkamp, 1998) for the initial velocity v_0. If we write for the estimated initial velocity v_r

$$v_r = R(t) \cdot v(t) \tag{15}$$

we calculate $R(t)$ by minimizing the mean error $\left\langle (v_0 - v_r)^2 \right\rangle$:

$$R(t) = \frac{e^{-\gamma t}}{e^{-2\gamma t} + \sigma^2 / 2\gamma {v_0}^2} \tag{16}$$

This gives the Tikhonov regularization with $\sigma^2 / 2\gamma {v_0}^2 = Var(v)/{v_0}^2$ as regularization parameter. The inverse square root of the regularization parameter $v_0 / \sqrt{Var(v)}$ can be interpreted as signal-to-noise-ration (SNR) of $v(t)$.

Example: kicked harmonic osciallator

For modeling of acoustic waves one needs in addition to the dissipation also an oscillating term. For pure oscillation without damping we have no loss of information (Fig. 7).

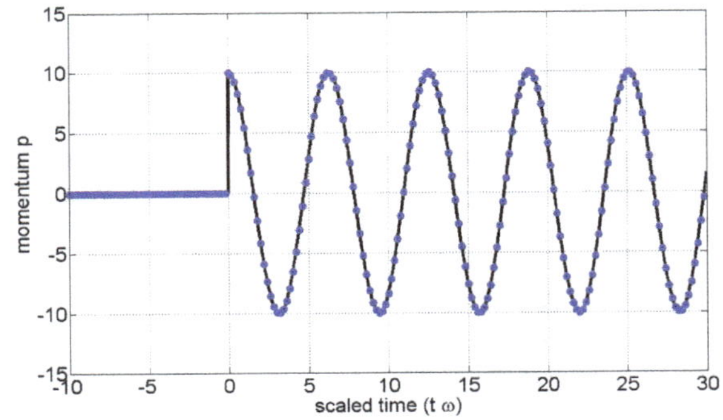

Fig. 7. Points on a sample path of the kicked harmonic oscillator without damping. The momentum used to kick the oscillator can be reconstructed without loss of information at any time after the kick if ω is known

The stochastically damped harmonic oscillator combines the oscillatory and the diffusive behavior and therefore it is a good starting point to model attenuated acoustic waves. The equations of motion are (using the momentum p instead of the velocity v):

$$\frac{dx(t)}{dt} = \frac{1}{m} p(t) \tag{17}$$

$$\frac{dp(t)}{dt} = -\frac{\gamma}{m} \cdot p(t) - m\omega_0{}^2 x + \sigma\, \eta(t) \tag{18}$$

These equations can be combined using a two dimensional random vector $\mathbf{\alpha} = (x, p)$ and were solved already by (Chandrasekhar, 1943) for definite initial conditions $x(0)$ and $p(0)$. Again we have changed the initial conditions to an oscillator with zero mean values kicked by an initial momentum p_0. In Fig. 8 the damping is chosen to be $\gamma = m\omega_0/3$.
Using the fluctuation-dissipation theorem $\sigma^2/2\gamma = kT$ one obtains for the distribution function

$$f(x,p,t) = \frac{1}{2\pi \frac{kT}{\omega_0}} e^{-\frac{1}{2mkT}(p-\overline{p}(t))^2 - \frac{1}{2kT} m\omega_0{}^2 (x-\overline{x}(t))^2}, \tag{19}$$

where $\overline{x}(t)$ and $\overline{p}(t)$ are the solutions of the ordinary (non-stochastic) damped harmonic oscillator. Then one gets for the information loss from eq. (9)

$$S(t) = -\frac{1}{T}\left(\frac{1}{2} m\omega_0{}^2 \overline{x}(t)^2 + \frac{\overline{p}(t)^2}{2m}\right) = -\frac{1}{T}(E_{pot} + E_{kin}), \tag{20}$$

which is equal to the entropy from thermodynamics, where $E_{pot}+E_{kin}$ is the total energy of the harmonic oscillator (sum of the potential and kinetic energy). This fact confirms again that the fluctuation-dissipation theorem for the damped harmonic oscillator $\sigma^2/2\gamma = k_B T$ can be derived from the equity of entropy production and information loss.

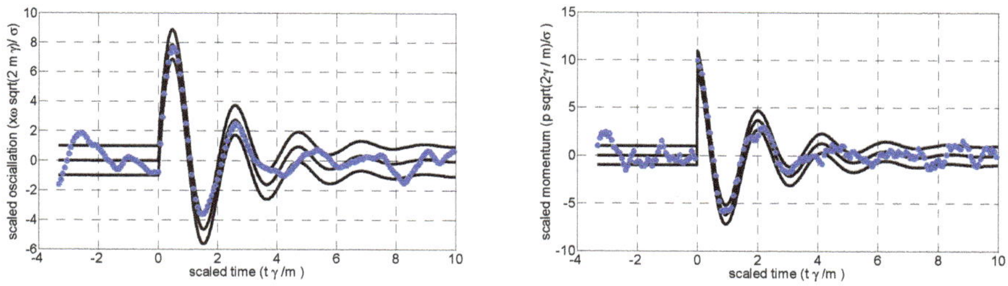

Fig. 8. Points on a sample path of the normalized kicked damped harmonic oscillator (eq. (17) and (18)). The solid lines represent the mean, and mean ± standard deviation of the scaled oscillation (left) and momentum (right). At the time t=0 a value of p_0=10 has been added to the scaled momentum. After some time the information about the value of p_0 gets more and more lost due to the fluctuations

For the mean value of the momentum $\overline{p}(t)$ one gets from eq. (7):

$$\overline{p}(t) = p_0[\cos(\omega t) - \frac{\gamma}{2m}\frac{\sin(\omega t)}{\omega}]e^{-\frac{\gamma t}{2m}} \tag{21}$$

Compared to the Ornstein-Uhlenbeck process (eq. (11)) the mean value has not only an exponential decay in time but also an oscillation with a frequency $\omega = \sqrt{\omega_0^2 - \gamma^2/(4m^2)}$. As mentioned above (Fig. 7) the oscillating term does not change the entropy and no information is lost. In the average only the exponential decay causes production of entropy and the information about the value of p_0 gets more and more lost due to the fluctuations. The entropy production which is proportional to the total energy is shown in Fig. 9.

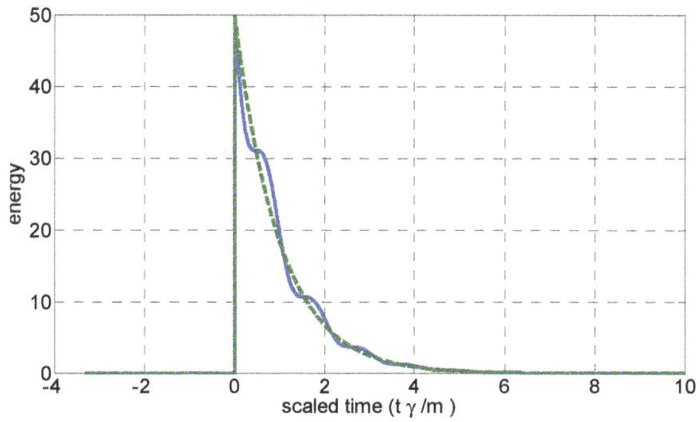

Fig. 9. Total energy of the damped harmonic oscillator (solid line) shows in the average an exponential decay with the time constant γ / m (dashed line)

Attenuated acoustic wave as a stochastic process

In photoacoustic imaging, the laser pulse at a time t = 0 generates an initial pressure distribution $p_0(\mathbf{r})$ (see section 2). For numerical calculations we use a discrete space $\mathbf{r} = \mathbf{r}_j$ (*j*=1,..,*N*), where $\mathbf{r}_j$ are *N* points on a cubic lattice with a spacing of Δr within the sample volume V. At a time t the pressure distribution $p_j(t)$ can be represented by a Fourier series (Barret et al. 1995), including the time $t = 0$ with the initial pressure distribution:

$$p_j(t) = p(\mathbf{r}_j, t) = \sum_{k=1}^{N} \hat{p}_k(t)\varphi_k(\mathbf{r}_j) \quad with \;\; \varphi_k(\mathbf{r}) = e^{2\pi i \boldsymbol{\rho}_k \cdot \mathbf{r}} \cdot D(\mathbf{r}) \tag{22}$$

$D(\mathbf{r})$ is a support function which is one within the sample volume V and zero outside. $\boldsymbol{\rho}_k$ are integer points on an infinite 3D lattices. From eq. (1) we get (if acoustic attenuation can be neglected):

$$\hat{p}_k(t) = \cos(\omega_k t)\hat{p}_k(0) \quad with \;\; \omega_k^{\,2} = 4\pi^2 c^2 \boldsymbol{\rho}_k^{\,2}. \tag{23}$$

Therefore we have only an oscillating term as shown in Fig. 7, but in higher dimensions, and no information is lost.

For an attenuated acoustic wave instead of the wave equation (1) we have used the Stokes equation (2) or the wave equation (3), giving for the spatial Fourier components $\hat{p}_k(t)$ not only an oscillating term but also an exponential decay:

$$\hat{p}_k(t) = [A_k \sin(\omega_k t) + B_k \cos(\omega_k t)]\, e^{-\lambda_k t} \tag{24}$$

where the phase factors A_k and B_k can be derived from the initial conditions and ω_k and λ_k are functions of $|\boldsymbol{\rho}_k|$. Like for the damped harmonic oscillator this exponential decay causes the loss of information and can be modeled by a Gauss-Markov process. The information content in Fourier space is the same as in real space (Fig. 10). Therefore we can describe the information loss in the attenuated acoustic wave by the same model as for the damped harmonic oscillator, only in higher dimensions, as for each wave-vector-index k an oscillator is needed.

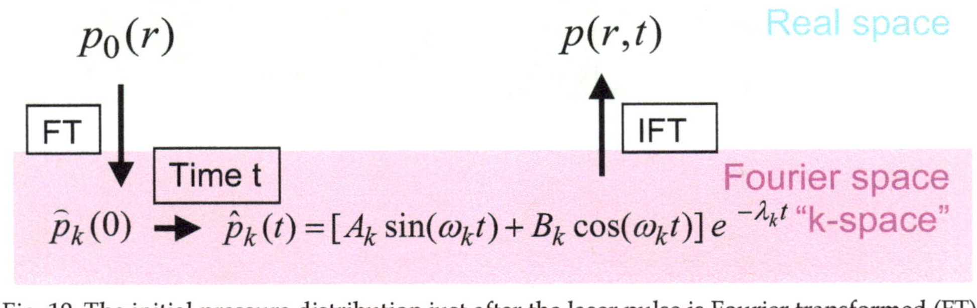

Fig. 10. The initial pressure distribution just after the laser pulse is Fourier transformed (FT). The time evolution of the Fourier series coefficients can be described similar to the mean value of a stochatsically damped harmonic oscillator. The pressure distribution after a time t is then calculated by an inverse Fourier transform (IFT)

One dimensional (1D) example: photoacoustic signal of a 0.2 mm thin layer in glycerin:

The effect of acoustic attenuation for the reconstructed image is similar in 1D, 2D, and 3D (Burgholzer et al., 2010a and 2010b). As in 1D the reconstructed image is just the shifted measured signal, the effect of attenuation and of its compensation can be directly seen. The photoacoustic signal of a 0.2 mm thin absorbing layer in glycerin is calculated by using the scheme of Fig. 10 after a time of 4.5 microseconds. In glycerin the acoustic pressure can be described well by the Stokes' equation with a relaxation time of 244 picoseconds (Shutilov, 1988).

For an attenuated acoustic wave instead of the wave equation (1) we have used the Stokes equation (2) or the wave equation (3), giving for the spatial Fourier components $\hat{p}_k(t)$ not only an oscillating term but also an exponential decay:

Putting eq. (24) into the Stokes equation (2) one gets:

$$\omega_k^{\ 2} = 4\pi^2 c^2 \rho_k^{\ 2}(1 - \frac{1}{4} c^2 \rho_k^{\ 2} \tau^2) \text{ and } \lambda_k = \pi c^2 \rho_k^{\ 2} \tau \quad (25)$$

The initial pressure distribution is a one dimensional square pulse corresponding to an absorbing layer of thickness 2a with a = 0.1 mm. In Fourier space we get:

$$\hat{p}_k(0) = \frac{1}{\pi k} e^{-ikr} \sin(ka) \, with \, k = 2\pi\rho_k \quad (26)$$

The calculated signal after a time of 4.5 microseconds is shown in Fig. 11 (dashed line). The dashed dotted line shows the signal without attenuation (no relaxation time) and therefore represents the ideal reconstructed image.

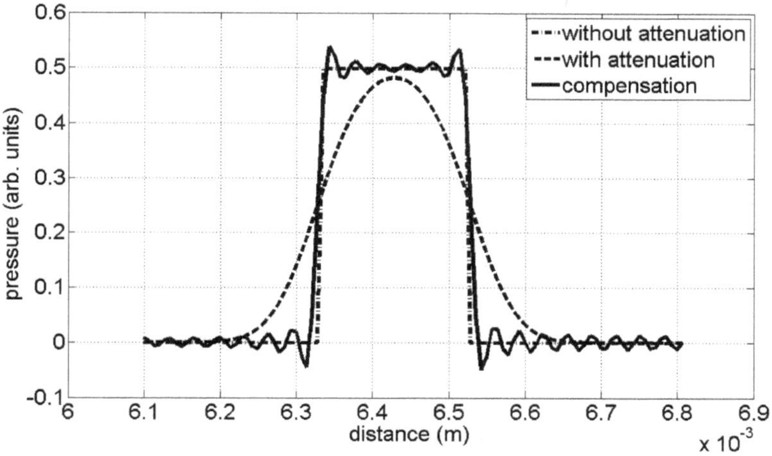

Fig. 11. Simulation result as an example for Stoke's equation: it is shown how an initial square pulse would look after a time of 4.5 microseconds in glycerine (τ=244ps) (dashed line). The dashed dotted line shows the pulse without attenuation (τ=0), which correspond to the reconstruction of the initial pressure pulse, and the solid line is the SVD reconstruction of the initial pressure pulse (see text)

For compensation of the acoustic attenuation we use the time reversed process of the Gauss-Markov process from Fig. 10. Similar to the stochastically damped oscillator the entropy production for the acoustic wave is set equal to the loss of information from the Gauss-

Markov process with standard deviation s_k and is approximated by an exponential decay in time (Fig. 9):

$$\Delta S(t) \approx \frac{1}{2} k_B \sum_k \frac{1}{s_k^2} \exp(-2\lambda_k t)\hat{p}_k(0)^2 \tag{27}$$

$$\Delta S(t) = \frac{\Delta Q}{T} \approx \frac{1}{2c_0^2 \rho_0 T} \sum_k \frac{1}{s_k^2} \exp(-2\lambda_k t)\hat{p}_k(0)^2 \tag{28}$$

which gives a variance of $\hat{p}_k(t)$ in the spatial Fourier space ("k-space") independent from k:

$$s_k^2 = k_B c_0^2 \rho_0 T \tag{29}$$

In real space this gives the thermodynamic fluctuations. Or the other way around, the thermodynamic fluctuations have a quantity such that the information loss of the Gauss-Markov process is equal to the dissipated heat divided by the temperature. Using the thermodynamic fluctuations for a detector size of 1 cm^2 and the SVD regularization from section 4 the compensated $\hat{p}_k(t)$ is calculated. By applying a subsequent inverse Fourier transform (IFT) one gets the compensated pressure profile in real space (solid line in Fig. 11).

6. Summary and conclusions

Acoustic attenuation is modeled as a stochastic process: this helps to understand how thermodynamic entropy production and the decrease of information, which is "transported" in the acoustic wave, are closely connected on a microscopic scale. This theoretical insight enables to answer an important question in photoacoustic imaging: what is the highest possible compensation of attenuation and therefore the best spatial resolution one can achieve?

We could show that for thermal fluctuations the information loss of the reconstructed image is equal to the entropy production due to attenuation of the acoustic wave. Therefore it is sufficient to calculate the entropy production from the macroscopic mean values and it is not necessary to take the fluctuations of the pressure into account.

The size and locations of detectors in photoacoustic imaging should be optimized to get the best resolution and sensitivity. Up to now in such models it is not assumed that the pressure is a random variable, which favors small point like detectors. Taking thermal fluctuations and other noise into account will help to get a more realistic model for detectors and the reconstructed images from measured signals with these detectors.

7. Acknowledgments

This work has been supported by the Christian Doppler Research Association, by the Federal Ministry of Economy, Family and Youth, by the Austrian Science Fund (FWF) project numbers S10503-N20 and TRP102-N20, by the European Regional Development Fund (EFRE) in the framework of the EU-program Regio 13, and the federal state Upper Austria.

8. References

Barret, H. H., Denny, J. L., Wagner, R. F. & Myers, K. J. (1995). Objective assessment of image quality II, *J. Opt. Soc. Am.* A 12(5) 834-852

Bauer-Marschallinger, J., Berer, T., Roitner, H., Grün, H., Reitinger, B. & Burgholzer, P. (2011). Ultrasonic attenuation of biomaterials for compensation in photoacoustic imaging, *Proc. SPIE* 7899, 789931

Bell, A. G. (1880). On the production and reproduction of sound by light: the photophone, *American Journal of Science* 20, 305–324

Berer, T., Hochreiner, A., Zamiri, S. and Burgholzer, P. (2010) Remote photoacoustic imaging on solid material using a two-wave mixing interferometer, *Opt. Lett.* 35, 4151-4153

Burgholzer, P., Hofer, C., Paltauf, G., Haltmeier, M. & Scherzer, O. (2005). Thermoacoustic tomography with integrating area and line detectors, *IEEE Trans. Ultrason. Ferroelectr. Freq. Control* 52, 1577–1583

Burgholzer, P., Grün, H., Haltmeier, M., Nuster, R. & Paltauf, G. (2007). Compensation of acoustic attenuation for high resolution photoacoustic imaging with line detectors, *Proc. of SPIE* Vol. 6437 643724-1

Burgholzer, P., Roitner, H., Bauer-Marschallinger, J. & Paltauf, G. (2010a). Image Reconstruction in Photoacoustic Tomography using Integrating Detectors accounting for Frequency-Dependent Attenuation, *Proc. of SPIE* Vol. 7564 756423-1

Burgholzer, P., Berer, T., Gruen, H., Roitner, H., Bauer-Marschallinger, J., Nuster, R. & Paltauf, G. (2010b). Photoacoustic Tomography using Integrating Line Detectors Invited, *Journal of Physics: Conference Series*, 214(1).

Callen, H. B. & Welton, T. A. (1951), *Phys. Rev.* 83, 34

Callen, H. B. & Greene, R. F. (1952). *Phys. Rev.* 86, 702

Chandrasekhar, S. (1943). Stochastic Problems in Physics and Astronomy, *Reviews of Modern Physics* 15, 1-89

De Groot, S.R. & Mazur, P. (1984) *Non-Equilibrium Thermodynamics,* Dover Publications, Inc., New York

Gillespie, D. T. (1992). *Markov Processes*, Academic Press, New York

Greene, R. F. & Callen, H. B. (1952). *Phys. Rev.* 88, 1387

Grün, H., Berer, T., Nuster, R., Paltauf, G. & Burgholzer, P. (2010). Three-dimensional photoacoustic imaging using fiber-based line detectors, *Journal of Biomedical Optics*, 15(2), 021306-1 - 021306-8

Haltmeier, M., Scherzer, O., Burgholzer, P. & Paltauf, G. (2004). Thermoacoustic computed tomography with large planar receivers, *InverseProbl.* 20, 1663–1673

Hansen, P. C. (1987). *Rank-deficient and discrete ill-posed problems: Numerical aspects of linear inversion*, SIAM, Philadelphia

Honerkamp, J. (1998). *Statistical Physics*, Springer-Verlag

Johnson, J. B. (1928). Thermal Agitation of Electricity in Conductors, *Phy. Rev.* 32, 97

Kiesel, S., Peters, K., Hassan, T. & Kowalsky, M. (2007). Behavior of intrinsic polymer optical fibre sensor for large-strain, *Meas. Sci.Technol.* 18, 3144–3154

Landau, L. D. & Lifshitz, E. M. (1980). *Statistical physics, Part 1*, Pergamon Press, Oxford

Langevin, P. (1908). Sur la théorie du mouvement brownien, Compets rendus Académie des Sciences (Paris) 146, 530-533. Translation by Anthony Gythiel, published in American Journal of Physics 65, 1079-1081 (1997)

La Rivière, P. J., Zhang, J. & Anastasio, M. A. (2006). Image reconstruction in optoacoustic tomography for dispersive acoustic media, *Optics Letters*. 31(6), 781-783

Lemons, D. S. (2002). An Introduction to Stochatic Processes in Physics, The Johns Hopkins University Press

Nachman, A. I., Smith III, J. F. & Waag, R. C. (1990). An equation for acoustic propagation in inhomogeneous media with relaxation losses, *J.Acoust.Soc.Am.* 88, 1584-1595

Nuster, R., Gratt, S., Passler, K., Grün, H., Berer, T., Burgholzer, P. & Paltauf, G. (2009). Comparison of optical and piezoelectric integration line detectors, in *Biomedical Optics: Photons Plus Ultrasound: Imagingand Sensing 2009*, edited by A. A. Oraevsky and L. H. Wang, *Proc. SPIE* 7177, 71770T

Nyquist, H. (1928). *Phys. Rev.* 32, 110

Paltauf, G., Nuster, R., Haltmeier, M. & Burgholzer, P. (2006). Photoacoustic tomography using a Mach-Zehnder interferometer as acoustic line detector, *Appl. Opt.* 46, 3352–3358

Paltauf, G., Nuster, R., Passler, K., Haltmeier, M. & Burgholzer, P. (2008). Optimizing Image Resolution in Three-Dimensional Photoacoustic Tomography With Line Detectors, in *Biomedical Optics: Photons Plus Ultrasound: Imaging and Sensing 2008, Proc SPIE,* 6856

Roitner, H. & Burgholzer, P. (2011). Efficient modeling and compensation of ultrasound attenuation losses in photoacoustic imaging, *Inverse Problems* 27, 015003

Roitner, H., Bauer-Marschallinger, J., Berer, T. & Burgholzer, P. (2011). Experimental Evaluation of ultrasound attenuation losses in photoacoustic imaging, submitted to *J.Acoust.Soc.Am.*

Royer, D. & Dieulesaint, E. (2000). *Elastic Waves in Solids I*, Springer

Shutilov, V. A. (1988). *Fundamental Physics of Ultrasound*, Gordon and Breach Science Publishers

Stokes, G. G. (1845). On the theories of the internal friction of fluids in motion, and of the equilibrium and motion of elastic solids, *Trans. Cambridge Philos. Soc.* 8, 287-319

Szabo, T. L. (1994). Time domain wave equations for lossy media obeying a frequency power law, *J.Acoust. Soc. Am.* 96, 491–500

Treeby, B. E. & Cox, B. T. (2010). Modeling power law absorption and dispersion for acoustic propagation using the fractional Laplacian, *J.Acoust. Soc. Am.* 127 , 2741-2748

Treeby, B.E. & Cox, B. T. (2010b). k-wave: Matlab toolbox for the simulation and reconstruction of photoacoustic wave fields, *J. Biomed. Optics* 15(2), 021314-1 - 021314-12

Treeby, B.E., Zhang, E. Z. & Cox, B. T. (2010c). Photoacoustic tomography in absorbing acoustic media using time reversal, *Inverse Problems* 26, 115003

Trefethen, L. M. (2000). *Spectral Methods in Matlab*, SIAM, Philadelphia

Uhlenbeck, G. E. & Ornstein, L. S. (1930). On the Theory of Brownian Motion, *Phys. Rev.* 36, 823-41

Wang, L. V. (2008). Prospects of photoacoustic tomography, *Medical Physics* 35, 5758–5767

Widrow, B. & Kollar, I. (2008). *Quantization noise*, Cambridge Books, Cambridge

Xu, M. & Wang, L. V. (2006). Photoacoustic imaging in biomedicine, *Review of Scientific Instruments*. 77(4), 041101-041122

Zhang, E. Z., Laufer, J. G., Pedley, R. B. & Beard, P. C. (2009). In vivo high-resolution 3D photoacoustic imaging of superficial vascular anatomy, *Phys. Med. Biol.* 54, 1035–1046

9

Evaluation Method for Anisotropic Drilling Characteristics of the Formation by Using Acoustic Wave Information

Deli Gao and Qifeng Pan
China University of Petroleum at Beijing, Beijing
China

1. Introduction

In drilling engineering, we must have solid understanding of the underground geological environment which is not only complicated and diversified but also someway concealed. Thus, in order to find an effective method of predicting it, a long-term research and practice must be required. In drilling engineering for oil & gas, problems including borehole deviation & its control, wellbore instability & its control, influence directly drilling quality & efficiency of a deep or complicated well for exploration and production of oil & gas fields. For instance, because of the complicated surface and underground conditions as well as the depth (over 5000 m) of oil & gas reservoirs in western China, such unstable factors such as hole deviation and instability often encountered with each other in deep drilling engineering. Because we did not have access to the geological parameters of the formations to be drilled including the rock drillability anisotropy and so on, huge economic loss had been caused and the steps to explore and produce oil & gas in western China had been seriously restricted. Hence, there are many researches and development programs to do for the right cognition and scientific evaluation of the geological environments, and for the further study of mechanism of the drilling process instability, and so on. The solution of these problems is the key to improve the performance of drilling & HSSE (health, safety, security, environment) and lower well construction cost.

The factors influencing the instability can be sorted into subjective category and objective category. In the objective category, there are the types of geological structure and in-situ stress, rock anisotropy, porosity, permeability, lithology, pressures, and mineral components, as well as rock strength and weak layer of the formation to be drilled, and so on. In the subjective category, there are the performance of down hole drilling system, the drilling parameters (weight and torque on bit, etc.), the drilling fluid performances (water loss, viscosity, rheological property and density) and its hydration on shale, the direction and open time of wellbore, the erosion and surge pressure of drilling fluid on the hole wall, the interaction between drillstring and hole wall. Thereby, in researches on the instability, the factors from the both categories should be taken into comprehensive consideration.

Whether in vertical drilling or in directional drilling, it is always a complicated academic and technological problem how to control the well trajectory exactly along the designed track to reach the underground targets. In rotary drilling, the forming of wellbore & its trajectory is the result of the rock-bit interaction. In this interaction, the drill bit anisotropy and its mechanical behavior (i.e. the drill bit force and tilt angle) are important factors that can directly affect the well trajectory. The mechanical behavior depends on by the bottom hole assembly (BHA) analysis. Accordingly, principal factors influencing the well trajectory generally contain BHA, drill bit, operating parameters in drilling, drilled wellbore configuration and the formations to be drilled. Of which the BHA, drill bit and operating parameters in drilling are the factors that can be artificially controlled, and the formation property (such as rock drillability and its anisotropy) is the objective factor which can not be changed by us. The trajectory can be predicted before drilling and also can be determined after drilling through surveys and calculations. Besides, the drilled wellbore will not only generate a strong reaction on the drill bit force and the drillstring deflection, but also will exert an influence on the anisotropic drilling characteristics of the formation. Due to the above-complicated factors, the hole deviation is always inevitable, which may seriously influence the wellbore quality and the drilling performance.

The well trajectory control is the process which forces drill bit to break through formations along the designed track forward by applying reasonable techniques. The anisotropic drilling characteristics of the drill bit & the formation and their interaction effects are the factors which will cause a direct influence on the well trajectory control. Thereby, it is a complicated scientific and technological problem for us how to make the cognition, evaluation and utilization of anisotropic drilling characteristics of the formation, as well as the prediction & control of mechanical action of the drill bit on the formations.

Rock drillability anisotropy of the formation to be drilled has significant effects on the well trajectory control so that it is very important to evaluate it. Definitions of rock drillability anisotropy and acoustic wave anisotropy of the formation to be drilled are presented in this chapter. The acoustic velocities and the drillability parameters of some rock samples from Chinese Continental Scientific Drilling (CCSD) are respectively measured with the testing device of rock drillability and the ultrasonic testing system in laboratory. Thus, their drillability anisotropy and acoustic wave anisotropy are respectively calculated and discussed in detail by using the experimental data. Based on the experiments and calculations, the correlations between drillability anisotropy and acoustic wave anisotropy of the rock samples are illustrated through regression analysis. What's more, the correlation of rock drillability in directions perpendicular to and parallel to the bedding plane of core samples is studied by means of mathematical statistics. Thus, a mathematic model is established for predicting rock drillability in direction parallel to the formation bedding plane by using rock drillability in direction perpendicular to the formation bedding plane with the well logging or seismic data. The inversion method for rock anisotropy parameters (ε,δ) is presented by using well logging information and the acoustic wave velocity in direction perpendicular to the bedding plane of the formation is calculated by using acoustic wave velocity in any direction of the bedding plane. Then, rock drillability in direction perpendicular to the bedding plane of the formation can be calculated by using acoustic wave velocity in the same direction. Thus, rock drillability anisotropy and anisotropic drilling characteristics of the formation can be evaluated by using the acoustic wave information based on well logging data. The evaluation method has been examined by case study based on oilfield data in west China.

2. Anisotropic drilling characteristics of the formation

Although many theories have been proposed to explain the hole deviation since the 1950s (Gao et al, 1994), it is only the rock drillability anisotropy theory (Lubinski & Woods, 1953) that was recognized by petroleum engineers and widely applied to petroleum engineering because it can be used to quantify the anisotropic drilling characteristics of the formation and to explain properly the actual cases of hole deviation encountered in drilling engineering. The theory suggested that since values of rock drillability are not always the same in the directions perpendicular and parallel to the bedding plane of the formation, the formation will bring the bit a considerable force, which may likely cause changes on the original drilling direction and hole deviation.

The orthotropic or the transversely isotropic formations are the typical formations encountered frequently in drilling engineering. The anisotropic effects of the formations (rock drillability) on the well trajectory must be considered in hole deviation control and directional drilling. Based on the rock-bit interaction model, the formation force is defined and modeled in this section to describe quantitatively anisotropic drilling characteristics of the formations to be drilled.

2.1 Definition of rock drillability anisotropy

Because of rock drillability anisotropy, the real drilling direction does not coincide with the resultant force direction of the drill bit (supposed that it is isotropic) on bottom hole. Besides calculating the drill bit force by BHA (bottom hole assembly) analysis, rock drillability anisotropy of the formation must be considered in hole deviation control.

The formation studied here is typical orthotropic one, and the transversely isotropic formation discussed previously is regarded as its particular case. Let $\vec{\mathbf{e}}_d$, $\vec{\mathbf{e}}_u$ and $\vec{\mathbf{e}}_s$ represent unit vectors in the directions of inner normal, up-dip and strike of the formation respectively, as shown in Fig.1.There are different physical properties along different directions of them. γ in Fig.1 represents dip angle of the formation to be drilled. Rock drillability anisotropy of the formation can be expressed by rock drillability anisotropy index. If the components of penetration rate of the drill bit (isotropic) along inner normal, up-dip and strike of the orthotropic formation are noted as R_{dip}, R_{str} and R_{n} respectively, correspondingly the net applied forces are F_{dip}, F_{str} and F_{n} respectively, the rock drillability can be defined as:

$$D_{\text{n}} = \frac{R_{\text{n}}}{F_{\text{n}}}\ ,\ D_{\text{dip}} = \frac{R_{\text{dip}}}{F_{\text{dip}}}\ ,\ D_{\text{str}} = \frac{R_{\text{str}}}{F_{\text{str}}} \tag{1}$$

Rock drillability anisotropy of the orthotropic formation may be represented by two indexes (I_{r1} and I_{r2}) which are defined as:

$$I_{\text{r1}} = \frac{D_{\text{dip}}}{D_{\text{n}}}\ ,\ I_{\text{r2}} = \frac{D_{\text{str}}}{D_{\text{n}}} \tag{2}$$

Dip angle and strike of the formation can be obtained from the analysis of well logging and geological structure survey. The values of I_{r1} and I_{r2} for the orthotropic formation can be evaluated by the experimental analysis or using the acoustic wave information.

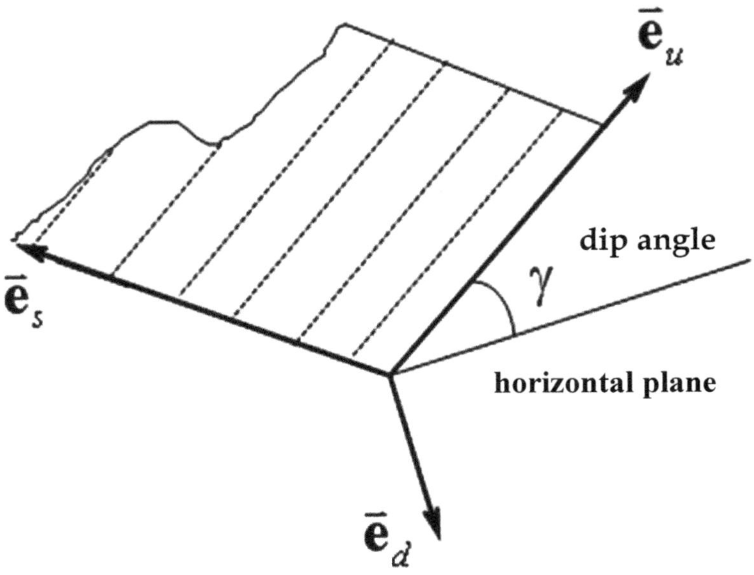

Fig. 1. Descartes coordinates for the formation geometry

2.2 The formation force

Assumed that the drill bit is isotropic for eliminating the effects of its tilt angle on hole deviation, the effects of the orthotropic formation on hole deviation can be presented by the formation force analysis. The two parameter equations related to the formation forces can be derived from the rock-bit interaction model (Gao & Liu, 1989):

$$\begin{cases} G_\alpha = \dfrac{t_{22}t_{13} - t_{12}t_{23}}{t_{11}t_{22} - t_{12}t_{21}} \\ G_\phi = \dfrac{t_{11}t_{23} - t_{21}t_{13}}{t_{11}t_{22} - t_{12}t_{21}} \end{cases} \tag{3}$$

Where G_α and G_ϕ are called as the building angle parameter (positive for building up the inclination of well trajectory) and the drifting azimuth parameter (positive for left walking of well trajectory) of the formation respectively , and the t_{ij} $(i, j=1,2,3)$ can be expressed as (Gao & Liu, 1990):

$$\begin{cases} t_{ij} = I_{r1}\delta_{ij} + (1 - I_{r1})a_{ij} + (I_{r2} - I_{r1})c_{ij} \\ \delta_{ij} = \begin{cases} 0, i \neq j \\ 1, i = j \end{cases} \end{cases} \tag{4}$$

where $a_{ij} = a_{ji}$, $c_{ij} = c_{ji}$ $(i, j=1, 2, 3)$ can be calculated by the following equations:

$$\left.\begin{aligned}
a_{11} &= (\sin\alpha\cos\gamma - \cos\alpha\sin\gamma\cos\Delta\varphi)^2 \\
a_{12} &= (\cos\alpha\sin\gamma\cos\Delta\varphi - \sin\alpha\cos\gamma)\sin\gamma\sin\Delta\varphi \\
a_{13} &= (\cos\alpha\sin\gamma\cos\Delta\varphi - \sin\alpha\cos\gamma)(\sin\alpha\sin\gamma\cos\Delta\varphi + \cos\alpha\cos\gamma) \\
a_{21} &= (\cos\alpha\sin\gamma\cos\Delta\varphi - \sin\alpha\cos\gamma)\sin\gamma\sin\Delta\varphi \\
a_{12} &= a_{21} \\
a_{22} &= (\sin\gamma\sin\Delta\varphi)^2 \\
a_{23} &= \sin\gamma\sin\Delta\varphi(\sin\alpha\sin\gamma\cos\Delta\varphi + \cos\alpha\cos\gamma) \\
a_{31} &= a_{13} \\
a_{32} &= a_{23} \\
a_{33} &= (\sin\alpha\sin\gamma\cos\Delta\varphi + \cos\alpha\cos\gamma)^2
\end{aligned}\right\} \tag{5}$$

$$\left.\begin{aligned}
c_{11} &= (\sin\Delta\varphi\cos\alpha)^2 \\
c_{12} &= -\cos\Delta\varphi\sin\Delta\varphi\cos\alpha \\
c_{13} &= (\sin\Delta\varphi)^2\sin\alpha\cos\alpha \\
c_{21} &= c_{12} \\
c_{22} &= (\cos\Delta\varphi)^2 \\
c_{23} &= -\cos\Delta\varphi\sin\Delta\varphi\sin\alpha \\
c_{31} &= c_{13} \\
c_{32} &= c_{23} \\
c_{33} &= (\sin\Delta\varphi\sin\alpha)^2
\end{aligned}\right\} \tag{6}$$

Where $\Delta\varphi = \varphi - \psi$; φ and α are respectively azimuth and inclination of well trajectory on the bottom hole; γ and ψ are respectively dip angle and up dip azimuth of the formation to be drilled.

It is obviously that the values of G_α and G_ϕ are not only controlled by rock drillability anisotropy of the formation, but also affected by the formation geometry and the well trajectory. Therefore, G_α and G_ϕ can be used to describe the anisotropic drilling characteristics of the formation to be drilled. Thus, the formation force can be mathematically defined as:

$$\begin{cases} GF_\alpha = G_\alpha W_{ob} \\ GF_\phi = G_\phi W_{ob} \end{cases} \tag{7}$$

Where GF_α and GF_ϕ are called as the inclination force (positive for building up the inclination) and the azimuth force (positive for decreasing the azimuth) of the formation respectively, and W_{ob} is weight on bit. It should be pointed out that both GF_α and GF_ϕ are only an equivalent expression of anisotropic drilling characteristics of the formation and they are completely different from the mechanical action forces of the drill bit on the formation. Rock drillability anisotropy of the formation is the internal cause of the generations of GF_α and GF_ϕ , while weight on bit is the its external cause.

2.3 G_α and G_φ of the transversely isotropic formation

By using equations (5) and (6) and making $I_{r1} = I_{r2} = I_r$, equation (3) can be simplified as the following expressions of G_α and G_φ for the transversely isotropic formation:

$$G_\alpha = \frac{(1-I_r)(\cos\alpha\sin\gamma\cos\Delta\varphi - \sin\alpha\cos\gamma)(\cos\alpha\cos\gamma + \sin\alpha\sin\gamma\cos\Delta\varphi)}{I_r + (1-I_r)\left[(\sin\gamma\sin\Delta\varphi)^2 + (\cos\alpha\sin\gamma\cos\Delta\varphi - \sin\alpha\cos\gamma)^2\right]} \tag{8}$$

$$G_\varphi = \frac{(1-I_r)\sin\gamma\sin\Delta\varphi(\cos\alpha\cos\gamma + \sin\alpha\sin\gamma\cos\Delta\varphi)}{I_r + (1-I_r)\left[(\sin\gamma\sin\Delta\varphi)^2 + (\cos\alpha\sin\gamma\cos\Delta\varphi - \sin\alpha\cos\gamma)^2\right]} \tag{9}$$

Where all the symbols here express the same meanings as the previous ones.

3. Experiments on rock anisotropy

Evaluation of rock drillability anisotropy is necessary for hole deviation control in drilling engineering. Many efforts have been made to evaluate rock drillability of the formation through the core testing, the inverse calculation and the acoustic wave. Proposed in this section is an alternative solution by using the acoustic wave to evaluate rock drillability anisotropy of the formation. First, a correlation between the P-wave velocity anisotropy coefficient and the rock drillability anisotropy index of the formation which are calculated according to the core testing data in laboratory, is established by means of mathematical statistics. Then, a mathematical model is obtained for predicting the rock drillability anisotropy index by using the P-wave velocity anisotropy coefficient. Thus, rock drillability anisotropy of the formation can be evaluated conveniently by using the well logging or seismic data (Gao & Pan, 2006).

3.1 Rock drillability anisotropy

3.1.1 Definition

The transversely isotropic formation is a typical anisotropic formation, whose anisotropy can be expressed by a rock drillability anisotropy index:

$$I_r = \frac{D_h}{D_v} \tag{10}$$

where $D_v = V_v/F_v$ and $D_h = V_h/F_h$ are respectively rock drillability parameters in the directions perpendicular and parallel to the bedding plane of the transversely isotropic formation; V_v & F_v and V_h & F_h are the corresponding components of the penetration rate & the net applied force of the isotropic bit to the formation.

When the rock drillability is tested in laboratory using the core samples, the weight on the bit and the rotary speed are constant so that rock drillability anisotropy index of the transversely isotropic formation can also be expressed as:

$$I_r = \frac{T_v}{T_h} \tag{11}$$

where T_v and T_h are two parameters representing the drilling time (seconds) in directions perpendicular and parallel to bedding plane of the core samples respectively. The standard definition of rock drillability can be expressed by the following equation (Yin, 1989):

$$K_d = \log_2 T \tag{12}$$

where K_d is the rock drillability and T the drilling time. Taking two sides of equation (11) into logarithm to the base 2, we can obtain the following equations:

$$\log_2 I_r = \log_2 T_v - \log_2 T_h = K_{dv} - K_{dh} = -\Delta K_d \tag{13}$$

$$I_r = 2^{-\Delta K_d} \tag{14}$$

3.1.2 Rock samples

Fourteen core samples used in laboratory came from the measured depth interval of 48m~1027 m of the well KZ-1 for scientific drilling in China, which were supplied by the Engineering Center for Chinese Continental Scientific Drilling (CCSD). In the directions perpendicular and parallel to the bedding plane, these core samples were cut into shapes of cube or cuboid and their surfaces of both ends were polished and kept parallel to each other, with an error of less than 0.2 mm. Then, the machined samples were put into an oven with a temperature of 105-110ºC and roasted for 24 h. Finally, all of the samples can be used for the testing of rock drillability after cooling down to room temperature.

3.1.3 Testing method

The rock drillability can be measured with a device for testing the rock drillability (shown in Fig.2). During the measurement, some weight is applied on the micro-bit by the function of a hydraulic pressure tank with the fixed poises, so that the weight on the micro-bit is kept at a constant value. The measured depth to be drilled to is set with the standard indicator, and the drilling time is logged with a stopwatch. Both the roller bit (bit of this kind has three rotating cones and each cone will rotate on its own axis during drilling) drillability and the PDC (the acronym of Polycrystalline Diamond Compact) bit drillability can be tested with the above-mentioned instrument, which is of the following standard data.

The diameter of the micro-bit is 31.75 mm.

Weight is 90±20 N on the roller bit and 500±20 N on the PDC bit.

The rotary speed is 55±1 r/min.

The total depth to be drilled to is 2.6 mm for the roller bit with a pre-drilled depth of 0.2 mm and 4 mm for the PDC bit with a pre-drilled depth of 1.0 mm.

During testing the rock drillability, the micro-bit is often checked so that each of the worn micro-bits should be replaced in time to ensure the testing accuracy. The testing points of drilling time for each tested side of a rock sample should be gained as many as possible and their average value is taken as the test value of the side. The grade value of each side drillability of the rock sample can be calculated by equation (16) with the test data of drilling time for each side of the rock sample.

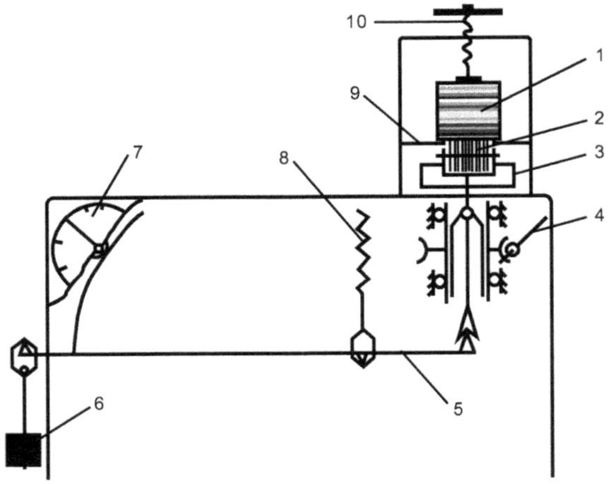

Fig. 2. Testing device for rock drillability(Note: 1. Rock sample; 2. micro-bit; 3. cutting tray; 4. turbine rod; 5. lever; 6. weight; 7. meter for measuring depth; 8. bar with thread for adjusting lever; 9. worktable; 10. compaction bar with thread)

3.1.4 Experimental result and analysis

Some testing results of rock drillability for the 14 core samples from CCSD are obtained in laboratory and shown in Table 1 and Table 2.

No. of the cores from CCSD	Measured depth, m	Rock drillability with the roller bit (K_{dRB})		Rock drillability anisotropy index
		Perpendicular to the bedding plane	Parallel to the bedding plane	
9	48	6.03	6.12	0.94
38	145	9.18	9.86	0.62
57	197	10.29	10.79	0.71
104	305	11.11	11.39	0.82
143	400	8.21	8.17	1.03
179	504	8.70	8.99	0.82
218	607	9.25	10.29	0.49
252	698	10.64	10.78	0.91
281	775	8.78	9.21	0.74
288	795	10.17	8.22	3.86
304	834	7.92	8.57	0.64
340	925	8.89	9.08	0.88
363	998	9.15	10.20	0.48
373	1027	8.12	10.21	0.23

Table 1. Experimental results of rock drillability with the roller bit

No. of the cores from CCSD	Measured depth, m	Rock drillability with the PDC bit (K_{dPDC})		Rock drillability anisotropy index
		Perpendicular to the bedding plane	Parallel to the bedding plane	
9	48	4.55	4.34	1.16
38	145	8.57	10.65	0.24
57	197	9.89	10.06	0.89
104	305	10.78	10.90	0.92
143	400	7.82	7.40	1.34
179	504	8.48	8.47	1.01
218	607	8.61	9.14	0.69
252	698	9.52	9.86	0.79
281	775	8.22	9.03	0.57
288	795	9.55	7.31	4.72
304	834	6.06	7.71	0.32
340	925	8.14	8.95	0.57
363	998	8.18	8.64	0.73
373	1027	7.93	8.77	0.56

Table 2. Experimental results of rock drillability with the PDC bit

It is observed clearly from Table 1 and Table 2 that the rock samples from CCSD have the anisotropic characteristics in the rock drillability. The rock drillability perpendicular to the bedding plan is different from that parallel to the bedding plane, whether it is for the roller bit or for the PDC bit. For the roller bit, indices of drillability anisotropy of the rock samples are ranged from 0.23 to 0.94, except the anisotropy indices of rock samples of 143# and 288#, which are 1.03 and 3.86 respectively. The case is similar to the PDC bit; indices of drillability anisotropy of the rock samples are between 0.24 and 0.92, except the anisotropy indices of rock samples of 9#, 143#, 179# and 288#, corresponding to 1.16, 1.34, 1.01 and 4.72, respectively. Generally, the rock drillability perpendicular to the bedding plan is less than that parallel to the bedding plane, so that the formation can be penetrated more easily in the direction perpendicular to the bedding plane.

3.2 Acoustic anisotropy of rock sample

3.2.1 Definition

It is supposed that the formation is the transversely isotropic, and thus the acoustic anisotropy of the formation rock can be expressed by an acoustic anisotropy index (I_v):

$$I_v = V_{av} / V_{ah} \tag{15}$$

where V_{av} and V_{ah} are the acoustic velocities in rock along the directions perpendicular and parallel to the bedding plane of the formation respectively.

3.2.2 Testing method

With the method of making the ultrasonic pulse penetrating through a rock sample, the acoustic velocities V_{av} and V_{ah} can be measured in laboratory. The ultrasonic testing system used in laboratory is shown in Fig. 3, in which the ultrasonic transducers can provide a frequency of 0.5 MHz and the butter and honey can be used as its coupling media. The pulse generator can generate electric pulses with a strength range of 1-300 V. The width and iteration frequency of the electric pulse can be adjusted and controlled. During testing, the signal generator makes an electric pulse signal which will touch off the emission end of the energy exchanger to generate ultrasonic pulses. The ultrasonic pulses (acoustic waves) propagating through the rock sample are incepted by the reception end of the energy exchanger. Finally, the propagation time and the signal strength of the ultrasonic pulses (acoustic waves) through the rock sample are logged by a digital memory oscillograph.

In order to reduce the errors from the artificial operations, the emission end of the energy exchanger is aimed at its reception end as accurately as possible during testing. Before each test, the ultrasonic testing system should be calibrated using the aluminum rod to ensure the accuracy of the test results. Testing for each point of a rock sample is conducted for three times in the actual testing. The average value of the test data of three times for each point is taken as a final test result for the point of a rock sample. With the test data, the acoustic velocity may be calculated by the following equation:

$$V = \frac{l}{t - t_0} \tag{16}$$

where V is the acoustic velocity; l is length of the rock sample, mm; t is propagation time of the acoustic wave, μs; and t_0 is delayed time of the testing system, μs.

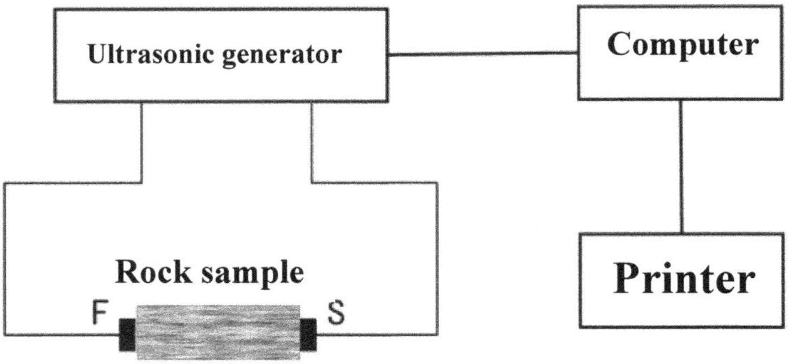

Fig. 3. The ultrasonic testing system

3.2.3 Experimental result and analysis

Some ultrasonic test results of the 14 core samples from CCSD are logged with the above test method and with the ultrasonic testing system in laboratory, and the rock acoustic velocities shown in Table 3 can be calculated by equation (16).

It can be obviously observed from Table 3 that the rock samples from CCSD are of the rock acoustic anisotropy. The rock acoustic velocity perpendicular to the bedding plan is different from that parallel to the bedding plane. Based on the acoustic velocity data in Table 3, the acoustic anisotropy of the rock samples can be calculated by equation (15). The acoustic anisotropy indices of the rock samples are between 0.85 and 0.98, except the 363# and 373#, which are 0.77 and 0.76 respectively. For the test of the rock samples from CCSD, the rock acoustic velocity perpendicular to the bedding plan is less than that parallel to the bedding plane, as shown in Table 3. The main reason for this difference is that there are many fractures with different scales in the rock sample. When the acoustic wave penetrates through the fractures, the fractures cause a loss of the pulse energy so as to make the acoustic velocity reduce more quickly, on the other hand, the pulse energy is dissipated in the process of propagation. According to some progress in geophysics (Patrick & Richard, 1984), the fractures can play a role in guiding the wave when the elastic wave has propagated in the direction parallel to the bedding plane of the rock sample, and play a role in obstructing the wave when the elastic wave has propagated in the direction perpendicular to the bedding plane. Therefore, the propagation of the acoustic wave penetrating through the rock sample is probably controlled by such a kind of geophysical mechanism.

No. of the core from CCSD	Measured depth, m	P-wave velocities of the rock samples, m/s		Acoustic anisotropy index of the rock sample
		Perpendicular to bedding plane (V_{av})	Parallel to bedding plane (V_{ah})	
9	48	4387	4457	0.98
38	145	4442	5120	0.87
57	197	6365	6826	0.93
104	305	5431	5714	0.95
143	400	4410	4928	0.89
179	504	4568	4744	0.96
218	607	4177	4671	0.89
252	698	5805	5990	0.97
281	775	4776	5516	0.87
288	795	5142	5453	0.94
304	834	4392	5129	0.86
340	925	4325	5096	0.85
363	998	3816	4928	0.77
373	1027	3761	4930	0.76

Table 3. Experimental results of acoustic velocities of the rock samples

3.3 Correlations between I_r and I_v

With the experimental data in table 1 to table 3 and the corresponding calculations, it can be found that the rock drillability anisotropy is inherently related to the acoustic anisotropy of the rock samples. Therefore, exponential function, logarithmic function, polynomial function, and linear function, are used to make a regression analysis of the data obtained by experiments. With the matching & extrapolating effects of these regression functions comprehensively considered, exponential function is finally selected as the regression model of correlation between I_r and I_v. The results of regression calculations for the correlations are listed in Table 4.

In Table 4, I_{rRB} is denoted as the drillability anisotropy index of the rock sample with a roller bit, I_{vp} as the acoustic anisotropy index of P-wave through the rock sample, ΔK_{dRB} as the rock drillability difference between both directions perpendicular and parallel to the bedding plane of the rock sample with a roller bit, calculated by equation (13), and I_{rPDC} as the drillability anisotropy index of the rock sample with a PDC bit.

Regression functions		$I_{rRB} = e^{(a+bI_{vp})}$	$\Delta K_{dRB} = e^{(a+bI_{vp})}$	$I_{rPDC} = e^{(a+bI_{vp})}$
a	Value	−3.418	8.129	−4.286
	Standard error	0.840	1.893	1.491
	t-ratio	−4.0679	4.295	−2.875
	Prob(t)	0.00226	0.00157	0.02068
b	Value	3.401	-9.993	4.366
	Standard error	0.914	2.361	1.661
	t-ratio	3.721	−4.232	2.628
	Prob.(t)	0.00397	0.00174	0.03026
Correlation coefficient R		0.793	0.832	0.710
F-ratio		16.92	22.41	8.09
Prob.(F)		0.0021	0.0008	0.0217

Table 4. Results of the regression calculations

4. Evaluation method based on acoustic wave information

Many studies have been made to evaluate the rock drillability anisotropy with the core testing method (Gao & Pan, 2006) and the inversion method (Gao et al, 1994). However, as for the core testing method, its result may not reflect the actual rock drillability anisotropy since the experimental conditions are different from the downhole conditions. Moreover, the profile of rock drillability anisotropy along the hole depth can not be established because of the limitation of the core samples. The inversion method needs to work with a bottom hole assembly (BHA) analysis program and some parameters in the inversion model are not easy to obtain so that its applications are limited to some extent. Thus, the evaluation method will be presented in this section so as to predict rock drillability anisotropy of the formation by using the acoustic wave information (Gao et al, 2008).

4.1 Acoustic wave velocity of the formation

The formation studied here is the transversely isotropic formation which is frequently encountered in drilling for oil & gas. Experimental investigation shows that layered rock has the transversely isotropic characteristics.

4.1.1 Phase velocity in the transversely isotropic formation

For the transversely isotropic media, Hooke's law can be written as

$$\begin{pmatrix} \sigma_{xx} \\ \sigma_{yy} \\ \sigma_{zz} \\ \sigma_{yz} \\ \sigma_{xz} \\ \sigma_{xy} \end{pmatrix} = \begin{pmatrix} C_{11} & C_{11}-2C_{66} & C_{13} & 0 & 0 & 0 \\ C_{11}-2C_{66} & C_{11} & C_{13} & 0 & 0 & 0 \\ C_{13} & C_{13} & C_{33} & 0 & 0 & 0 \\ 0 & 0 & 0 & C_{44} & 0 & 0 \\ 0 & 0 & 0 & 0 & C_{44} & 0 \\ 0 & 0 & 0 & 0 & 0 & C_{66} \end{pmatrix} \begin{pmatrix} \varepsilon_{xx} \\ \varepsilon_{yy} \\ \varepsilon_{zz} \\ \varepsilon_{yz} \\ \varepsilon_{xz} \\ \varepsilon_{xy} \end{pmatrix} \tag{17}$$

Elastodynamic equation of the elastic media can be obtained from the textbook and expressed as

$$\begin{cases} \dfrac{\partial \sigma_{xx}}{\partial x} + \dfrac{\partial \sigma_{yx}}{\partial y} + \dfrac{\partial \sigma_{zx}}{\partial z} + X - \rho \dfrac{\partial^2 u}{\partial t^2} = 0 \\ \dfrac{\partial \sigma_{yy}}{\partial y} + \dfrac{\partial \sigma_{zy}}{\partial z} + \dfrac{\partial \sigma_{xy}}{\partial x} + Y - \rho \dfrac{\partial^2 v}{\partial t^2} = 0 \\ \dfrac{\partial \sigma_{zz}}{\partial z} + \dfrac{\partial \sigma_{xz}}{\partial x} + \dfrac{\partial \sigma_{yz}}{\partial y} + Z - \rho \dfrac{\partial^2 w}{\partial t^2} = 0 \end{cases} \tag{18}$$

where X, Y, and Z are respectively the body force in directions of x, y and z (Xu, 2011). u, v and w are the corresponding displacements. ρ is the density of the elastic media, g/cm^3. Substituting equation (17) into equation (18) and solving with geometric equations without considering body force, we can get the following wave equation :

$$\begin{cases} \rho \dfrac{\partial^2 u}{\partial t^2} = C_{11} \dfrac{\partial^2 u}{\partial x^2} + C_{66} \dfrac{\partial^2 u}{\partial y^2} + C_{44} \dfrac{\partial^2 u}{\partial z^2} + (C_{11} - C_{66}) \dfrac{\partial^2 v}{\partial x \partial y} + (C_{13} + C_{44}) \dfrac{\partial^2 w}{\partial x \partial z} \\ \rho \dfrac{\partial^2 v}{\partial t^2} = C_{66} \dfrac{\partial^2 v}{\partial x^2} + C_{22} \dfrac{\partial^2 v}{\partial y^2} + C_{44} \dfrac{\partial^2 v}{\partial z^2} + (C_{11} - C_{66}) \dfrac{\partial^2 u}{\partial x \partial y} + (C_{13} + C_{44}) \dfrac{\partial^2 w}{\partial y \partial z} \\ \rho \dfrac{\partial^2 w}{\partial t^2} = C_{44} \dfrac{\partial^2 w}{\partial x^2} + C_{44} \dfrac{\partial^2 w}{\partial y^2} + C_{33} \dfrac{\partial^2 w}{\partial z^2} + (C_{13} + C_{44}) \dfrac{\partial^2 u}{\partial x \partial z} + (C_{13} + C_{44}) \dfrac{\partial^2 v}{\partial y \partial z} \end{cases} \tag{19}$$

Because of the symmetry of the stress and strain in the direction normal to z direction, the wave equation can be simplified to two dimensions without any loss of generality. In the plane of y=0(that is the xz plane), the wave equation (19) can be written as

$$\begin{cases} \rho \dfrac{\partial^2 u}{\partial t^2} = C_{11} \dfrac{\partial^2 u}{\partial x^2} + C_{44} \dfrac{\partial^2 u}{\partial z^2} + (C_{13} + C_{44}) \dfrac{\partial^2 w}{\partial x \partial z} \\ \rho \dfrac{\partial^2 v}{\partial t^2} = C_{66} \dfrac{\partial^2 v}{\partial x^2} + C_{44} \dfrac{\partial^2 v}{\partial z^2} \\ \rho \dfrac{\partial^2 w}{\partial t^2} = C_{44} \dfrac{\partial^2 w}{\partial x^2} + C_{33} \dfrac{\partial^2 w}{\partial z^2} + (C_{13} + C_{44}) \dfrac{\partial^2 u}{\partial x \partial z} \end{cases} \tag{20}$$

The solutions of equation (20) are

$$\rho v_{Pa}^2(\theta)=\frac{1}{2}[C_{33}+C_{44}+(C_{11}-C_{33})\sin^2\theta+D(\theta)] \tag{21}$$

$$\rho v_{SVa}^2(\theta)=\frac{1}{2}[C_{33}+C_{44}+(C_{11}-C_{33})\sin^2\theta-D(\theta)] \tag{22}$$

$$\rho v_{SHa}^2(\theta)=C_{66}\sin^2\theta+C_{44}\cos^2\theta \tag{23}$$

Where

$$D(\theta)=\{(C_{33}-C_{44})^2+2[2(C_{13}+C_{44})^2-(C_{33}-C_{44})(C_{11}+C_{33}-2C_{44})]\sin^2\theta+ \\ +[(C_{11}+C_{33}-2C_{44})^2-4(C_{13}+C_{44})^2]\sin^4\theta\}^{1/2},$$

where v_{Pa} is phase velocity of the P-wave; v_{SVa} is phase velocity of P-SV wave; v_{SHa} is phase velocity of SH-wave; θ is phase angle which is the angle between the wave front normal and the unique (vertical) axis as shown in Fig.4.

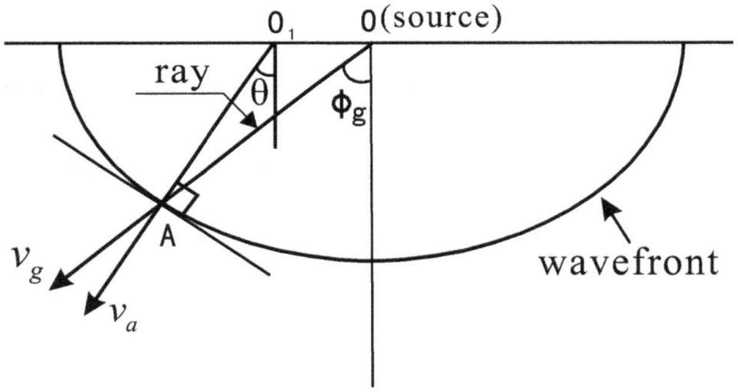

Fig. 4. Phase angle and group angle

It is defined that

$$\varepsilon=\frac{C_{11}-C_{33}}{2C_{33}} \tag{24}$$

$$\gamma=\frac{C_{66}-C_{44}}{2C_{44}} \tag{25}$$

$$\delta^*=\frac{1}{2C_{33}^2}[2(C_{13}+C_{44})^2-(C_{33}-C_{44})(C_{11}+C_{33}-2C_{44})] \tag{26}$$

and

$$\alpha_0=\sqrt{C_{33}/\rho}\ ,\ \beta_0=\sqrt{C_{44}/\rho} \tag{27}$$

where α_0 is the vertical P-wave velocity; β_0 is the vertical SV-wave velocity; ρ is rock density. ε , γ and δ^* are rock anisotropy parameters of the formation.
Substituting equation (24), (25), (26) and (27) into equation (21), (22) and (23), we can get

$$v_{Pa}^2(\theta) = \alpha_0^2[1+\varepsilon\sin^2\theta + D^*(\theta)], \tag{28}$$

$$v_{SVa}^2(\theta) = \beta_0^2[1+\frac{\alpha_0^2}{\beta_0^2}\varepsilon\sin^2\theta - \frac{\alpha_0^2}{\beta_0^2}D^*(\theta)] \tag{29}$$

$$v_{SHa}^2(\theta) = \beta_0^2[1+2\gamma\sin^2\theta] \tag{30}$$

where

$$D^*(\theta) = \frac{1}{2}(1-\frac{\beta_0^2}{\alpha_0^2})\{[1+\frac{4\delta^*}{(1-\beta_0^2/\alpha_0^2)^2}\sin^2\theta\cos^2\theta + \frac{4(1-\beta_0^2/\alpha_0^2+\varepsilon)\varepsilon}{(1-\beta_0^2/\alpha_0^2)^2}\sin^4\theta]^{1/2} - 1\} \tag{31}$$

Letting $\theta = \frac{\pi}{2}$ and substituting it into equations (28), (29), (30) and (31), we can get

$$\begin{cases} v_{Pa,90} = \alpha_0(1+2\varepsilon)^{0.5} \\ v_{SVa,90} = \beta_0 \\ v_{SHa,90} = \beta_0(1+2\gamma)^{0.5} \end{cases} \tag{32}$$

For the case of weak rock anisotropy (i.e. the quantity of ε, γ, θ and δ^* is small), expanding equation (31) in the Taylor series at fixed θ and neglecting the second order of small quantity can be simplified as

$$D^* \approx \frac{\delta^*}{(1-\beta_0^2/\alpha_0^2)}\sin^2\theta\cos^2\theta + \varepsilon\sin^4\theta \tag{33}$$

Using equations (33) in equation (28) and (29), expanding v_{Pa}, v_{SVa} and v_{SHa} in a Taylor series at the fixed θ and neglecting the second order of small quantity can be expressed as

$$v_{Pa}(\theta) = \alpha_0(1+\delta\sin^2\theta\cos^2\theta + \varepsilon\sin^4\theta) \tag{34}$$

$$v_{SVa}(\theta) = \beta_0[1+\frac{\alpha_0^2}{\beta_0^2}(\varepsilon-\delta)\sin^2\theta\cos^2\theta] \tag{35}$$

$$v_{SHa}(\theta) = \beta_0(1+\gamma\sin^2\theta) \tag{36}$$

Where

$$\delta = \frac{1}{2}\left[\varepsilon + \frac{\delta^*}{(1-\beta_0^2/\alpha_0^2)}\right].$$

4.1.2 Phase velocity and group velocity

The phase velocity is the velocity in the direction of the phase propagation vector, normal to the surface of constant phase, which is also called the wave front velocity since it is the propagation velocity of the wave front along the phase vector. The phase angle is formed between the direction of phase vector and the vertical axis. In contrast, the ray vector points always from the source to the considered point on the wave front. The energy propagates along the ray vector with the group velocity, while the group angle is formed between the propagation direction and the vertical axis. The difference between the phase angle and the ray angle is illustrated in Fig.4.

The relationship between the phase angle and the group angle can be expressed by the following equation:

$$\tan(\phi_g - \theta) = \frac{\frac{dv_a}{d\theta}}{v_a} \tag{37}$$

Where ϕ_g is the group angle; v_a is the phase velocity; θ is the phase angle. Expanding equation (37) leads to

$$\tan(\phi_g(\theta)) = (\tan\theta + \frac{1}{v_a}\frac{dv_a}{d\theta}) / (1 - \frac{\tan\theta}{v_a}\frac{dv_a}{d\theta}) \tag{38}$$

The group velocity (v_g) is related to the phase velocity (v_a) as shown by the following formula

$$v_g^2(\phi_g(\theta)) = v_a^2(\theta) + \left(\frac{dv_a}{d\theta}\right)^2 \tag{39}$$

Where v_g is the group velocity.

The following section makes the solution for the relationship between the group velocity and the phase velocity, the group angle and the phase angle of the P-wave at any angle.

Another rock anisotropy parameter δ is introduced and expressed as

$$\delta = \frac{1}{2}\left[\varepsilon + \frac{\delta^*}{(1-\beta_0^2/\alpha_0^2)}\right] = \frac{(C_{13}+C_{44})^2 - (C_{33}-C_{44})^2}{2C_{33}(C_{33}-C_{44})} \tag{40}$$

Letting $t = 1 - \beta_0^2/\alpha_0^2$, equation (38) can be rescaled as

$$\frac{v_{Pa}^2(\theta)}{\alpha_0^2} = 1 + \varepsilon\sin^2\theta + D(\theta) \tag{41}$$

Where

$$D(\theta) = \frac{1}{2}\sqrt{4(\varepsilon^2 + 2t\varepsilon - 2t\delta)\sin^4\theta + 4t(2\delta-\varepsilon)\sin^2\theta + t^2} - \frac{1}{2}t.$$

Making derivation to both sides of equation (41) , we can get

$$\frac{dv_{Pa}(\theta)}{d\theta}=\frac{\alpha_0^2\sin\theta\cos\theta}{v_{Pa}(\theta)R(\theta)}\left[2(\varepsilon^2+2\varepsilon t-2t\delta)\sin^2\theta+2t\delta-\varepsilon t+\varepsilon R(\theta)\right] \tag{42}$$

Where

$$R(\theta)=\left(4(\varepsilon^2+2t\varepsilon-2t\delta)\sin^4\theta+4t(2\delta-\varepsilon)\sin^2\theta+t^2\right)^{\frac{1}{2}}=2D(\theta)+t\,.$$

Substituting equations (41) and (42) into equation (38), we can get

$$\tan\phi_g=\frac{\left\{2\left[M_3(\theta)-M_2(\theta)-2M_1\right]\sin^2\theta-M_4(\theta)-2M_3(\theta)\right\}\tan\theta}{2\left[M_3(\theta)-M_2(\theta)\right]\sin^2\theta-M_4(\theta)} \tag{43}$$

Where

$$M_1=\varepsilon^2-2t\delta+2t\varepsilon\,;$$

$$M_2(\theta)=4t\delta+\varepsilon R(\theta)-2t\varepsilon\,;$$

$$M_3(\theta)=2t\delta+\varepsilon R(\theta)-t\varepsilon\,;$$

$$M_4(\theta)=t^2-tR(\theta)+2R(\theta)\,.$$

Substituting equation (41) and (42) into equation (39), we can get the following equation:

$$v_{Pg}\left(\phi_g(\theta)\right)=\frac{1}{v_{Pa}(\theta)R(\theta)}\left\{v_{Pa}^4(\theta)R^2(\theta)+\alpha_0^4\sin^2\theta\cos^2\theta\left[2M_1\sin^2\theta+M_3(\theta)\right]^2\right\}^{\frac{1}{2}} \tag{44}$$

Where v_{Pg} is the group velocity of the P-wave; v_{Pa} is the phase velocity of the P-wave.
In the case of $\theta=0^\circ$ and $\theta=90^\circ$, $\frac{dv_{Pa}(\theta)}{d\theta}=0$, the phase velocity is equal to the group velocity.Thus, when the group angle of the P-wave at the considered point of the formation is given, its phase angle can be calculated by using equation (43). Phase velocity and group velocity of the point can also be made out by using equation (41) and (44).

4.1.3 Methodology for determining rock anisotropy parameters

From the above discussion, we know that if rock anisotropy parameters δ, ε and γ, are known, the wave velocity at any direction can be calculated by using acoustic wave velocity perpendicular to the bedding of the formation. In other words, the acoustic wave velocity perpendicular to the bedding of the formation can be make out if the wave velocity at any direction and rock anisotropy parameters, δ, ε and γ, are known.
It is assumed that the formation to be drilled is transversely isotropic with symmetry axis perpendicular to the bedding of the formation and the formation properties do not change significantly from one well to another. Acoustic wave logging provides a way to measure the velocity of P-wave or S-wave in the formation (or slowness time). The schematic figure for measuring the velocity of P-wave or S-wave is illustrated in Fig.5.

In the figure 5, S_1 and S_2 are monopole sonic transducers. R_1, R_2, R_3 and R_4 are sonic receivers. When the sonic is transmitted from S_1, time difference between R_2 and R_4 is recorded. In the same way, when the sonic is transmitted from S_2, time difference between R_1 and R_3 is recorded. The average of time difference between R_2 and R_4 and time difference between R_1 and R_3 is the velocity in the formation measured.

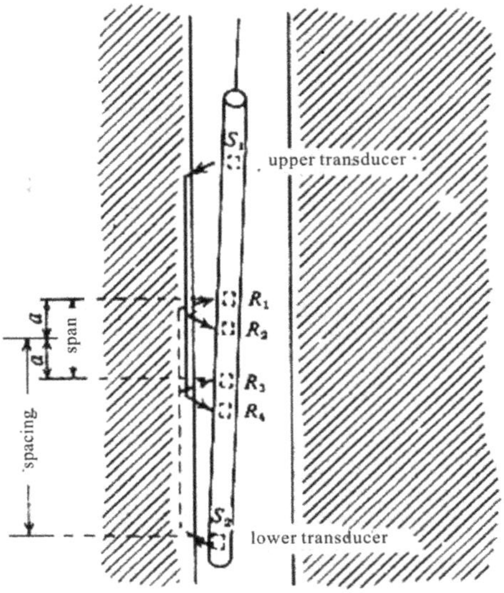

Fig. 5. The principle of acoustic wave logging

Since available S-wave velocity is limited in logging data, we restrict ourselves to take consideration of the P-wave only. The frequency of the acoustic wave logging is about 20kHz~25kHz , which has long wave length. Since a monopole sonic transducer has a mini-bulk, a monopole borehole sonic tool may be approximated by a point source in line with an array of point receivers. The group velocity surface is the response from a point source and so the monopole sonic tool response is approximated as a point source coupled with a series of point receivers in an infinite media (neglecting borehole effects). Therefore, we measure group velocity with borehole sonic tools.

Three parameters, the vertical P-wave velocity (α_0) and the anisotropy parameters ε and δ can be recovered using borehole sonic measurement at different angles relative to the axis of symmetry by following objective function:

$$\Delta v_P = \frac{1}{n}\sum_{i=1}^{n}\left[v_{Pmi}(\theta) - v_{Pci}(\theta)\right]^2 \tag{45}$$

where $v_{Pmi}(\theta)$ is the measured P-wave velocity; $v_{Pci}(\theta)$ is the P-wave velocity calculated by equation (32); n is the total number of the measured signals. The goal of the inversion is to find the optimization value of C_{11}, C_{13}, C_{33} and C_{44}, to minimize value of Δv_P , by which α_0, δ and ε can be calculated, as shown in Fig.6 (Gao et al, 2008).

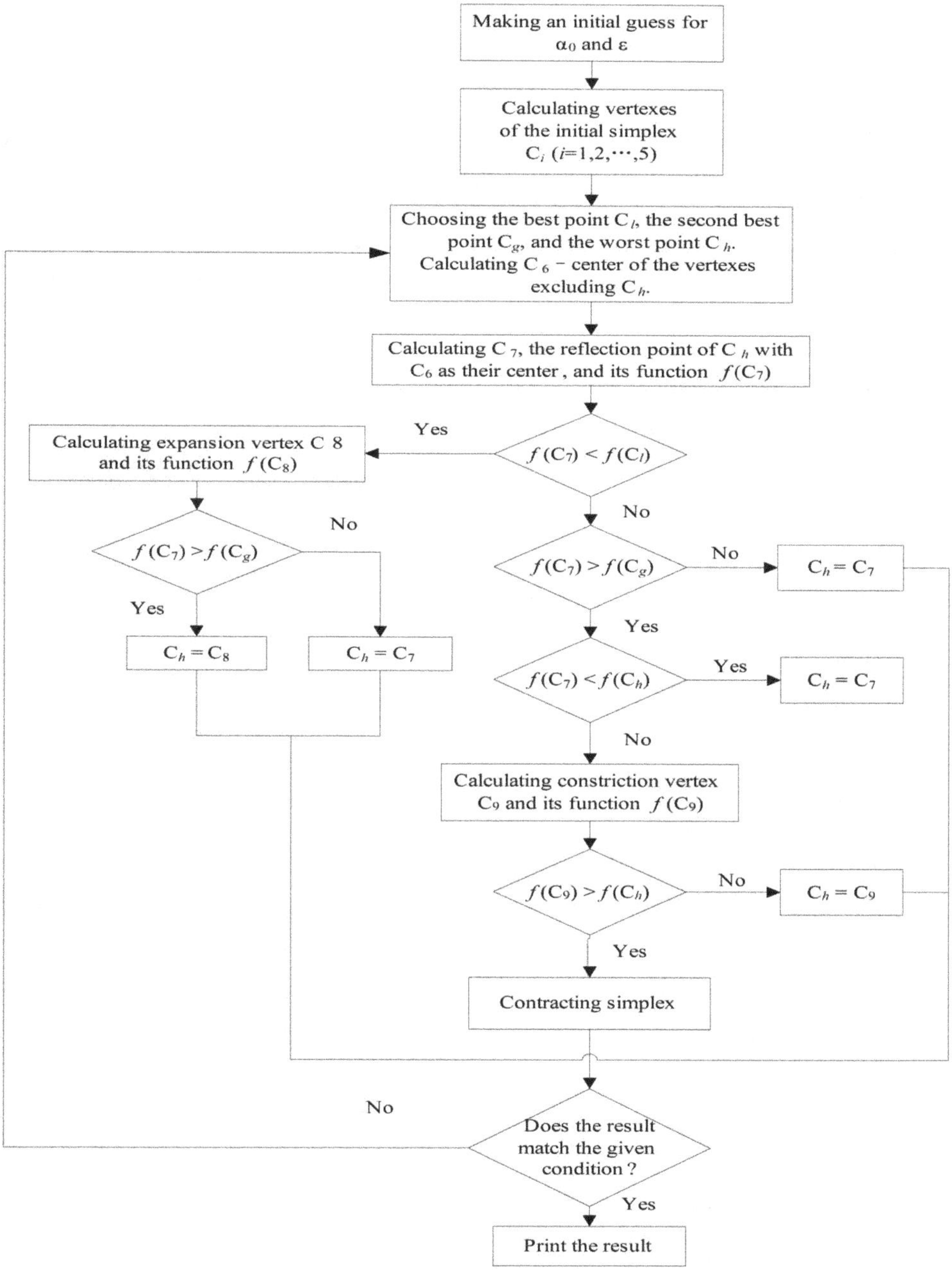

Fig. 6. Flow chart for the inversion calculations of rock anisotropy parameters

4.2 Prediction model of rock drillability anisotropy

Based on the previous section 3, a calculation model has been established to predict rock drillability anisotropy of the formation:

$$I_r = 2^{C_1 K_{dv} + C_2} \tag{46}$$

$$K_{dv} = C_3 + C_4 \ln(\Delta t) \tag{47}$$

Where K_{dv} is the rock drillability perpendicular to the bedding plane of the formation; Δt is the time interval of acoustic wave in the same direction, us/m; C_j (j=1,2,3,4) are the regression coefficients based on the experimental data and the survey data in drilling engineering. For example, by the regression analysis based on some oilfield data in west China, we can get such coefficients as C_1=0.05246, C_2=-0.76732, C_3=32.977, C_4=-4.950.

4.3 Evaluation method of rock drillability anisotropy based on acoustic wave

From equations (46) and (47), it is shown that the key point for the evaluation of rock drillability anisotropy is how to obtain the rock drillability perpendicular to the bedding plane of the formation which depends on the time interval of acoustic wave in the same direction. Thus, the evaluation of rock drillability anisotropy comes down to determine the time interval of acoustic wave perpendicular to the bedding plane of the formation.

Provided that the formation is of the transversely isotropy and has the symmetry axis perpendicular to the bedding plane of the formation, the angle between hole axis and the formation normal can be calculated by the following formula which is derived from transformation of the formation coordinates to the bottom hole coordinate.

$$\omega = \arccos\left[\cos\alpha\cos\beta - \sin\alpha\sin\beta\cos(\phi - \phi_f)\right] \tag{48}$$

where ω is the angle between hole axis and normal of the formation; α is hole inclination, degree or radian; ϕ is azimuth, degree or radian; β is stratigraphic dip, degree or radian; ϕ_f is azimuth of the formation tendency, degree or radian.

When rock anisotropy parameters of a hole section is known, its acoustic wave velocity perpendicular to the bedding plane of the formation can be calculated by the following procedures:

1. Calculating group angle according to stratigraphic dip angle & up dip direction, and inclination & azimuth of hole.
2. Making an initial guess for the acoustic wave velocity $v_{P,0}$.
3. Reading shear wave velocity from shear wave logging or calculating it by equation (46).
4. Calculating phase angle by equation (43).
5. Calculating phase velocity and group velocity of the P-wave by equation (41) and equation (44), respectively.
6. Comparing the P-wave group velocity with the measured velocity.
7. If group velocity of the P-wave matches the measured velocity, $v_{P,0}$ is what we find. Otherwise, we should repeat step 2 to step 7 until they are matched.

The flow chart for inversion of the acoustic wave velocity perpendicular to the bedding plane of the formation is shown in figure 7.

The rock drillability can be calculated by equation (47) after obtaining the time interval of the acoustic wave perpendicular to the bedding plane of the formation. Thus, the profile of rock drillability anisotropy index can be established by using equation (46).

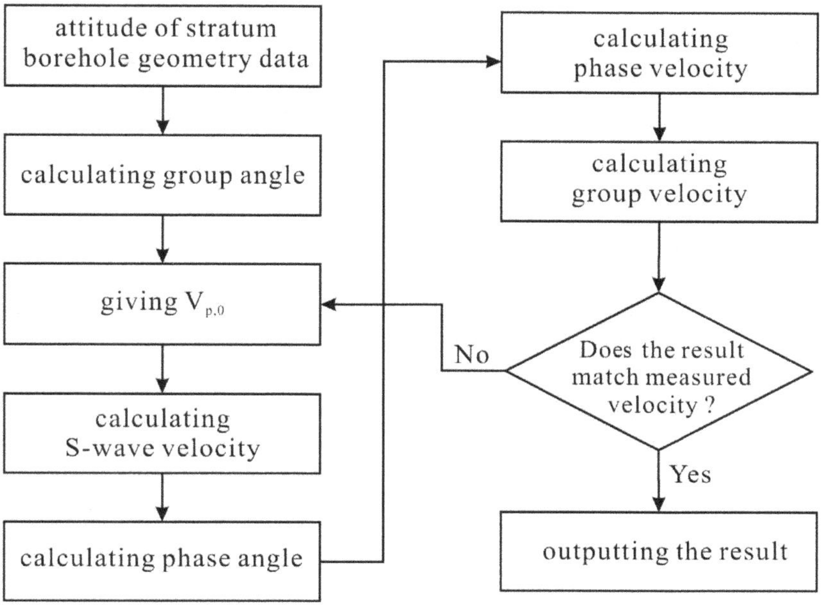

Fig. 7. Inversion method of the acoustic wave velocity perpendicular to the bedding plane

5. Case study

Based on some well logging data and drilling information from Qinghai oilfield in west China, the case study is presented in this section to verify the evaluation method for the anisotropic drilling characteristics of the formation to a certain extent.

Based on these data in table 5, rock drillability anisotropy of the fromation and its anisotropic drilling characteristics can be calculated by using the evaluation method described above. The inversion result of shale anisotropy parameters is shown in table 6.

Well Number	Well logging information
Well 5	gamma-ray, compensated acoustic wave and compensated density, inclinometer data and geologic stratification data
Well 6	gamma-ray, compensated acoustic wave and compensated density, inclinometer data and geologic stratification data
Well 7	gamma-ray, compensated acoustic wave and compensated density, inclinometer data and geologic stratification data, and some other records

Table 5. Well drilling & logging information from some completed wells at the Honggouzi conformation in Qinghai oilfield

$v_{P,0}$ (m/s)	ε	δ	root-mean-square (m/s)
4029.1	1.6833	1.6098	98.6153

Table 6. The inversion result of shale anisotropy parameters for the Honggouzi conformation

From the data in the table 6, we can see that the shale is of strong rock anisotropy. The acoustic wave front of the shale section is shown in figure 8.

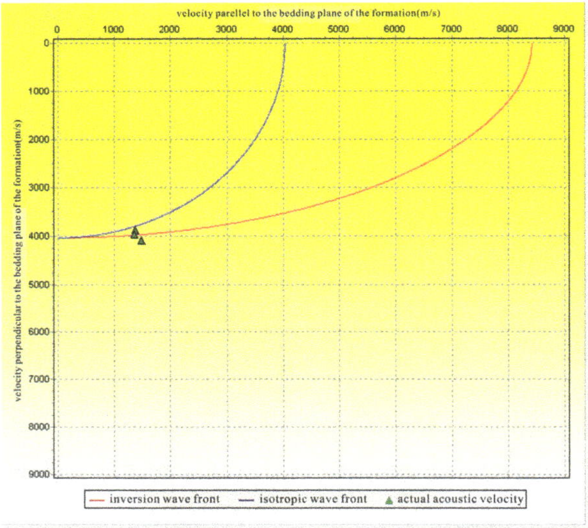

Fig. 8. Acoustic wave front of the shale section of the Honggouzi conformation

Based on the wave front of the shale shown in Fig.8, the rock anisotropy parameters of the formation at any measured depth, ε and δ, can be approximately calculated by the following equations:

$$\left.\begin{aligned} &\varepsilon = V_{sh}\varepsilon_c \\ &\delta = V_{sh}\delta_c \\ &V_{sh} = (2^{\Delta GR \cdot G_{cur}} - 1)/(2^{G_{cur}} - 1) \\ &\Delta GR = (GR - GR_{min})/(GR_{max} - GR_{min}) \end{aligned}\right\} \quad (49)$$

where ε and δ are rock anisotropy parameters of the formation at any measured depth; ε_c and δ_c are rock anisotropy parameters of the shale section; V_{sh} is the shale content, %; GR is gamma ray value; GR_{max} is the maximum value of gamma ray; GR_{min} is the minimum value of gamma ray; G_{cur} is the Hilchie index whose value is 3.7 for the Neogene Stratigraphy and 2 for old strata.

After obtaining the rock anisotropy parameters(ε and δ), we can calculate the acoustic wave perpendicular to the bedding plane of the formation by using the inversion method shown in Fig.7 and the well logging data of acoustic wave. Thus, the rock drillability anisotropy of the formation can be calculated by equations (46) & (47), and the corresponding anisotropic drilling characteristics can be evaluated by equations (8) & (9), as shown in figure 9.

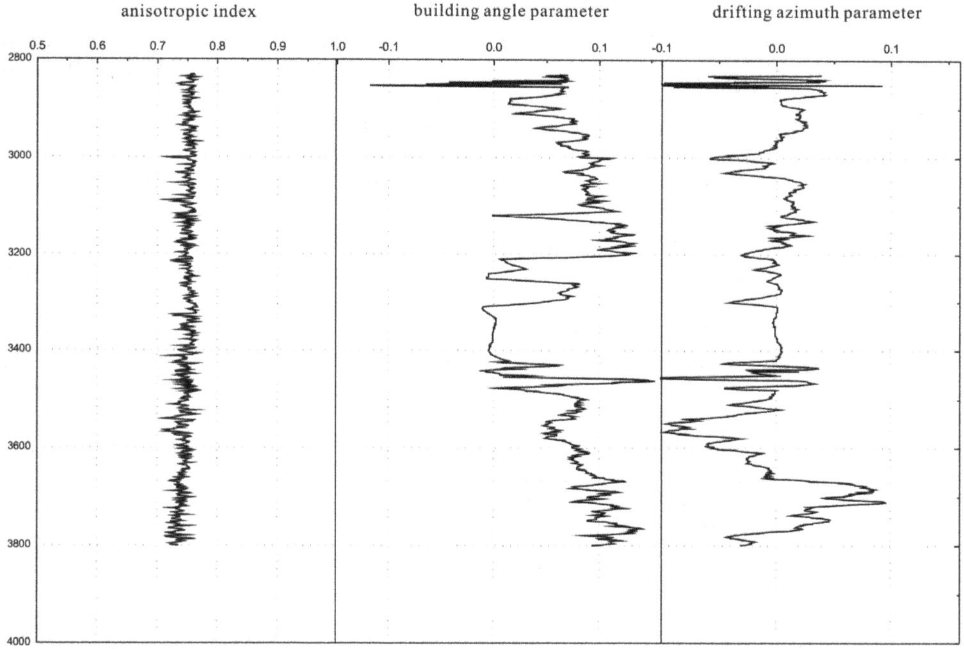

Fig. 9. The evaluation results of anisotropic drilling characteristics of the formation at the Honggouzi conformation in Qinghai oilfield

6. Conclusion

The orthotropic formation and the transversely isotropic formation are the typical formations encountered frequently in drilling engineering. Based on rock-bit interaction model, the two parameter equations have been derived for us to calculate the anisotropic drilling characteristics of them as soon as rock drillability anisotropy of the formations is evaluated quantificationally by using the oilfield data.

The correlation between rock drillability anisotropy and acoustic wave anisotropy of the formation can be matched to each other by an exponential function which is of the best extrapolative performance and relativity. Coefficients in the model are various for different formations to be drilled.

To a certain extent, the research results presented here have shown a new way for us to evaluate conveniently rock drillability anisotropy of the formation by using well logging or seismic data. Case study shows that this evaluation method is better for applications of rock drillability anisotropy of the formation in drilling engineering.

7. Acknowledgment

The authors are grateful for the financial support from the national research project (Grant No. 2010CB226703) and the supply of many core samples by CCSD.

8. References

Gao, D.; Dai, D. & Pan, Q. (2008). Evaluation of deflecting characteristics of anisotropy formation by using well-log information. *ACTA PETROLEI SINICA*, Vol.29, No.6, (December 2008), pp. 927-932, ISSN 0253-2697

Gao, D. & Pan, Q.(2006). Experimental study of rock drill-ability anisotropy by acoustic velocity. *Petroleum Science*, Vol. 3, No.1, (March 2006), pp.50-55, ISSN 1672-5107

Gao, D. (1995). Predicting and scanning of wellbore trajectory in horizontal well using advanced model. *Proceedings of the Fifth International Conference on Petroleum Engineering Held in Beijing*, China, SPE 29982, pp.297-308, 14-17 Nov. 1995

Gao, D.; Liu, X. & Xu B.(Dec. 1994). *Prediction and Control of Well Trajectory*, Petroleum University Press, ISBN 7-5636-0584-3/TE·95, Dongying, China

Gao, D. & Liu, X.(1990). Anisotropic drilling characteristics of the typical formations. *Journal of the University of petroleum, China*, Vol.14, No.5, (Oct. 1990), pp.1-8, ISSN 1000-7393

Gao, D. & Liu, X.(1989). A new model of rock-bit interaction. *Oil Drilling and Production Technology*, Vol.11, No.5, (Oct. 1989), pp.23-28, ISSN 1000-7393

Lubinski, A. & Woods,H.(1953). Factors affecting the angle of inclination and dog-legging in rotary bore holes. DPP, 1953: 222-242

Patrick, J. & Richard L.(1984). An experimental test of P-wave anisotropy in stratified media. *Geophysics*, Vol.49,No.4, (1984),pp. 374-378

Xu, B.(Jan. 2011). *Concise Elasticity and Plasticity*, Higher Education Press, ISBN 978-7-04-030725-2, Beijing, China

Yin H.(1989). Study of formation anisotropy-rock drillability. *Oil Drilling and Production Technology*, Vol.11, No.1, (Feb. 1989),pp.15-22, ISSN 1000-7393

10

Low Frequency Acoustic Devices for Viscoelastic Complex Media Characterization

Georges Nassar
Université de Lille –Nord de France
France

1. Introduction

The evolution in consumer expectations in terms of quality and safety in the agro-industry has led to the need to develop new methods of investigating product quality and the processes involved. Many fields of production still rely too much on the know-how of the operators who, with their experience acquired over time, have become key players in the company. In addition, the manufacturing of quality food products frequently relies on artisanal know-how that is difficult to industrialise and often synonymous of high production losses, therefore prohibitive costs. In contrast, so as to limit costs, the industrial production process is often associated with poorer quality. The objective evaluation of product quality involves the development of methods and sensors adapted to the product or the manufacturing process.

Indeed, beneath an apparent simplicity, agro-industry products have complex physical properties linked to elasticity, viscosity and plasticity. One of the major difficulties lies in the complexity of the processes which depend on numerous physical parameters. The matter is subjected to numerous mechanical, thermal or chemical treatments thus migrating towards viscoelastic or even plastic properties that are more difficult to quantify.

The originality of the approach adopted consists in the study and set up of an ultrasonic measuring device associated with its electronic environment in order to reply to a specific need due to the complexity of the physico-chemical phenomena involved. A global approach to this problem is very tricky as the physical properties of the media evolve significantly throughout processing.

We thus focused on the development of sensors and methods of characterisation dedicated to different phases of the industrial processes. Two very closely linked aspects were therefore studied targeting product characterisation and process control.

Work has been carried out to develop acoustic and ultrasonic instrumentation designed to monitor the change in state of the matter (liquid-gel transition and product cohesion), then to monitor the evolution of its elastic properties. The process control applications concern the development of a very low frequency, non-destructive monitoring method to reply to the specificities of the physical properties of the matter.

In this document, we report the scientific approach highlighting the design of the ultrasonic sources which dispenses with classic design through the choice of specific resonance modes for the sensors. Their design aims at promoting low frequency resonance in a relatively small scale composite structure. This sensor technology was adapted according to the frequency chosen to study the change in physical state of the media and to monitor the evolution of the acoustic properties of products that are often heterogeneous.

Several approaches were used to optimise this technology: an analytical approach to determine the sensor's first vibratory mode which was consolidated by a numerical study, then confirmed and validated experimentally.

The application is based on two points:

- The physical or physico-chemical phenomenon linked to the transition phase and the sol-gel transition in the dairy field; the opaqueness and the fragility of this type of gel justifies the importance of quantifying the metrological parameters such as measurement accuracy, stability over time and mechanics of the sensor implemented.
- The interaction of the sensor with its environment in a process. A bivariate study (sensor/propagation medium) was carried out in order to select the required geometry to ensure that the sensor is adapted to its environment: loaded metallic plates subjected to mechanical stress and heated to a temperature of around 100°C (plate heat exchanger) or strongly absorbent media (fermenting bread dough...). This part of the study consisted in finding a good compromise between the geometry of the sensor, its location in the overall system and the required sensitivity.

2. Context

Many biochemical industrial activities involve very complex physicochemical phenomena which enable products to be processed. These products often go through a variety of states during processing.

Processing uses energy from chemical reactions (e.g. enzymatic), thermal or mechanical energy, or even a combination of these different forms of energy. Physical modifications can also occur (incorporation of air in the matter…). All these forms of energy are often combined within the same process and it is difficult to quantify the contribution of each in the product processing phenomena.

Due to the complexity of the processes (several processing stages, multiplicity of forms of energy…) and the products (viscoelastic matter, visco-elasto-plastic…), associated with the legitimate concern of not interfering with the process, few measurements have been carried out during the various stages of processing. The temperature, pressure and flow are often monitored during processing even though they are not necessarily correlated to the desired properties of the product under development.

This is why we chose to develop acoustic sensors adapted to the constraints imposed by either the product or the process.

Among the essential parameters sought-after, rheological measurements are often determining in terms of the consumers' perception of the qualities of the end product such as texture, viscosity, elasticity.... Several techniques of investigation exist but the majority are laboratory applications and are difficult to adapt for in-line controls. The difficulty thus arises of using multiple techniques to obtain a more complete characterization of the process and the interpretation of the data obtained with these analysis techniques.

Furthermore, the quality control of the processed products also involves evaluating the performance of the process. The temperature, pressure and flow are of course part of the characteristics measured for the control and/or closed-loop control of the various stages of product processing. However, in the case of complex processes they are difficult to correlate to the final properties of the product.

The processes can also evolve over time. This is the case with heat exchangers for which the performance varies over time due to fouling. Only preventive maintenance leading to additional production costs can ensure stable performances of the process over time. The development of sensors integrated in the process to provide information on the evolution of the performance remains essential.

This work presents a selection of studies which have led to the development of low frequency acoustic sensors specifically adapted to monitor changes in the physical state of complex media and the process: fragile gel, highly heterogeneous or highly absorbent media, media with complex rheological behaviour...

Several cases were studied:

- A low frequency acoustic sensor adapted to the characterization of complex products using an omni-directional source in the case of media undergoing a change in physical state;
  - continuous homogeneous medium: sol-gel transition,
  - complex heterogeneous medium: transition from a suspension of particles in a liquid to a cohesive visco-elasto-plastic solid.
- A very low frequency acoustic sensor used to monitor the response of a medium subjected to mechanical vibrations. Such technology is designed to study the processing phenomena of a highly absorbent product such as bread dough during fermentation.

Finally, the identification of the needs and constraints imposed by certain environments (temperature, hygiene, attenuation...) have led to the combination of these types of technology to monitor a process (e.g. fouling of plate heat exchangers, search for an optimum point in the kneading phase...). By taking into account the coupling of the sensor with its environment this technique can, in certain cases, exploit the noise emitted by the process itself, as in kneading for example.

In this work, we chose to illustrate the potential of low frequency acoustic methods on applications from the agri-foodstuffs sector. These same states can also be found in the pharmaceutical and cosmetics industries, the aviation industry, the medical field as well as in material chemistry.

The methodology implemented can be divided into several phases:

- Analysis of the product and/or process
- Definition and/or optimisation of the appropriate method and a sensor meeting the various constraints :
  - Analytical study
  - Numerical modelling
- Experimental validation of the sensor
- Validation of the application

3. Sensors suitable for studying media of which the physical properties evolve over time

3.1 Monitoring changes in the physical state of the matter

The analysis of the different stages in the formation of macromolecular networks is of major importance, since understanding the structure and properties (physical or chemical) of gels requires the understanding of the process of organization. In many physical, chemical or biological processes, the union of small separate elements to form aggregates of different sizes and further macroscopic phases makes connectivity an essential characteristic of this

type of process. Many models have been proposed to explain the phenomenon of aggregation. The most important ones are those of Flory (Flory, 1953), Stockmayer (Stockmayer, 1943), Case (Case, 1960), Gupta (Gupta et al., 1979), Eichinger (Eichinger, 1981), Allsopp (Allsopp, 1981) and San Biagio (San Biagio et al., 1990). In most cases the phenomenon is described by the classical theory as a "particular case of percolation" and the two-dimensional growth of the network according to Caylay's tree. Other studies including those of De Gennes (De Gennes, 1989) and Stauffer (Stauffer, 1981, 1985) describe the phenomenon of random aggregation and the problems of percolation and gelation. However, the different characteristics of the macromolecular chain-making system can be evaluated according to Clerc (Clerc et al., 1983), using for example a Monte-Carlo simulation, predicting the influence of the characteristics of the starting solution and the gelation conditions on the structure and the arrangement of the masses.
In fact, the gelation process is a transition from an entirely soluble system to a heterogeneous two-phase system: composed of an insoluble entity (infinite-size macromolecule) and a soluble phase. This transition is accompanied by radical changes in some physical properties of the medium. Below the gelation "point", the viscosity of the medium increases and the medium ceases to flow by developing an elasticity.
To study this phenomenon, several physical measurement techniques exist i.e. optical, thermal, rheological and acoustic (Nassar, 1997). However, sampling and sensitivity to a limited range of physical properties are often drawbacks. Consequently, different techniques are required to explore an entire process with the difficulty of bringing together the heterogeneous data provided by these techniques. This is, for example, the case of optical methods which are penalized by the opacity of the substances analyzed as well as the size of the molecules formed in relation to the wavelength. Thermal methods are insensitive to the mechanical characteristics of the medium. The fragility of some gels (milk gel) limits rheological techniques. In many cases, several analytical techniques exist, but they are only used in the laboratory.

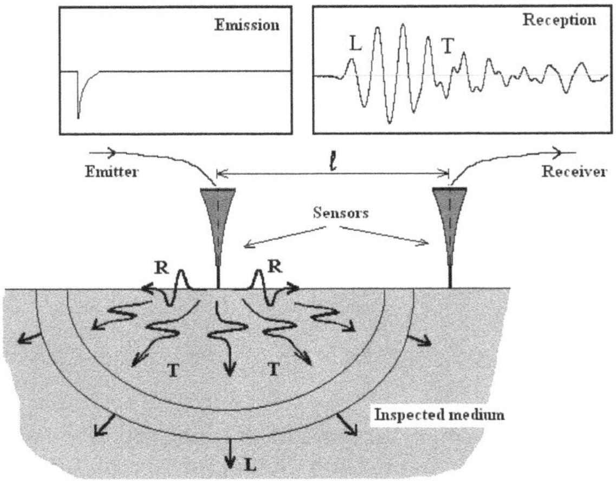

Fig. 1. Basic principle

To develop further instrumentation in order to understand and to quantify the modification process of media in real conditions, a low-frequency ultrasonic technique using sensors with

extremely pointed ends that act as point sources was examined. The application of this technology using two near-field coupled sensors to explore the relationship between the physical properties measured during the evolution of the time of flight of the wave and the structural changes during matter formation (Figure 1) was investigated.
The sensors were near-field coupled through the medium to be characterised. Such disposition privileges the Signal/Noise ratio and avoids the loss of acoustic pressure which is inversely proportional to the ray of the spherical wave. In a metal or ceramic solid all the waves are generated simultaneously, but in the media we are concerned with, the dominating longitudinal wave is the fastest and is relatively simple to exploit.
The advantage of these sensors is that they can be adapted to the measurement configuration envisaged according to the nature of the wave and the appropriate resonance mode.

3.2 Study of a low-frequency ultrasonic device

The aim of the study was to define and develop optimal ultrasonic instrumentation to understand the phenomenon and quantify the viscoelastic properties of changing media.
The usual ultrasonic characterization techniques are generally based on the use of a resonant piezoelectric transducer in thickness mode. As the resonant frequency of a transducer is inversely proportional to its size, it is greater for low frequencies around 100 KHz. Some researchers like Degertekin (Degertekin & Khury-Yakub, 1996a, 1996b, 1996c), Shuyu (Shuyu, 1996, 1997) and Nikolovski (Nikolovski & Royer, 1997) used this physical principle, but associated a tapered volume with the ceramic components to concentrate the mechanical energy.
The aim of this part of the work was to obtain a low frequency acoustic point source to generate a spherical wave in the medium. To do this a different procedure from that traditionally used in classic sensor design was implemented. A new technique was used which consisted in setting in resonance the entire mechanical structure of a reduced-size unit through the contact of an extremely pointed end with the material to be analyzed. In order to behave like an acoustic point source, the size of the point was smaller than the wavelength in the medium.
The first part presents a theoretical analysis of low-frequency ultrasonic resonators, beaming a spherical wave in the medium. The choice of a triangular shaped resonator and its mechanical behavior will be assessed and the study completed by a numerical approach based on the application of the finite elements method to characterize all the resonator vibration modes and visualize the corresponding distortions when the structure is excited. As the analytical results were in good agreement with the numerical results, they were applied to the whole triangular-shaped sensor to validate the findings experimentally. The resonance mode frequencies determined by the numerical calculation were then correlated with the electrical impedance measurements.

3.3 Study and design

3.3.1 Analytical approach

For a possible analytical analysis, the structure of a standard ultrasonic sensor is based on a simple triangle shape (Figure 2).
The propagation of longitudinal waves in the triangular part of the sensor was studied to determine the resonance frequency of the elongation mode and the velocity amplification

ratio between the ends. The analysis is based on an extension of Ensminger's (Ensminger, 1960) theory.
According to figure 2, the x section is written:

$$S = e \cdot \ell(x) \text{ Then } S = e \cdot \ell_1 (x_1 + x)/x_1 \tag{1}$$

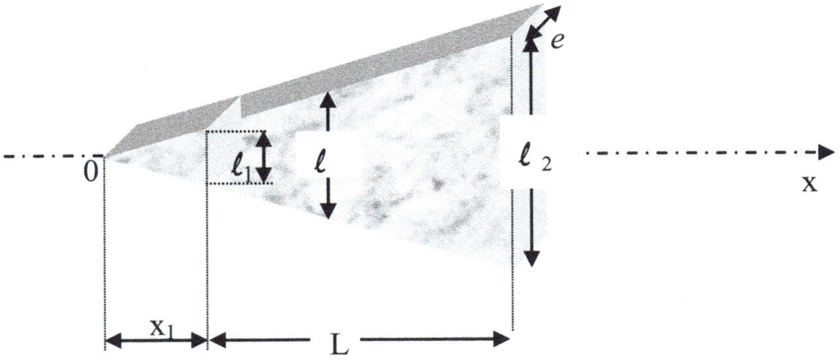

Fig. 2. Basic analytical shape

Ensminger studied the propagation of a wave in extensional mode in a cone with no loss of which the lateral dimensions were short in comparison with the length. In the case of a triangular shape, this equation takes the following form:

$$\frac{\partial^2 v}{\partial x^2} + \frac{1}{(x_1 + x)} \frac{\partial v}{\partial x} + \frac{\omega^2}{c^2} v = 0 \tag{2}$$

Where:
v is the velocity of the particles, ω is the pulsation and c is the velocity of the longitudinal wave in the material making up the vibratory element.
On the basis of the dimensions given in figure 3, the solution to this differential equation leads to an approximate velocity amplification ratio between the two extremities (Nassar, 1997); $|v(0)/v(L)| = 1/0.46 = 2.16$ for a resonance frequency: $f = 60$ KHz.

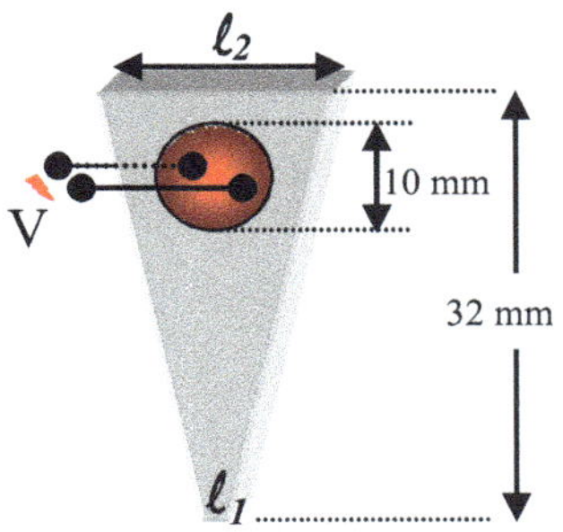

Fig. 3. Basic shape. Triangular sensor of thickness $e = 1$ mm, $L = 32$ mm, $\ell_1 = 2$ mm and $\ell_2 = 16$ mm

3.3.2 Numerical approach

While an analytical study can only take into consideration one particular mode of vibration of the triangular part of the sensor, a numerical study based on the finite elements method can determine all the vibrating modes of these parts as well as those of the realised sensor.
For a real structure; whole sensor included a binding rod and a triangular truncated part (Figure 4), the displacement differential equations were solved with a continuous regime, taking into account the boundary conditions at the surfaces. The materials were defined by Young's modulus E, Poisson's coefficient ν and density ρ. The results presented below were applied without loss and they were compared to the characteristics of the longitudinal mode determined by the analytical calculations. This comparison was also made for the triangular sensor which was studied as a whole.
For our study, the ANSYS analysis software was used. The sensors used were made essentially of piezoelectric material. A source of excitation was engraved in the general structure of the vibrating element (triangle part), providing mechanical continuity without any break (Figure 4a). This type of engraving was considered as it has been demonstrated (Nassar, 1997) that for the same longitudinal mode, the amplification ratio at the ends is **71** times bigger when there is one engraved source providing mechanical continuity with the vibrating element so that one source can impact the structure by gluing (Figure 3b).

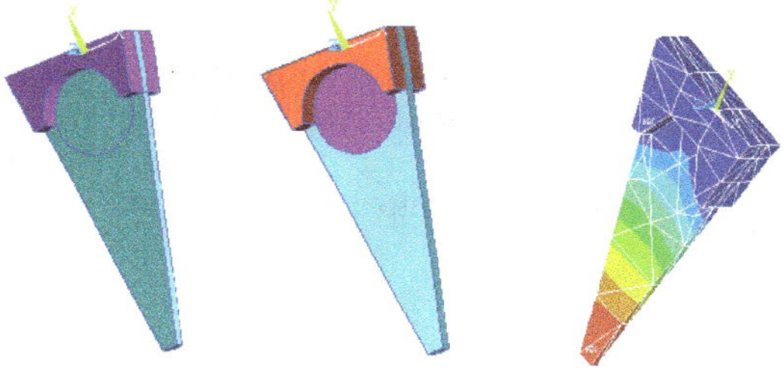

Fig. 4. From left to right: (a) engraved source; (b) embedded source; (c) elongation mode

Table 1 show the resonance frequencies and the vibration velocity transformation ratio (|v(0) /v(L)|) and table 2 present the difference rate of this ratio according to the position of the excitation source

	Longitudinal resonance frequency Hz	Velocity ratio v(0)/v(L) at the extremities
Analytical calculation	60132	2.16
Numerical calculation	**57611**	2.32
Electrical impedance measurements	59 kHz	

Table 1. Comparative study. Analytical and numerical approach

sensors	Embedded source	Engraved source
Frequency (Hz)	57729	**57611**
Standardised velocity ratio	$1.4\ 10^{-2}$	1.00

Table 2. Impact of the nature and the location of the source

The results show a good correlation between the frequencies determined by the calculations and those determined numerically or using impedance measurements.
A significant increase in the amplitude of vibration was observed resulting from the design of the electrode on an active element.

3.4 Application for monitoring changes in state

3.4.1 Pointed sensors for sol-gel transition

The milk gelation can be considered as an aggregation of different sized molecules (Walstra& Vliet, 1986; Fox, 1989; Dalgleish, 1993). This model was explored for several reasons: the available knowledge, the experimental conditions that are known and relatively easy to conduct, the complex medium with the physical properties of liquid and gel states in close contact.
As the reaction progresses, the average mass of each aggregate increases and the number of molecules in the medium tested decreases. The aggregation process results in a giant macromolecule defining the gel.
This process was examined using two identical ultrasonic sensors near-field coupled through the medium to be characterized. The working frequency was 60 kHz.
Figure 5 presents a schematic diagram of the measuring device. The emitter is driven by a sharp electrical pulse lasting 15 μs. These conditions provide a longitudinal vibration mode at the end of the sensor which behaves like a point source. This phenomenon generates a divergent ultrasonic wave in the medium, one part of which was measured using a receiver located at a constant distance from the transmitter by the first the zero-crossing of the wave.
The propagation of the wave in the medium is more or less a compressional wave, as suggested by the time of flight corresponding to a velocity of 1600 m/s in reconstituted milk samples at 25 °C.

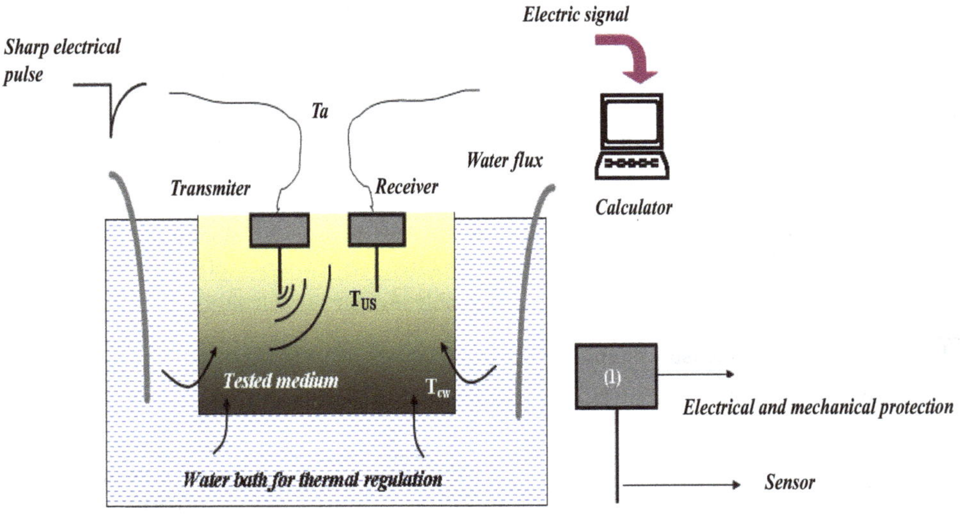

Fig. 5. Diagram of the measuring cell. Tus is the temperature of the product at "sensor level", Tcw is the temperature of the container walls and Ta is the ambient temperature

3.4.2 Ultrasonic monitoring of gelation: measurement of the variation in the time-of-flight of the wave

3.4.2.1 Measurement stability

The time of flight dt(ns) of the signal, measured in distilled temperature-controlled water at 30±0.1°C (reference medium) remains stable. The precision obtained was 1ns over a global reply time of 10µs, given a relative precision of 10^{-4} per measurement at the zero-crossing point

3.4.2.2 Measurement reliability in gelation process

The reliability of a measurement system resides in its reproducibility and its faculty to follow all the stages of the gelation process. In the standard conditions using 12 grams of skimmed milk powder dissolved in 100 ml of distilled water and in accordance with the literature (Noël & al., 1989), the sol-gel transition clotted between 15 and 16 min. At a regulated ambient temperature at 30 °C similar to the one of the medium under test, as shown in figure 6, the time of flight of the signal decreased indicating an increase in the mechanical resistance of the product. This variation had not reached a plateau value, indicating that the medium was still changing.

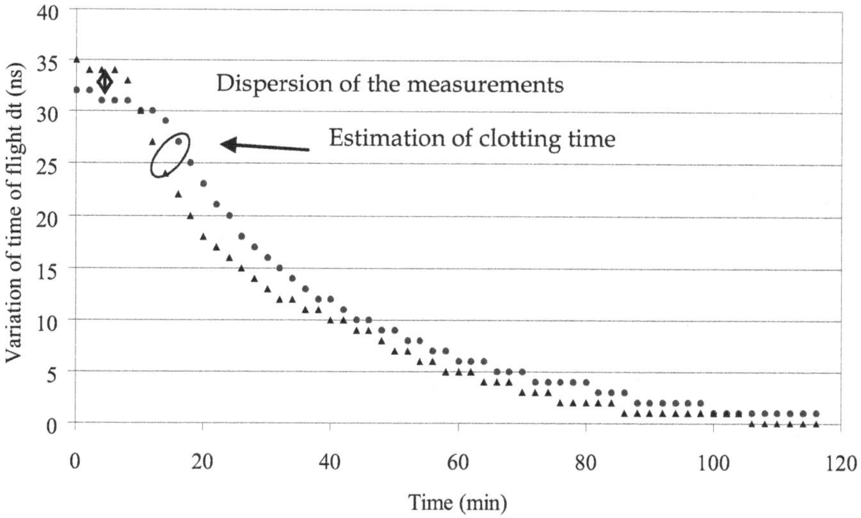

Fig. 6. Typical curves of the gelation process of two media prepared in the same conditions at a reaction temperature of 30.1 °C observed using the ultrasonic technique. The difference between the two curves provides a quantitative estimation of the global dispersion of the ultrasonic measurement due to the electronics and sample preparation

Figure 6 also provides a qualitative estimation of the reproducibility of the ultrasonic measurements. The curves show the progress of the action of the rennet in two media prepared in the same conditions. The maximum dispersion of the measurements was 5 ns due to the electronic parts and the milk reconstitution process.

3.4.2.3 Evolution of the molecular network

Figure 7 presents the variations in the time of flight resulting from the variation of the ultrasonic wave velocity during the milk gelation process at different ambient temperatures.

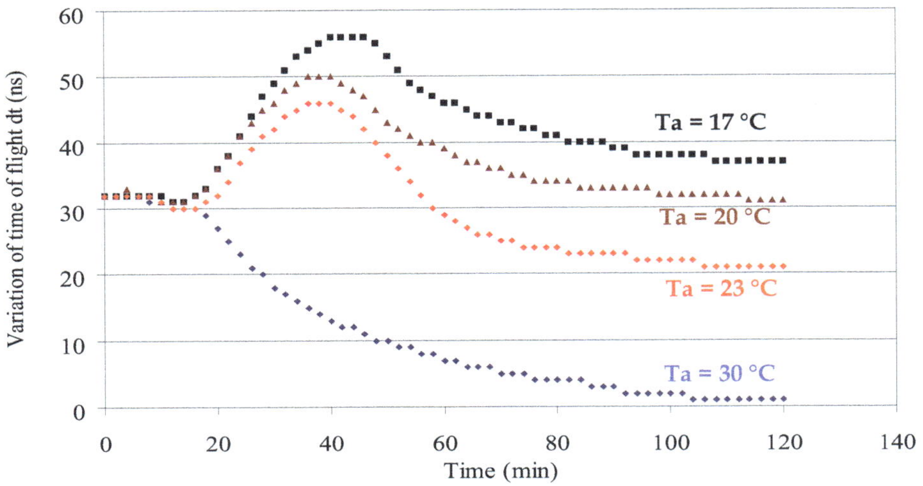

Fig. 7. Ultrasonic sensor responses during milk gelation at 30.1 °C, but at different ambient temperatures (from 17 °C to 30 °C)

The curve observed at an ambient temperature of 30 °C, similar to the temperature of the milk, was similar to those obtained by measurements using other physical methods (McMahon & Brown, 1984). When the ambient temperature was significantly different from the product temperature, ultrasonic curves showed a specific pattern.

According to Dalgleish (Dalgleish, 1982), gelation process was defined like a transition phenomenon in which a soluble suspension made of macromolecules (liquid phase) becomes insoluble when a giant mass forms. We can assume that basic macromolecules are synthesized by linking monomers (building units) via covalent bonds. This chemical reaction is due to the presence of functional groups of the monomers that are able to form chemical bonds with other functional groups of the monomers (Mercier & Marechal, 1993). The network formation occurs if the functionality of the units is greater than two (Mercier & Marechal, 1993). As the reaction progresses, the conversion status of the system is characterized by the connectivity rate p (Flory, 1953; Stockmayer, 1943). Below the gelation threshold, the viscosity of the medium increases as the connectivity rate p approaches the critical advancement rate pc. This phenomenon is known as a critical connectivity transition. Above the threshold, the medium ceases to flow and the gel phase develops some elasticity. This phenomenon introduces structural changes in the physical properties and more particularly in the mechanical behavior of the medium, thus resulting in the transition from a liquid state to a viscoelastic solid state.

To illustrate this process schematically, let us consider an initial solution containing units that can link together. At the beginning, the medium behaves like a viscous solution in a sol phase due to the presence of a single type of finite-size masses: in this case, p is low (Figure

8b). When p increases, bigger and bigger masses are formed (Figure 8c). For a certain critical value of p, pc, a giant chain appears (continuous connectivity of the space from one side to the other $A \leftrightarrow B$; Figure 8d) defining the gel point. Above the threshold pc, the medium in the gel phase has the macroscopic behavior of a viscoelastic solid. For $p = 1$ all the units belong to the giant mass (Figure 8e).

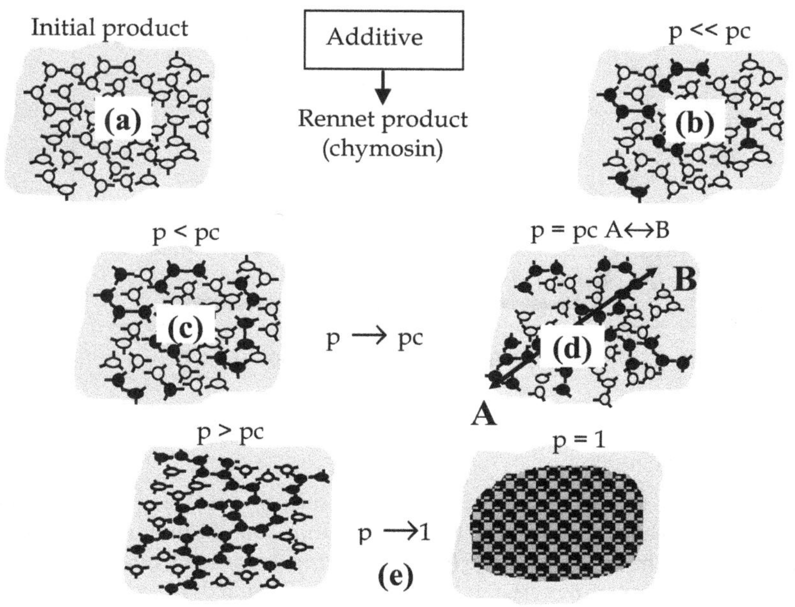

Fig. 8. Schematic representation of the sol-gel transition. a) Initial phase, b) Suspension of molecules of finite sizes, c) Agglomeration and formation of macromolecules of large masses, d) Critical connection phase: p = pc, e) Network continuity connection to give a single giant macromolecule, the gel

As the properties of a gelling medium are proportional to the reaction progress, it is possible to represent the behavior of the viscoelasticity in terms of the connectivity rate p according to the following cases (Figure 8):

- For p < pc, the system is a liquid whose viscosity increases as the gel point approaches.
- For p = pc, elastic behavior appears.
- For p > pc, the medium becomes a solid gel whose elasticity increases with p.

In order to relate the theoretical aspect to the experimental results, the following curve (Figure 9) has been divided into five different stages. The phenomenon describes the gelation process when the ambient temperature was different from the product temperature. It was undetectable when these temperatures were the same.

According to figure 9, "stage (a)" could be interpreted as a proteolytic phase characterized by the appearance of two polypeptides resulting from the effect of the rennet product. This stage is followed by the formation of aggregates of finite size (Figure 8b). This molecular reorganization might be related to the change of slope of the curve (stage b) reflecting a decrease in the time of flight, which means an increase in the velocity in the sol medium.

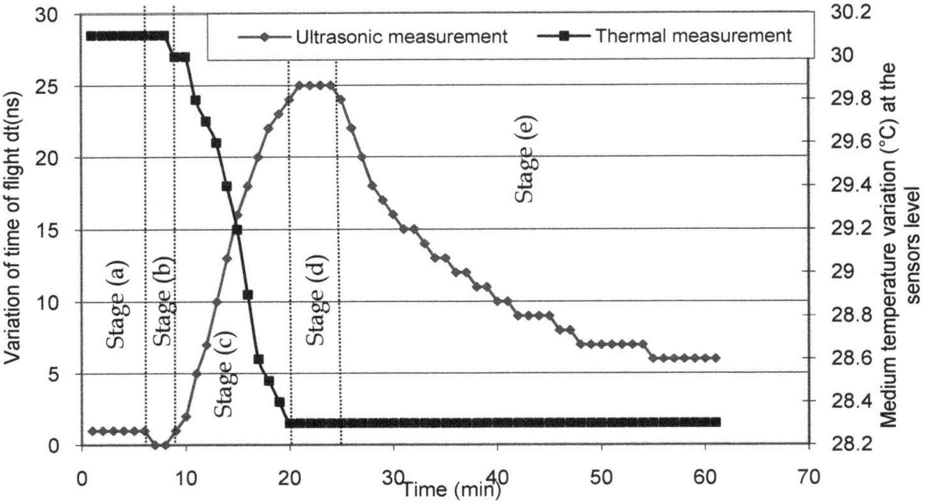

Fig. 9. The specific ultrasonic response during milk gelation when the ambient temperature was different from the temperature of the milk

The temperature inside the medium remained constant during this stage and resulted from the propagation of heat by pure and free convection. The formation of more or less voluminous masses in the medium (stage (c)) induced the transition from a viscous state to a viscoelastic state, slowing down the free convection. This led to a slight temperature decrease in the medium at the level of the sensors (T_{US} in Fig. 4) to reach a new equilibrium where heat was mainly transmitted through the container walls (regulated temperature: T_{cw}°C), by conduction. The changes in the medium during this stage could be interpreted in the following manner:

1. Due to a thermal conduction phenomenon, a slight temperature gradient appears in the medium, between the container walls (T_{cw} = 30.1°C) and the center of the vat, at the location of the measuring point; T_{US}°C (following on the ambient temperature). The temperature decrease induced an increase of the time of flight.
2. The time of flight decreased as the reaction progressed. This decrease can be attributed to the development of an elastic modulus resulting from the formation of macromolecules, changing the medium from a viscous liquid state to a viscoelastic solid state. The phenomenon was expressed physically by the evolution of the connectivity rate p towards its critical value pc .

These two phenomena make "stage (c)" a competition between :

1. An increase of the time of flight resulting from a decrease in temperature.
2. A decrease in the time of flight resulting from the appearance of an elastic component in the changing medium.

The connectivity rate p, reached a critical value pc in "stage (d)", at the maximum of the curve when the existence can be assumed of a giant macromolecular chain linking the two extreme sides of the considered space (Figure 8d). During this stage, the mechanical aspect

of the medium could dominate the remaining part of the reaction, due to the weak thermal variation (≈ 2 °C) resulting from the difference in temperature between the medium and the environment (ambient temperature).
During "stage (e), the gel strengthened. The gel was stronger when the ambient temperature was very close to the temperature medium (Figure 7). This phenomenon can be shown experimentally by an increase in Δdt(ns) resulting from a decrease in time of flight, whereas theoretically it was explained by the evolution of the connectivity rate p towards 1 following the establishment of continuous connections of finite size masses on the giant molecule, the gel.

3.4.3 Monitoring the formation dynamics of the cohesion forces in a fractionated medium: measurement of the variation in the wave amplitude

A key step often met in agro-industry processes is the formation of the matrix of the final product. In cheese-making, this phase involves the cohesion of the elements making up the medium. Generally, it is the conversion of the matter from a heterogeneous state (made up of overlapping grains) to a homogenous state. In this particular case, the cohesion of the curd grains, essential step in the process, varies according to the process conditions as well as the enzymatic and bacterial activities in the medium. It is thus necessary to take these into account in the description of the cohesion.

Analysis of the medium during the cohesion process

During draining, very different physical states are involved in the conversion of the curd grains from a heterogeneous medium to a more homogenous medium. It is therefore difficult to describe the interaction between the ultrasonic wave and the curd grains during draining using just one physical model. So, for a better evaluation of the different phases in the processing of the medium:

1. From moulding and for a very short period of time the grains are touching and cohesive links begin to form between the contact surfaces, thus forming a skeleton containing connected porosities through which the whey continues to drain (Figure 10a).
2. When the whey evacuation channels become blocked (Figure 10b), this phase is described by the multilayer model by Brekovskikh (Brekovskikh, 1980). It is equivalent to a material made up of layers of grains and whey of which the thickness is equivalent to the size of the grains as well as whey evacuation channels. The layers of whey become thinner and thinner until they disappear (Figure 10c), producing a homogenous medium (final phase). This approximation is valid insofar as the main signal beam is confined to a narrow area of the medium, which is the case in our measurements. Indeed, the zone of interest is comparable to the size of the grains (Figure 11).

The evolution of the medium throughout the entire draining phase in the mould was described using these two models: the outflow of the whey and the cohesion of the grains.

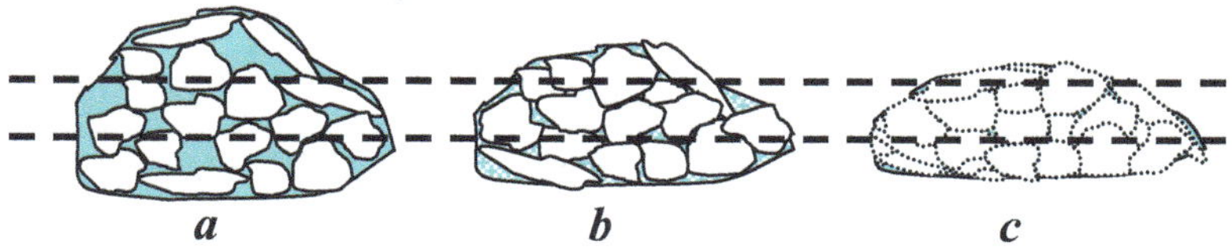

Fig. 10. Evolution of the medium over time

Fig. 11. Estimation of the propagation zone of the main signal beam

Heated to a constant temperature of 35 °C, the experimental mould was instrumented with a transmitter and two ultrasonic receivers spread out so as to integrate the signals transmitted through two paths presenting enough interfaces between the grains. This thus reduced measurement dispersion linked to the random number of interfaces (Figure 12).

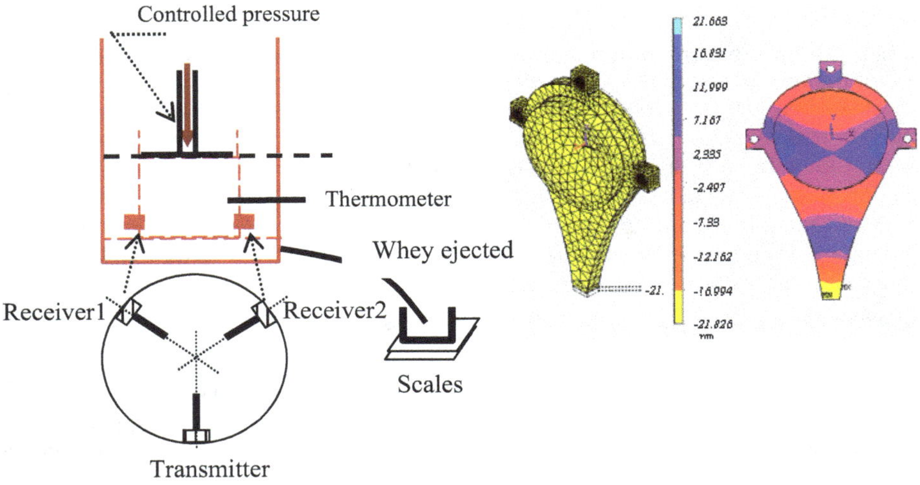

Fig. 12. On the left: The experimental measuring device with a horizontal cross-section of the mould at sensor level and on the right: omni-directional, optimised sensor in elongation mode at 246 kHz

Figure 13 shows that the amplitude of the ultrasonic signal is a parameter that is sensitive to variations in the properties of the medium under investigation.

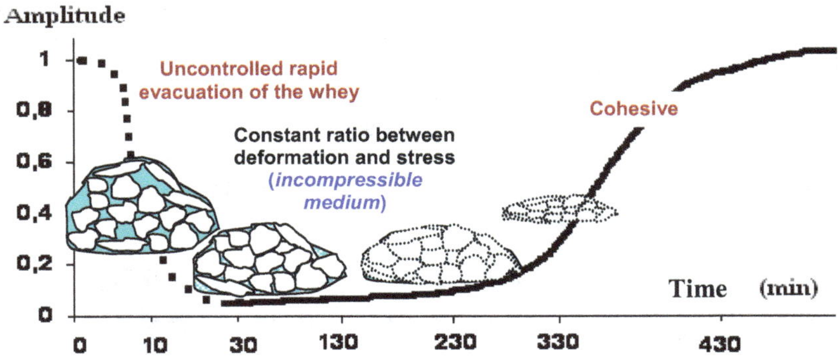

Fig. 13. Typical curve reflecting the cohesion phenomenon as seen by the variation in the ultrasonic amplitude

3.5 Case of highly absorbent matter

The characterisation of media using ultrasounds is often limited by the heterogeneous nature of the matrix which can, in the case of cosmetic, pharmaceutical and agro-food products, be viscoelastic and heterogeneous (foam or emulsion for example). Wave attenuation in such media is mainly due to viscous absorption and scattering from heterogeneities. The higher the frequency the greater the attenuation, hence the necessity to analyse the media using low frequencies in order to characterise the evolving matter.

The search for a compromise between the analysis frequency and the volume of the medium to be characterised led us to propose specific sensor geometries associated with specific excitation conditions.

In order to manage this constraint, a very low frequency acoustic technique was adapted so as to communicate sufficient energy to a particularly absorbent sample. This was achieved by mechanical excitation caused by a shock. An electrical image of this excitation is obtained using a second identical sensor used as the synchronisation reference (Figure 14).

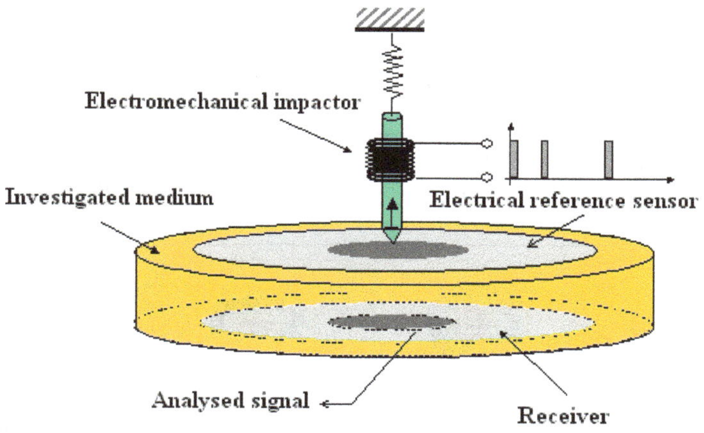

Fig. 14. Schematic diagram

3.5.1 Principle of the sensor proposed

The sensor proposed is illustrated in figure 15. In the shape of a thin disc, its structure is made up of a ring with an embedded piezoelectric disc.

A structure like this offers the advantage of being able to work at resonances lower than those of the piezoelectric disc and thus several resonance modes can be used of which the main ones are flexion and radial modes. This type of sensor also offers a large area of contact with the medium studied, which, in the case of soft and aerated materials, can be advantageous.

The resonance modes of a circular structure, notably those of a disc or a thin ring, have been studied for many years by several authors (Aggarwal, 1952a, 1952b; Moseley, 1960; Vogel & Skinner, 1965; Leissa, 1969; Blevins, 1979; Irie & al., 1984; Lee & Singh, 1994) . The main modes of resonance of a disc are radial modes in the disc plane and flexion modes outside the disc plane (Tables 3, 4 & 5). The tables show a good correlation between the theoretical, numerical and experimental analyses.

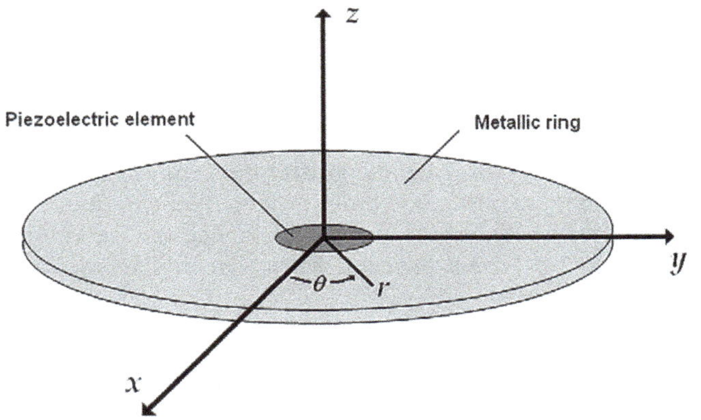

Fig. 15. Set-up of the sensor proposed

Flexion modes	(s, n)	Theoretical (Hz)	FEM (Hz)
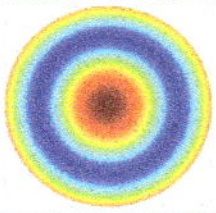	(1.0)	1798	1794
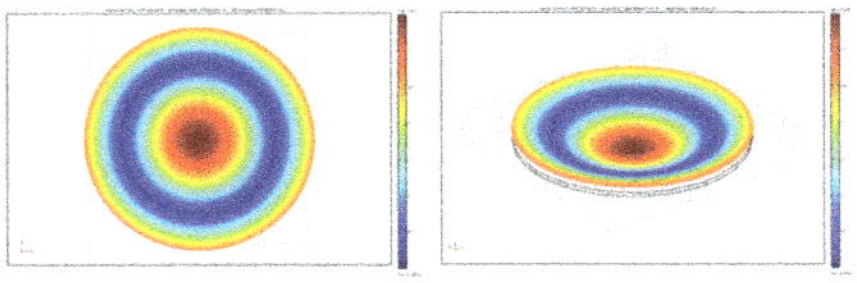			

Table 3. Illustration of the deformations. Amplitude perpendicular to the disc plane Flexion mode (1.0) for a frequency of 1.8 kHz (theoretical and using finite elements)

Radial modes	Analytical calculation (kHz)	FEM (kHz)
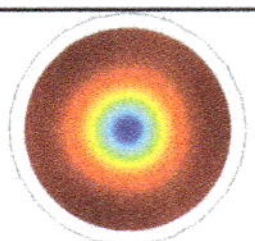	35.5	35.5
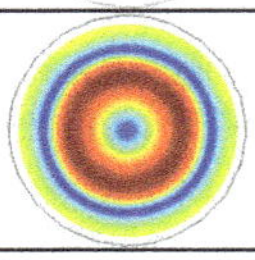	92.6	92.7

Table 4. The resonance frequencies of the first two radial modes for a free aluminium disc (R=5cm; h=2mm) obtained analytically and numerically using finite elements

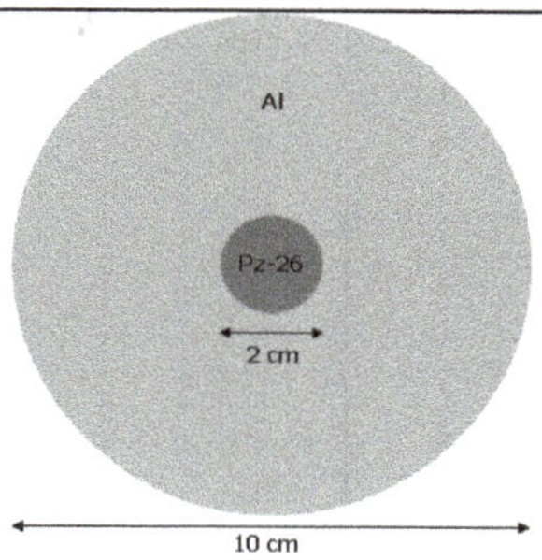 Radial modes	Frequency (kHz)
1	35.9
2	90
Flexion modes	
(1.0)	1.6
(2.0)	6.5

Table 5. Resonance frequencies of the first radial and flexion modes for the composite sensor

3.5.2 Application for monitoring fermenting bread dough

The objective of this application was to establish the links between the product evolution kinetics and the acoustic characteristics measured.

From a practical point of view, impulse excitation was used in this system. The excitation was obtained by a controlled mechanical impact (rod of an electromagnet), thus exciting the disc used for the synchronisation. The vibration induced in the dough is received by a receiver disc identical to that of the synchronisation disc (Figure 16).

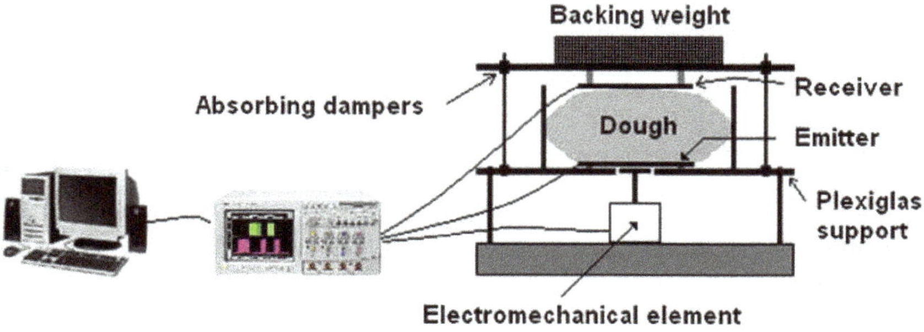

Fig. 16. Experimental measuring device

A metrological study of the measuring device carried out using standard samples (for example a pocket of water at 25°C) showed that the standard deviation of the amplitude and the velocity was approximately 2%. Signal acquisition was carried out over 3 hrs.

3.5.3 Dynamic monitoring of the fermentation process of bread dough

After controlled kneading of the dough, the measurement chamber was placed in an enclosure in order to control the temperature and humidity. The acoustic values studied were the variation of the time-of-flight and the wave amplitude on reception.

Figure 17 shows the variations in these two values. It can be noted that the critical points and phases appear simultaneously on the two curves.

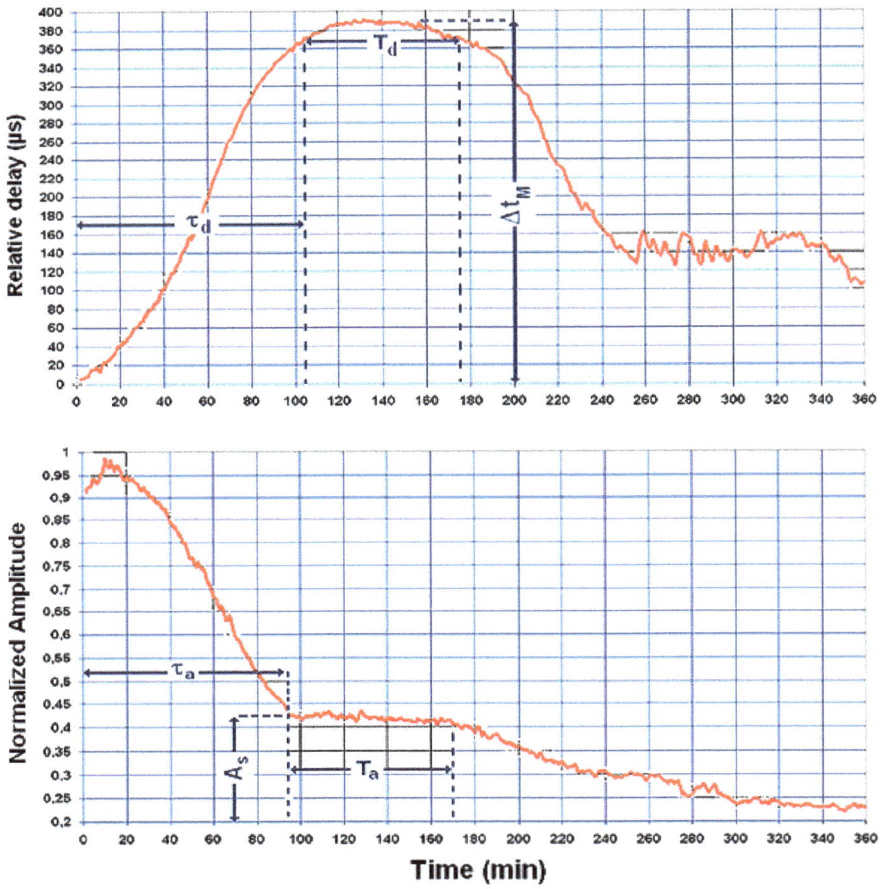

Fig. 17. Evolution of the standardised amplitude and the relative signal delay on reception during the fermentation phase

Where:

- τ_r is the time necessary to reach a relatively stable zone,
- T_r reflects the period of stability during which the relative delay reaches its maximum and remains relatively constant,
- Δt_M is the maximum relative delay. It is linked to the gas fraction contained in the dough and therefore the extensibility of the latter.
- τ_a is the period during which the amplitude of the signal decreases before reaching a plateau,
- T_a is the period of stability of the amplitude,
- A_S is defined as being the amplitude of the signal during the period of stability.

A repeatability study was carried out to estimate the dispersion of the parameters (delay and amplitude). Several tests were performed under the same operating conditions. The standard deviation of the measurements of these parameters was around 3%.
Table 6 summarises the variations in the characteristic parameters observed on the curves according to the evolution in the temperature

	20°	27°	34°
$\tau_r\,(min)$	165	105	60
$T_r\,(min)$	145	70	55
$\Delta t_M\,(\mu s)$	380	385	374
$\tau_a\,(min)$	160	95	55
$T_a\,(min)$	130	75	50
$A_S\,(\%)$	40	43	44

Table 6. Parameters relating to the variation in temperature

It can be noted that the maximum relative delay is relatively constant (approximately 380µs) for the three products made under the same operating conditions. This parameter seems to be independent of the temperature, which is in agreement with the hypothesis that it varies according to the gas fraction contained in the matter and the elastic properties of the matrix.

4. Acoustic sensor for in-line monitoring of a manufacturing process

In certain industrial processes it is often difficult to access useful information in real-time due to the conditions imposed on the mechanical and thermal parameters, pressure, hygiene..., conditions which require a specific installation of the sensor with regard to its environment. The difficulty thus arises of an integration taking into account both the process constraints and the acoustic constraints. This is the case of a plate heat exchanger which can be considered as a typical example in this category (Figure 18).

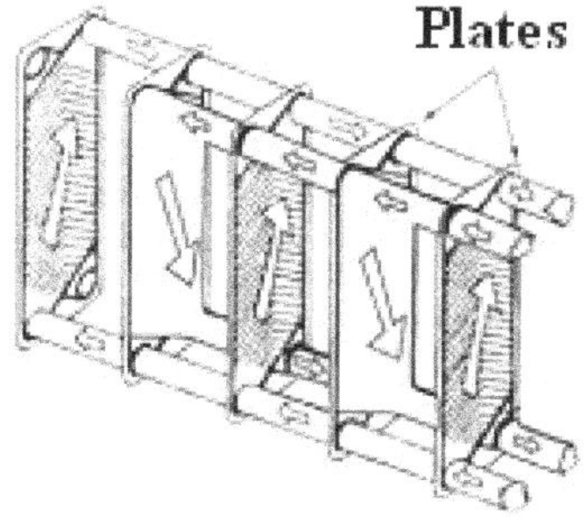

Fig. 18. Standard plate heat exchanger

4.1 Sensor selection criteria

For the exchanger, the sensor selected is not cumbersome and is sensitive over a temperature range reaching over 100°C (Figure 19). The excitation and synchronisation

modes remain the same as the previous case (disc sensor). The principle of the measurement is to excite a vibration mode in one or several plate exchangers and to analyse the evolution under the effect of fouling by measuring the response of the plates using a receiver.
A bivariate system-sensor study enabled the geometry of the latter to be defined over the same vibration frequency range as the system (exchanger).

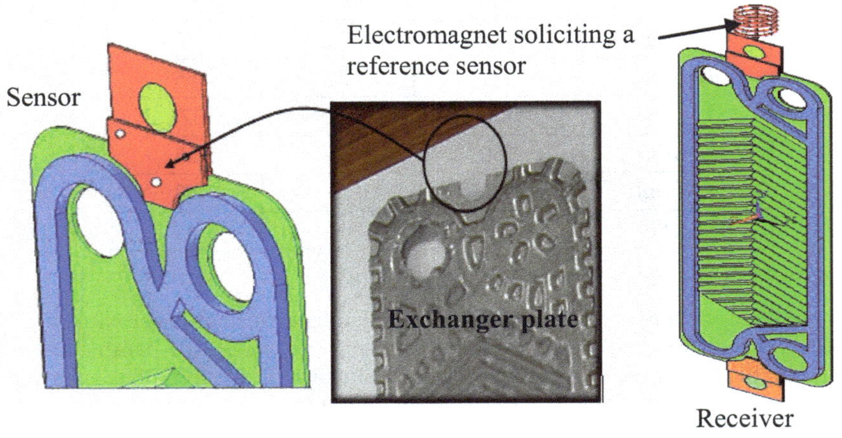

Fig. 19. Positioning of the sensors on an exchanger plate

4.1.1 Sensor excitation mode

In order to monitor the evolution of the damping of the plate modes due to fouling of the exchanger, it is necessary to excite these modes with enough energy to preserve the signal-noise ratio (of the signal received) after going through the exchanger.
A mechanical shock is the only way of producing enough energy for local excitation.
The frequency response obtained by modal analysis in the absence of structural constraints is given in the first column in table 7. This column gathers the different modes specific to the structure studied. Some correspond to simple, longitudinal or transversal displacements, others to more complex displacements (flexions, torsions...).

Mode	Frequency (Hz) - numerical	Frequency (Hz) - experimental
1	1683	1586
2	2387	2894
3	3557	3639
4	5422	5330
5	5734	6639
6	7417	7390

Table 7. The first 6 modes specific to the sensor

The second column shows the modal frequencies obtained from the analysis of the impedance of the sensor mounted on a heat exchanger.
The mean standard deviation between the frequencies obtained by modal analysis and those obtained experimentally is 5 %. The good correlation between these results indicates that the numerical modelling provides a good estimation of the resonance frequencies of the sensor.

4.1.2 Excitation by mechanical shock: estimation of the frequency range

The mechanical excitation in question is ensured via the core of an electromagnet.
As an indication, figures 20a and 20b show the temporal and frequency responses of the sensor.

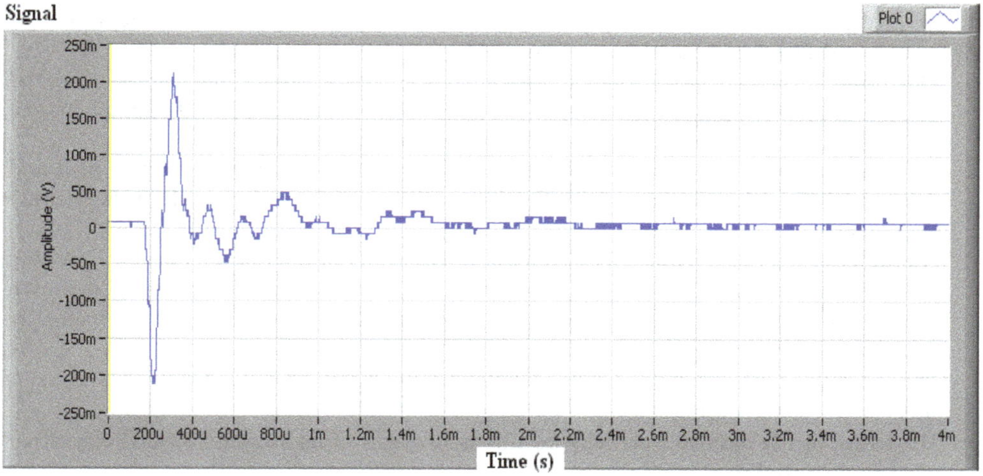

Fig. 20a. Temporal response of a mechanical shock

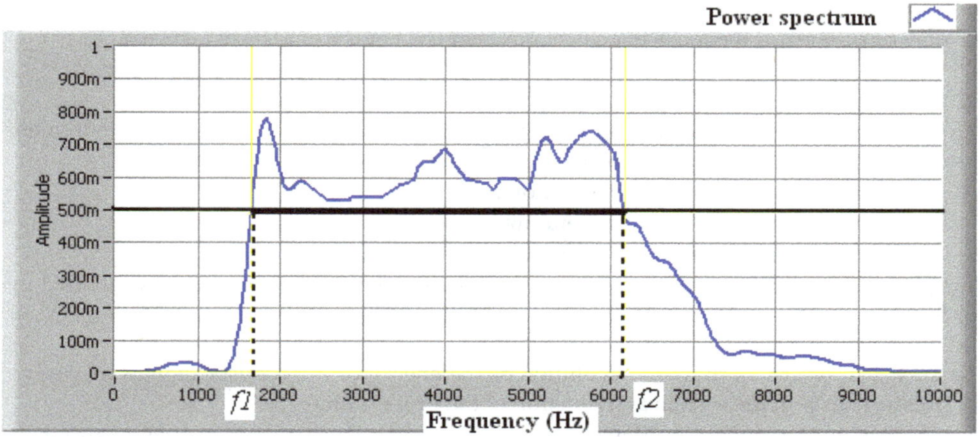

Fig. 20b. Spectral response associated with the shock

The curves show the temporal response and the frequency range of the sensor following a stress induced by a mechanical shock of short duration. The experiments carried out on the overall system (sensor & exchanger) in real configuration show that the temporal response is maximum 4 ms and its frequency response is around a central frequency of approximately 4 kHz.

4.2 Application

4.2.1 Fouling mechanism

Heat exchanger fouling is a dynamic process. The phenomenon continues to evolve, generally until equilibrium is reached or cleaning is required. The period of fouling can vary from a few hours to several months.

Müller (Müller-Steinhagen & Middis, 1989) looked at five stages in the process of the appearance and development of particulate fouling:

- The initiation, which corresponds to the time necessary before fouling, can be observed on a clean surface. The duration depends on the nature of the deposit, the initial state of the surface (material, roughness) and the temperature of the wall.
- The denaturing of the product (protein, organic matter...) under the effect of heat and the surrounding parameters (pH...), their aggregation and transport within the vicinity of the wall.
- The adhesion of the particles transported to the wall, controlled by surface adhesion forces (Van der Waals, electrostatic...) and cohesion of the deposit. It has been shown that the particles can adhere to a clean surface or adhere to other particles already deposited.
- The dislodging of deposited particles, caused by hydrodynamic forces which exert shear stress on the deposit.
- The aging of the deposit over time results in changes in its structure which can either weaken or consolidate it.

Generally, the initiation phase is rarely taken into account in particulate fouling models. The mechanisms that govern the deposit of particles are generally presented as being the transport of the particles to the surface, then the "adhesion" to the wall and finally the possible dislodging of the particles.

4.2.2 Results

Before studying the phenomenon of fouling, the metrological variation of the measurement system was taken into account according to the main technological parameters:

- Variation in temperature at constant flow.
- Variation in flow at constant temperature.
- Variation in viscosity at constant temperature and flow.

This phase is essential in order to separate the interferences of acoustic values generated by the fouling phenomenon from those linked to the technological conditions of the exchanger and its environment.

The curves in figure 21 show the evolution of the energy of the acoustic signals as well as the pressure drop in the system as a function of the process time.

The "Power" curve shows the damping effect linked to the load on the plate caused by fouling.

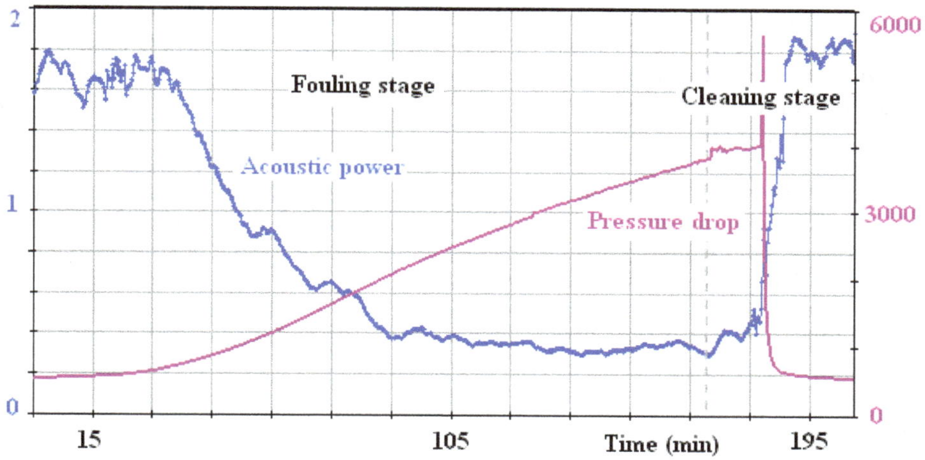

Fig. 21. Evolution of the power of the acoustic signal received during the fouling test and cleaning

In conclusion, this work concerned the monitoring of fouling using acoustics. By adopting a multi-stage experimental protocol we have been able to show that the variation in the acoustic signal can be used to predict variations in the pressure drop as well as the state of fouling in the plate heat exchanger under very specific operating conditions.
Finally, this study illustrates an example of a non-intrusive acoustic technique for the local monitoring in real time of the fouling of plate heat exchangers. The results show that it is possible to follow the relative kinetics of the state of fouling in each zone of the exchanger with the right choice and positioning of the sensors.

5. Conclusion

This chapter has proposed a synopsis of all the work that has led to the development of novel low frequency sensors. By using structural resonance modes excited by a transducer, these sensors present the advantage of having small sized sources with regard to the acoustic wavelength generated. These sensors are omni-directional but can nevertheless present significant contact areas with the medium to be characterised. This is the case for sensors developed for the characterisation of gels. The close contact of the elements set in resonance with the medium enables phenomena linked to changes in state to be monitored easily. Various applications have led us to develop sensors with very different geometries and which are optimised with the application in mind.
Indeed, for each need expressed, the approach consisted in optimising not only the geometry of the sensors but also their optimum position according to the problem posed. Three different cases were thus studied:

- identical near-field coupled sensors, through the medium to be characterised. They were used for monitoring the evolution of the ultrasonic values to characterise a sol-gel transition or the cohesion kinetics of a medium. For certain applications, the sensors are immersed in the medium. This direct immersion is essential for characterising fragile media.

- a low frequency receiver associated with an excitation of the medium via a mechanical shock in the case of very absorbent and scattering media. A second identical sensor is used for the synchronisation of the acquisitions thus reducing, by standardisation, the scattering of the values measured. The mechanical shock produces significant vibratory energy over a broad frequency range.
- finally, the sensors were coupled to heat exchanger plates in order to characterise fouling. This work has shown the interest of using acoustic sensors to monitor processes, providing an often local and dynamic response to the evolution of the performances of the process.

The work carried out provides a solid base of knowledge on ultrasound-complex media interactions. This knowledge could be put to good use in the development of sensors and integrated ultrasonic methods and their applications in the analysis and monitoring of local properties.

6. References

Aggarwal R. R., (1952a). Axially Symmetric Vibrations of a Finite Isotropic Disk. I, Journal of acoustical society of America, Vol. 24, N0. 5, pp. 463-467

Aggarwal R. R., (1952b). Axially Symmetric Vibrations of a Finite Isotropic Disk. II, Journal of acoustical society of America, Vol. 24, N°. 6, pp. 663-666

Allsopp, M. W. (1981). The developement and importance of suspension PVC morphology, Pure an applied chemistry, Vol. 53, pp. 449-465.

Blevins R. D., (1979). Formulas for natural frequency and mode shape, Van Nostrand Reinhold Co., ISBN 0-4422-0710-7, New York, USA

Brekhovskikh, L.M., (1980). Waves in layered media, Academic Press, ISBN 0-12-130560-0, New York, USA

Case, L. C. (1960). Molecular distributions in polycondensations involving unlike reactants. VII. Treatment of reactants involving nonindependent groups, Journal of polymer science, Vol. 48, pp. 27-35

Clerc, J. P. ; Giraud, G. ; Roussenq, J. ; Blanc, R. ; Carton, J.P. ; Guyon, E. ; Ottavi, H. & Stauffer, D. (1983). La Percolation: modèles, simulation analogiques et numériques, Annales de Physique, Vol. 8, Masson, Paris, France.

Dalgleish, D. G. (1982). Developments in Dairy Chemistry, edited by P. F. Fox (Applied Science, London,), Vol. 1, Chap. 5, ISBN 0-8533-4142-7, London, United kingdom

Dalgleish, D.G. (1993). Cheese: Chemistry, Physics and Microbiology, General Aspect, 2nd ed., Vol. 1, p. 69, Fox, P.F., Chapman & Hall, ISBN 0-1226-3652-X, London, United kingdom.

De Gennes, P. G. (1989). Scaling Concepts in Polymer Physics, Cornell University Press, Ithaca, ISBN 0-8014-1203-X, New York, USA

Degertekin, F. L. & Khury-Yakub, B.T. (1996). Hertzian contact transducers for non-destructive evaluation, Journal of acoustical society of America, Vol. 99, pp. 299-308

Degertekin, F. L. & Khury-Yakub, B.T. (1996). Lamb wave excitation by Hertzian contacts with applications in NDE. IEEE Transactions on Ultrasonics Ferroelectrics and Frequency, Vol. 44, N°. 4, pp. 769-778

Degertekin, F. L. & Khury-Yakub, B.T. (1996). Single mode lamb wave excitation in thin plates by Hertzian contacts, Applied physics letters, Vol. 69, N°. 2, pp. 146-148

Eichinger, B. E. (1981). Random elastic networks. I. Computer simulation of linked stars, Journal of chemical physics, Vol. 75, pp. 1964-1979

Ensminger, D. E. (1960). Solid cone in longitudinal half-wave resonance, Journal of acoustical society of America, Vol. 32, pp. 194-196

Flory, P. J. (1953). Principles of polymer chemistry, Cornell University Press, Ithaca & London, ISBN 0-8014-0134-8, New York, USA

Fox, P. F. (1989). Proteolysis during cheese manufacture and ripening. A review, Journal of dairy science, Vol. 72, pp. 1379-1385

Gupta, S. K.; Kumar A. & Bhargava, A. (1979). Molecular weight distribution and moments for condensation polymerization of monomers having reactivity different from their homologues, Polymer, Vol. 20, pp. 305-310

Irie T., Yamada G. & Muramoto Y., (1984). Natural frequencies of in-plane vibration of annular plates, Journal of sound and Vibration, Vol. 97, N°. 1, pp. 171-175

Lee M. & Singh R., (1994). Analytical formulations for annular disk sound radiation using structural modes, Journal of acoustical society of America, Vol. 95, N°. 6, pp. 3311-3323

Leissa A. W., (1969). Vibration of plates, NASA SP-160, U.S. Government Printing Office, Washington, D.C.

McMahon D. J. & Brown R. J., (1984). Enzymic coagulation of caseine micelles: a review, Journal of dairy science, Vol. 67, pp. 919-929

Mercier, J. P. & Marechal, E. (1993). Chimie des Polymères 1st ed. Presses polytechniques et universitaires romandes, Lausanne, Chap. 1, 3, 8. Lausanne, Swiss.

Moseley D. S., (1960). Contribution to the Theory of Radial Extensional Vibrations in Thin Disks, Journal of acoustical society of America,Vol. 32, N°. 8, pp. 991-995

Müller-Steinhagen H. & Middis J., (1989). Particulate fouling in plate heat exchangers, Heat Transfer Engineering, Vol. 10, N°. 4, pp. 30-36

Nassar, G. (1997). Etude et Optimisation d'un Dispositif Ultrasonore De Suivi en Ligne des propriétés viscoélastiques, Doctoral dissertation, Valenciennes University-France.

Nikolovski, J. P. & Royer, D. (1997). Local and selective detection of acoustic waves at the surface of a material", IEEE Ultrasonics Symposium, pp. 699-703, ISBN 0-7803-4153-8, Toronto, Ontario, Canada, October 5-8, 1997

Noël, Y. ; Flaud, P. & Quemada, D. (1989). Traitement Industriel des Fluides Alimentaires Non Newtoniens, Tome II, Actes du 2ème Colloque la Baule, La Baule, France, September 11-13, 1989, pp. 215-224.

San Biagio, P. L.; Bulone, D.; Emanuele, A.; Madonia, F.; Di Stefano, L.; Giacomazza, D.; Trapanese, M.; Palma-Vittorelli, M. B.; & Palma, M.U. (1990). Spinodal demixing, percolation and gelation of biosttural polymers, IUPAC 10th Int. Symp. on Polymer Networks, Vol. 40, pp. 33-44, Jerusalem, Israel, December, 1990

Shuyu, L. (1996). Study on the longitudinal-torsional composite mode exponential ultrasonic horns, Ultrasonics, Vol. 34, pp. 757-762

Shuyu, L. (1997). Study on the longitudinal-torsional composite vibration of a sectional exponential horn, Journal of acoustical society of America, Vol. 102, pp.1388-1393

Stauffer, D. (1981). Can percolation theory be applied to critical phenomena at gel point?, Pure an applied chemistry, Vol. 53, pp. 1479-1487

Stauffer, D. (1985). Introduction to Percolation Theory, Taylor & Francis Ltd., ISBN 0-7484-0253-5, London, United Kingdom

Stockmayer, W. H. (1943). Theory of molecular size distribution and gel formation in branched -chain polymers, Journal of chemical physics, Vol. 11, pp. 45-55

Vogel S. M. & Skinner D. W., (1965). Natural frequencies of transversely vibrating uniform annular plates, Journal of applied mechanics, Vol. 32, pp. 926-931

Walstra, P. & Vliet, V. (1986). The physical chemistry of curd making, Netherlands. milk dairy journal, Vol. 40, pp. 241-259

11

Modeling of Biological Interfacial Processes Using Thickness–Shear Mode Sensors

Ertan Ergezen et al.*
School of Biomedical Engineering, Health and Sciences, Drexel University, Philadelphia
USA

1. Introduction

Biological interfaces and accompanying interfacial processes constitute one of the most dynamic and expanding fields in science and technology such as biomaterials, tissue engineering, and biosensors. For example, in biomaterials, the bio-interfacial processes between biomaterials and surrounding tissue plays a crucial role in the biocompatibility of the layer (Werner, 2008). In tissue engineering, cellular adhesion plays an important role in the regulation of cell behavior, such as the control of growth and differentiation during development and the modulation of cell migration in wound healing, metastasis, and angiogenesis (Hong et al., 2006). Performance of a biosensor is highly dependent on interfacial processes involving the sensor sensing interface and a target analyte. Therefore, quantitative information on the novel and robust immobilization of detector molecules is one the most important aspects of the biosensor field (Kroger et al., 1998).
Thickness shear mode (TSM) sensors have been used in a variety of studies including interfacial biological processes, cells, tissue and properties of various proteins and their reaction (Cote et al., 2003). Phenomena such as cell adhesion (Soonjin et al., 2006.), superhydrophobicity (Sun et al., 2006, Roach et al., 2007), particle-surface interactions (Zhang et al.,2005), organic and inorganic particle manipulation (Desa et al., 2010) and rheological and interfacial properties of blood coagulation (Ergezen et al. 2007) were studied using TSM sensors. Due to the high interfacial sensitivity of TSM sensors, it has been shown that cell motility can be monitored by analyzing the noise of the TSM sensor response (Sapper et al., 2006). It has also been demonstrated that the number of motile sperm in a semen sample can be assessed in real-time using a flow-chamber integrated with a thickness shear mode sensor (Newton et al., 2007).

1.1 Quantification of Thickness Shear Mode (TSM) sensor response

The TSM sensor response is affected by the complex nature of the interface. Its response is influenced by the geometrical and material properties of the interacting surfaces such as surface roughness (Cho et al., 2007), hydrophobicity (Ayad and Torad, 2009), interfacial

* Johann Desa, Matias Hochman, Robert Weisbein Hart, Qiliang Zhang, Sun Kwoun, Piyush Shah and Ryszard Lec
School of Biomedical Engineering, Health and Sciences, Drexel University, Philadelphia
USA

slippage (Zhuang et al., 2008), coverage area (Johanssmann et al., 2008), sensitivity profile (Edvardsson et al., 2005) and penetration depth of the shear acoustic wave (Kunze et al., 2006).

Various theoretical models have been developed for quantitative characterization of the TSM sensor response to interfacial interactions. Nunalee et al (2006) developed model to predict of the TSM sensor response to a generalized viscoelastic material spreading at the sensor surface in a liquid medium. Cho et al (2007) created a model system to study the viscoelastic properties of two distinct layers, a layer of soft vesicles and a rigid bilayer. Urbakh and Daikhin (2007) developed a model to characterize the effect of surface morphology of non-uniform surface films on TSM sensor response in contact with liquid. Hovgaard et al (2007) have modeled TSM sensor data using an extension to Kevin-Voigt viscoelastic model for studying glucagon fibrillation at the solid-liquid interface. Kanazawa and Cho (2009) discussed the measurement methodologies and analytical models for characterizing macromolecular assembly dynamics.

The physical description based on a wave propagation concept in a one-dimensional approximation has been proven as the best model of thickness shear mode (TSM) sensors. The fundamentals have been published in several books (Rosenbaum, 1998). Martin et al. have (1994) applied this background to sensors by using Mason's equivalent circuit to describe the thickness shear mode sensor itself and transmission lines as well as lumped elements for viscoelastic coatings, semi-infinite liquids etc.. Follow-up papers have introduced a more straightforward definition of the elements of the BVD-model (Behling et al, 1998) as well as several additional approximations, e.g. based on perturbation theory, to derive less complex equations, have suggested a simplified notation to separate the mass from so-called nongravimetric effects, or have applied the transmission line model to several subsystems (Voinova et al, 2002) for demonstration of specific situations just to call some examples. More recent papers deal with deviations from the one-dimensional approximations, e.g. by introducing generalized parameters by deriving specific solutions e.g. for surface roughness or with discontinuity at boundaries.

TSM sensors combined with the theoretical models mentioned above were used to determine the properties of liquids (Lin et al., 1993), high protein concentration solutions (Saluja et al., 2005), and thin polymer films (Katz et al., 1996).

For viscoelastic layers, their mechanical impedance depends upon the density, thickness, and the complex shear modulus of the loading. Identification of the all the system parameters from the impedance measurements has been very challenging and uncertain without a priori knowledge of the thicknesses and/or some of the material properties (Lucklum et al. 1997).

Furthermore, Kwoun (2006) showed the beneficial features of the multi-resonance operation of the TSM (called as "multi-resonance thickness shear mode) sensor to study the formation of biological samples, specifically collagen and albumin, on the sensor surface. In this work, it was demonstrated that the different harmonic frequency clearly showed the different characteristics of mechanical properties, especially shear modulus, of the biological sample. Although this work was one of the pioneer studies to demonstrate the strengths of the MTSM measurement technique, it is limited as it is a semi-quantitative method. Exact values of mechanical properties of anisotropic collagen and albumin samples were not able to be defined due to complexity of the non-linear simultaneous equations of the model. An improved MTSM technique combined with an advanced data analysis technique was proposed by Ergezen et al (2010). A new approach merging the multi-harmonic thickness shear mode (MTSM) measurement technique and genetic algorithm-based data analysis technique has been used. This novel method was utilized to solve two unmet needs:

1. Identification of all four parameter by using the MTSM sensor's single harmonic response results in an under-determined problem. The MTSM sensor response enables the identification of two parameters by providing imaginary and real components of the mechanical impedance. In other words, there are fewer equations than the material/geometrical parameters of the interface, therefore, the stochastic method is the only approach that can address this problem mathematically. In this project it was shown that combination of the MTSM measurement technique and the genetic algorithm-based data analysis technique (called as MTSM/GA technique) was used to solve this under-determined problem. *It was reported for the first time, a novel approach that enables determining all four parameters, which define the response of the MTSM technique.*
2. Most of the biological interfaces constitute multi-layer structures. Multi-layer modeling of biological interfacial processes was proposed by several researchers and by us (Wegener et al., 1999, Ergezen et al., 2007). In contrast, there has been very limited (Lucklum et al., 2001) theoretical study and no experimental studies based on the MTSM sensor for quantitative characterization of multi-layer biological processes. *It was reported, for the first time, the most comprehensive theoretical and experimental study for quantitative characterization of multi-layer biological interfacial processes.*

A new approach merging the multi-harmonic thickness shear mode (MTSM) sensor and a data extraction technique based on stochastic global optimization procedure has been proposed. For this purpose, the MTSM/GA technique is being developed and calibrated with a polymer layer (having known properties). This was then used to estimate the properties of a protein layer with unknown properties adsorbed to the MTSM sensor surface. It was demonstrated that this new method has the potential to be a novel tool for quantitatively characterization of interfacial biological layers.

2. Theory

2.1 Multi-Harmonic Thickness Shear Mode (MTSM) sensor

Piezoelectric MTSM sensors transmit acoustic shear waves into a medium under test, and the waves interact with the medium. Shear waves monitor local properties of a medium in the vicinity of the sensor and of the medium/sensor interface (on the order of nm - µm); thus, they provide a very attractive technique to study interfacial processes. Measured parameters of acoustic waves are correlated with medium properties such as interfacial mass/density, viscosity, or elasticity changes taking place during chemical or biological processes.

The shear acoustic wave penetrates the medium over a very short distance. The square of the depth of penetration of an acoustic shear wave in MTSM sensor is related to medium viscosity, elasticity, density and the frequency of the wave (please see Appendix IA.) (Kwoun et al. 2006). Figure 1a shows the acoustic wave penetrating the adjacent medium and Figure 1b shows that the depth of penetration decreases at higher harmonic frequencies in a semi-infinite medium.

Therefore, by changing the frequency, one can control the distance at which the wave probes the medium. Multi-harmonic operation of MTSM sensor will enable to control the interrogating depth into the biological processes. Therefore it will provide a more in depth characterization of the biological interfacial processes. For example, it was suggested that cell adhesion on extra cellular matrix should be modeled as a multi-layered structure (Wegener et al. 2000). Therefore MTSM sensors can provide information about mechanical and structural properties of the biological processes from different depths (slicing the medium).

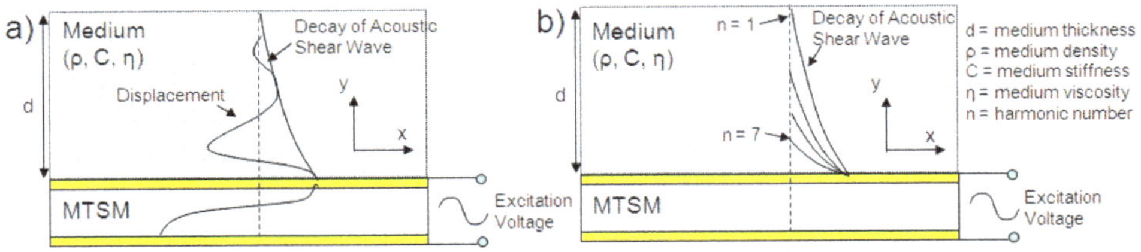

Fig. 1. a) Acoustic wave penetrating into the medium b) depth of penetration decreases at higher harmonic frequencies

It should be noted that it was assumed that the medium is semi-infinite and the mechanical properties are not frequency dependent in fig. 1.

2.2 Electrical response of MTSM sensor

The MTSM sensor is a piezoelectric-based sensor which has the property that an applied alternating voltage (AC) induces mechanical shear strain and vice versa. By exciting the sensor with AC voltage, standing acoustic waves are produced within the sensor, and the sensor behaves as a resonator. The electrical response of the MTSM sensor in air over a wide frequency range is shown in figure 2, where S_{21} is the magnitude response of the MTSM sensor ($|S_{21}|$=20log(100/(100+Z_t)), Z_t=total electromechanical impedance of the MTSM sensor (Rosenbaum 1998). As an example, the magnitude and phase responses of MTSM sensor are presented at the first (5 MHz), third (15 MHz), fifth (25 MHz) and seventh (35 MHz) harmonics in air.

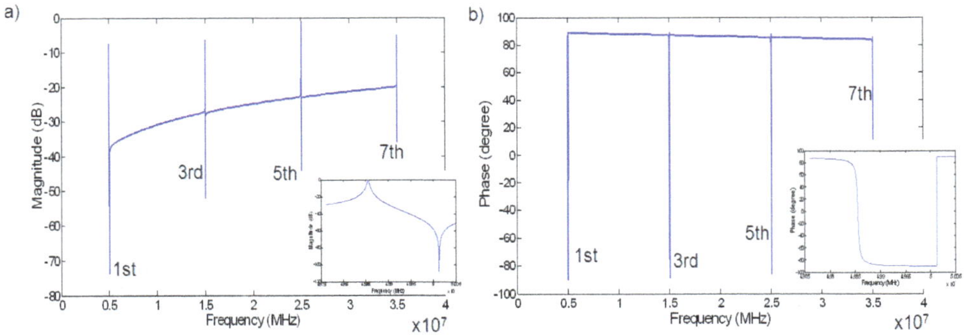

Fig. 2. A typical a) frequency vs. magnitude response and b) frequency vs. phase response characteristic and the associated resonance harmonics for the MTSM sensor, spanning a wide frequency range (5 MHz to 35 MHz). (Insets) Magnified view of magnitude and phase response at 5 MHz

An example of the MTSM's magnitude response in the vicinity of the fundamental resonant frequency is given below (figure 3a). When the TSM sensor is loaded with a biological media, there will be a shift in resonant frequency and a decrease in the magnitude. These changes can be correlated with changes in the mechanical and geometrical properties of the medium such as thickness, viscosity, density and stiffness. Depending on the changes at the interface of the sensor surface-medium interface, a positive and/or negative shift can be seen in the frequency response (Figure 3b).

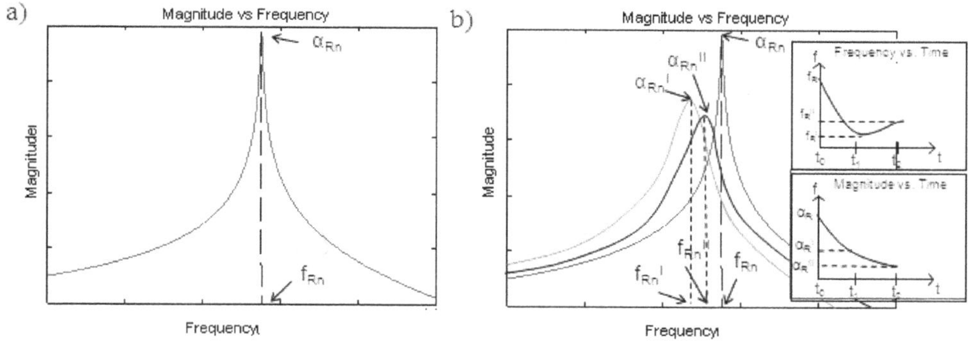

Fig. 3. (a)Demonstration of a typical qualitative frequency-dependent response curve for the MTSM sensor in the vicinity of the resonant frequency; n = harmonic number, α_{Rn}=Initial maximum magnitude, f_{Rn}=Initial resonant frequency, (b) In the case of both positive and negative frequency shifts throughout the experiment, α_{Rn}^{I}, α_{Rn}^{II}=Instantaneous maximum magnitudes of loaded MTSM sensor at time t_1 and t_2 respectively, f_{Rn}^{I}, f_{Rn}^{II} =Instantaneous resonant frequencies of the loaded MTSM sensor at time t_1 and t_2 respectively (Inlet) resonant frequency and magnitude are monitored as a function of time

2.3 MTSM/GA data processing technique

This section will be structured in the following manner; first, the general structure of a genetic algorithm will be explained. Second, advantages of genetic algorithm over other techniques will be discussed. Finally, implementation of MTSM-GA technique for determination of material parameters will be explained.

Principles of operation of a genetic algorithm (GA)

Basic definitions of GA terms are defined in Appendix IB. Genetic algorithm (GA) is based on the genetic processes of biological organisms (figure 4). GA works with a population of individuals, each representing a possible solution to a given problem. Each individual is assigned a fitness score according to how good a solution to the problem it is. The highly-fit individuals are given opportunities to reproduce, by cross breeding with other individuals in the population. This produces new individuals as offspring, which share some features taken from each parent.

Comparison of GA to other data processing techniques

Complex models are ubiquitous in many applications in the fields of engineering and science. Their solution often requires a global search approach. Therefore the objective of optimization techniques is to find the globally best solution of models, in the possible presence of multiple local optima. Conventional optimization and search techniques include; (1) gradient-based local optimization method, (2) random search, (3) stochastic hill climbing, (4) simulated annealing, (5) symbolic artificial intelligence and (6) genetic algorithms. The detailed information on each technique and comparisons to Genetic Algorithms (GA) are already explained by Depa and Sivanandam (2008). Here, the aim is not to analyze these techniques in detail but to show the suitability of GA as a parameter estimation algorithm. As discussed by Depa and Sivanandam, some of the advantages of GA over other techniques are: (1) it is good for multi-mode problems, (2) it is resistant to becoming trapped in local optima, (3) it performs well in large-scale optimization problems, (4) it handles large, poorly understood search spaces easily. These advantages match with the requirements for an optimization technique to be applied in this application. Therefore GA was chosen as an optimization technique and successfully combined with the MTSM technique.

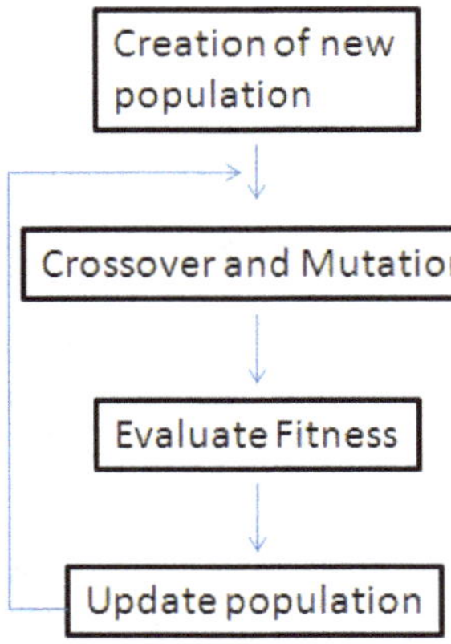

Fig. 4. Flow chart of a genetic algorithm

Structure of the MTSM/GA technique

The structure of MTSM-GA technique is presented in figure 5. As seen from the figure, there are two inputs to the GA, namely; range of variables and MTSM sensor response. GA outputs the determined values of the variables by using GA functions such as crossover, mutation and fitness evaluation. In the following sections, initially, the inputs to the GA will be explained. Then the structure of GA and its internal functions will be presented.

MTSM sensor response

The first input to the GA is the MTSM sensor response. Both magnitude and phase responses were continuously monitored during the experiments (see materials and methods section). Then the specific points on these responses such as resonant frequency, maximum magnitude, minimum phase, frequency at minimum phase, and phase at maximum magnitude were input to GA for calculating the fitness score for each individual. The changes in these target points were calibrated with the diwater/glycerin changes.

Selection of the ranges for variables

The next step of the technique is to set the ranges for the variables (chromosomes). These ranges represent the bounded space within which the GA will search for solutions. The ranges should be reasonable for each parameter in order to determine accurate solutions. For example, for a Newtonian liquid the stiffness is 0, therefore one should not set the range to be between 1e5 N/m^2 and 1e7 N/m^2. If this were done the algorithm will not converge to a solution because of the inappropriate choice of ranges.

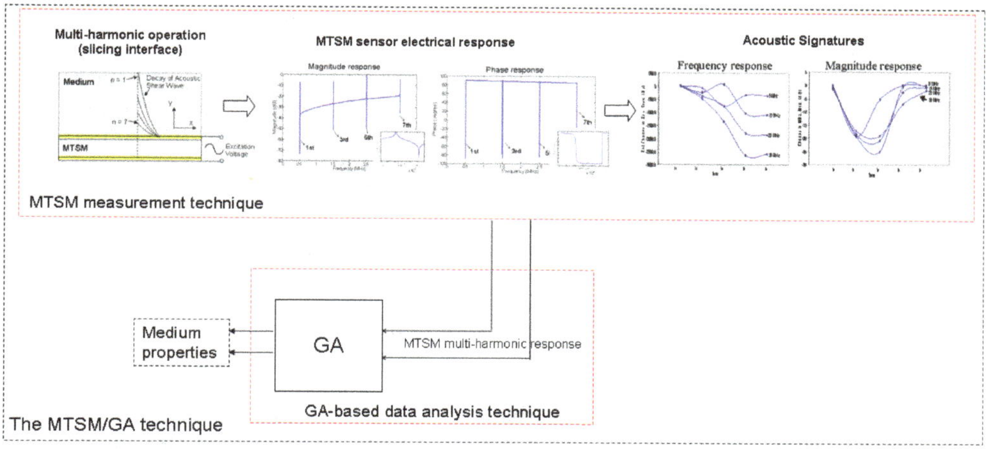

Fig. 5. Basic structure of MTSM/GA technique

As shown by Kwoun (2006), the viscoelastic materials can be divided in to four regimes, namely; liquid like, soft rubber, hard rubber and solid like. As seen from table 1, the viscosity values might change between 0.001 and 0.1 kg/m.s and stiffness value changes between 0 - 1e9 N/m^2 . Typical range of density values for a polymer was determined to be between 1000 - 1400 kg/m^3.

Phase	η (kg/m.s)	C (N/m^2)
Liquid like	0.001 - 0.01	0-1e5
Soft Rubber	0.01 - 0.1	0-1e5
Hard Rubber	0.01 - 0.1	1e5 - 1e7
Solid Like	- 0.1	1e7 - 1e9

Table 1. Four regimes of a viscoelastic system

Genetic Algorithm and its internal functions

This section will be divided into three sections. First, the GA's main parameters such as number of populations, crossovers, mutation rates and genes per chromosome will be analyzed. Then the fitness function of the GA will be explained. Finally, the technique combination of sub-spacing and zooming to determine the values for four variables will be presented.

Selection of GA parameters

Different combinations of the GA parameters were evaluated. Here, the combination that gives the best result is presented. Each variable was represented by a binary chromosome that contains 16 genes. A random population of 100 individuals was generated. Tournament selection was implemented for selection of individuals for mutation and crossover. In order to carry out the crossovers the entire population is divided into groups of 5 individuals each, these groups are randomly selected. From each group, the individual with the highest fitness together with another individual of this group are selected for crossover. The two selected individuals are the parents and yield two offspring. Both the parents and the offspring pass to the next generation. This idea was implemented in order to reduce the selection pressure.

The crossover between the parents is a simple one meaning that a random crossover point is selected and two kids' genome are formed with the left and right genes of the crossover point of each parent. A relatively high mutation probability (0.5) is present in order to avoid local minimum, otherwise all the individuals might end up having the same genome and this genome corresponding to a not optimal solution. Also elitism was implemented to assure that the best individual of a generation survives to the next generation. This ensures that the algorithm keeps the best solution until a better one is found.

Fitness function

One of the most important parts of a genetic algorithm is the fitness function. The fitness function must reflect the relevant measures to be optimized. This function evaluates the function being searched for the set of parameters of each member of the population. The output of the fitness function is a vector that contains the fitness for each member of the population. This vector helps in the selection of individual for generating new offspring or individuals that will be included in the new generated population.

The approach used, in this study to model biolayers on a MTSM sensor, is Mason's transmission line model (please see Appendix C). This model is a one-dimensional model that describes the electrical characteristics of an acoustic structure wherein, each layer of load can be represented as a T-network of impedances.

Once the initial population is created the algorithm randomly generates a population (includes 100 individuals) chosen from the ranges of the variables (the section titled "selection of the ranges for variables"). Then each individual was input to fitness function (transmission line model). The error between the model (transmission line model) and the experimental results were compared by using the following equation:

$$fit\_func = \frac{100}{1+(\sqrt{(\alpha_{Re}-\alpha_{Rt})^2}+\sqrt{(f_{Re}-f_{Rt})^2}+\sqrt{(P_{Me}-P_{Mt})^2}+\sqrt{(f_{Me}-f_{Mt})^2}+\sqrt{(\alpha_{ARe}-\alpha_{ARt})^2}+\sqrt{(f_{ARe}-f_{ARt})^2})}$$

The denominator of this function represents the difference between the model and the experimental data (we use the plus one in order to avoid the eventual division by zero). In this project, rather than fitting the whole magnitude and phase curve, certain points such as α_R = maximum magnitude, f_R = resonant frequency, P_M = minimum phase, f_M = resonant frequency at minimum phase, α_{AR} = minimum magnitude, f_{AR} = anti-resonant frequency has been compared between the model and the experimental results. Subscript "e" indicates experimental results and subscript "t" stands for theoretical model. This function is monotonously increasing with the kindness of the solution provided by the genetic algorithm. The algorithm was terminated at after 500 generations.

Set-up of the Genetic Algorithm

Acoustic impedance seen at the sensor/film interface is derived from transmission line theory (Martin and Frye 1991). Surface mechanical impedance is related to density and thickness of the film, and complex modulus (= $G^I + jG^{II}$). Therefore there are four independent variables to define the surface acoustic impedance. The MTSM sensor response contributes two parameters by providing real and imaginary part of mechanical impedance. Hence using single harmonic response results in an under-determined problem. Genetic optimization technique has been applied to under-determined problems to obtain approximate solutions with satisfactory accuracy (Wang and Dhawan, 2008). Here genetic algorithm has been improved by combining sub-space and zooming techniques. It was shown that this combination provides very good approximation with less than 1% error.

First, sub-spacing method was applied. This method gives a quick idea of where the solution can be and also it decreases algorithm running time dramatically (Garaia and Chaudhurib, 2007). Therefore the solution space was divided in 10 sub-spaces. Genetic algorithm was run 5 times in each subspace. Each subspace's convergence performance was evaluated. The sub-space with the best fitness score was considered to be the candidate solution space. It was observed that the candidate sub-space had a distinct convergence performance compared to the others. This method dramatically increased the efficiency of GA by eliminating the irrelevant solution spaces.

Secondly, GA was run 100 times (this number was chosen to have 95% confidence level and 10% confidence interval statistically). The termination criterion for each run was 500 generations. After 100 runs, it was observed that, for two out of four variables, observed points having a uniform distribution (skewness < 0.5) were accumulating around one number in a narrow range (in ±20% of candidate solution point). The average value of the observed points was also equal or very close (<5%) to solution (theoretically shown). Therefore GA was always able to converge to "the most likely" values for two out of four variables after these two steps (from our observations, mostly stiffness and thickness, and sometimes, viscosity and thickness). It was shown theoretically that one can always put these numbers, and calculate the other two variables with the error of less than <15% at this step.

Then zooming method was applied to reduce the search space around the candidate optimum solution point. Several zooming methods have been developed for different applications (Ndiritu and Daniel, 2001, Kwon et al. 2003). In this project, the GA was run 30 times, and then the new range was set to be between maximum and minimum numbers of the 30 points. This zooming continued until the error was less than 1% for all variables. This error was achieved after 6 zooming.

These results showed that the MTMS/GA technique combined with sub-spacing and zooming methods can be applied successfully to approximate the solution with good accuracy for this under-determined problem.

3. Materials and methods

The MTSM/GA technique first experimentally tested with the polymer SU8-2002 layer spin coated on sensor surface. The determined properties of the layer were compared with the values obtained from literature. The technique was then applied to obtain the mechanical and geometrical properties of a protein layer adsorbed on gold layer. The methods and chemicals used in the experiments are described below.

a. Deposition of the thin polymer film

The SU 8-2002 (MicroChem) polymer solution was spin coated on MTSM sensor by using the following procedure. First, the gold electrode surface of TSM sensors was cleaned using

Piranha solution (one part of 30% H2O2 in three parts H2SO4). After 2 min exposure time, the sensors were rinsed with distilled water. The surface was dried in a stream of nitrogen gas. The SU 8 - 2002 sample was dispensed on MTSM sensor surface and sensors were spin coated for 40 seconds. The sensors were then soft baked for 1 min at 95 ^{o}C. The SU 8-2002 films were exposed to UV light for 4 seconds under 25 mJ/cm^2. This was followed by 1 min hard baking on hot plate at 95 ^{o}C.

b. Antibody adsorption on MTSM sensor surface

The reference measurements were taken for air and phosphate buffer saline (PBS). Next, the sensors were exposed to rabit-immunoglobulin G (IgG) (50 µg/ml) suspended in diwater (Fisher Scientific, pH: 5.34, Cat No: 25—555-CM) for 50 minutes to allow IgG coating of the sensor surface by adsorption.

c. Characterization of geometrical properties of the thin film

The thicknesses of the SU 8 - 2002 films were determined by using optical profilometer (Zygo Inc. Model #: NV6200). For the thickness measurements, a very small portion of MTSM sensor surface was not exposed to UV light. After the films were developed, the SU 8-2002 layer was removed from this portion. To obtained different thicknesses of film layer, 1:1 solution of SU8-2002 and cyclopentanone (Acros Organics) was prepared.
The surface topography of the film layer was measured using atomic force microscopy (AFM). The prepared samples were placed on a glass slide installed on the atomic force microscope (Bioscope; Veeco), that was mounted on the inverted fluorescence microscope (TE2000; Nikon, Melville, N.Y.). Measurements were made using contact mode with a scan rate of 2 Hz.

d. Measurement system and MTSM sensor data analysis technique

A 14 mm diameter, 0.33 mm thick, 5 MHz quartz crystal with deposited 7 mm gold electrodes was placed in a custom fabricated brass sensor holder (ICM). The sensor holder was connected to a Network Analyzer (NA) (HP4395A). A LabView program on a personal computer was used to control the network analyzer and collect the data at 5, 15, 25 and 35 MHz. The experiments were done in room temperature (24°C±1°C). Magnitude and phase responses of MTSM sensor were monitored during the experiments (figure 6). The sampling rate was 30 seconds. Each experiment was repeated three times.

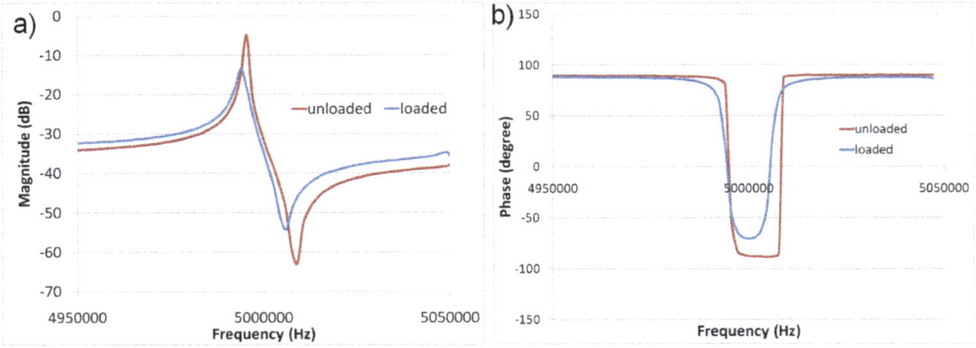

Fig. 6. a) Magnitude and b) phase responses of MTSM sensor

4. Results and discussions

Initially, two different thicknesses of SU8 2002 layers were spin coated on sensor surface and changes in the frequency and magnitude responses were monitored at 5, 15, 25 and 35 MHz. The thicknesses of the layers were measured by using optical profilometer (fig. 7a). The average thicknesses of the layers were 1920±25 nm and 770±50 nm respectively. Surface topography of the SU8 - 2002 layers was measured by using AFM (fig. 7b). The average roughness of the layer was 20 nm and no cracks on the surface were observed.

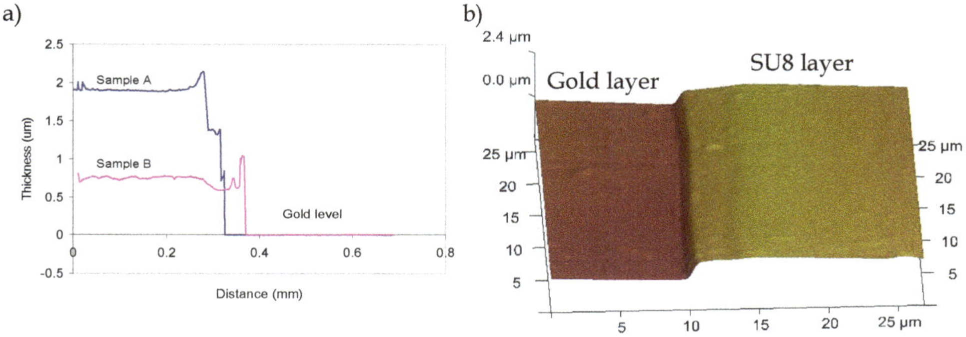

Fig. 7. A) Thickness measurements from optical profilometer sample a. SU8-2000 solution sample b. 1:1 dilution of SU8-2002 and cyclopentanone B) Surface topography of SU 8 layer

a. Determination of mechanical and geometrical properties of SU8 layer of 1.92 μm thickness

First set of experiments were performed by spin coating 2 μm thick SU 8 - 2002 layer on sensor surface. The MTSM/GA determined properties are presented in table 2. The average thickness of the polymer layer determined to range from 2080 nm to 2140 nm among the harmonics. Although these values are slightly higher than the value (1920±25 nm) obtained in control experiments, they are still in less than 10% experimental errors. The variation between the frequencies for density value was also very small, ranging from 1240 to 1253 kg/m³. These numbers correlate well with the literature value of 1200 kg/m³ (Jiang et al., 2003) for SU8.

MTSM Frequency	MTSM/GA Results		Profilometer	Jiang et al. [38]
(MHz)	d(nm)	ρ (kg/m³)	d (nm)	ρ (kg/m³)
5	2120±60	1253±10		
15	2140±50	1246±11		
25	2080±110	1240±50	1920±25	1200
35	2080±60	1240±28		

Table 2. Comparison density and thickness values of SU 8-2002 layer determined using MTSM/GA sensor at 5, 15, 25 and 35 MHz with profilometer and Jiang et al. (Jiang et al. 2003)

The frequency dependent shear modulus of SU 8-2002 layer obtained using the MTSM/GA is presented in table 3. Both loss and storage modulus varies with the operating frequency. These extracted values were compared with the values obtained by Jiang et al (2003) (table 3). Jiang et al calculated the shear modulus of SU8 layer by using the impedance-admittance characteristics of the equivalent circuit models of loaded and unperturbed TSM sensors operating at 9 MHz.

MTSM Frequency	MTSM/GA Results		Jiang et al.[38] (at 9 MHz)	
(MHz)	G^I (N/m²)	G^{II} (N/m²)	G^I (N/m²)	G^{II} (N/m²)
5	(4.55±2.12) x10^7	(1.89±0.26) x10^5	7.80e7	2.00e5
15	(2.33±0.18)x10^8	(1.00±0.03) x10^6		
25	(3.82±0.52)x10^8	(4.69±0.52) x10^6		
35	(5.81±0.71) x10^8	(6.49±0.18) x10^6		

Table 3. Comparison of determined G^I and G^{II} values of SU8 layer using MTSM/GA at 5, 15, 25 and 35 MHz and Jiang et al (2003)

As seen in table 3, the values obtained by Jiang et al. fall between the values obtained using the MTSM/GA method for 5 and 15 MHz. The small variation in the G^I and G^{II} may be due to difference in the film preparations. Alig et al. (1996) has shown that variations in film preparation methods can affect the mechanical properties of the polymer layers.

b. Determination of mechanical and geometrical properties of SU8 layer of 0.770 μm thickness

The second set of experiments was done with the ~770 nm thick SU 8-2002 layer on MTSM sensor. As seen from the table 4, the thickness of the layer determined using the MTSM/GA method correlates well with the expected thickness for each harmonic (less than 10% error). Furthermore the results vary only 10 nm between the harmonics. Similarly, determined values for density were consistent between the harmonics, which is around ~1200 kg/m³.

MTSM Frequency	MTSM/GA Results		Profilometer	Jiang et al.[39]
(MHz)	d(nm)	ρ (kg/m³)	d (nm)	ρ (kg/m³)
5	820±45	1180±40	770±50	1200
15	820±20	1190±30		
25	810±35	1190±50		
35	810±52	1213±35		

Table 4. Determined density and thickness values by MTSM/GA or 770 nm thick SU8 layer at 5, 15, 25 and 35 MHz

The initial losses before coating were -0.53 dB and -2.5 dB for 5 and 35 MHz respectively. The losses increase to -0.59 dB for 5 MHz and -4.18 dB for 35 MHz. As seen from these results, the losses remain relatively low when the thickness of the layer was decreased to 770 nm in contrast to the phenomenon observed when the film thickness was 2 μm. For 2 μm

film thickness, the losses increase to -1.9 dB and -11.5 dB at 5 MHz and 35 MHz respectively, while initial loses were similar to what observed for 770 nm film thickness.
The shear modulus values determined via the MTSM/GA technique are presented. Both loss and storage modulus were decreased compared to the values obtained when film thickness was 2 µm (figure 8). It has been shown that the scale effect on the mechanical properties of the polymers might be the reason for the decrease in the values (Liu et al, 2009, Luo et al, 2003).

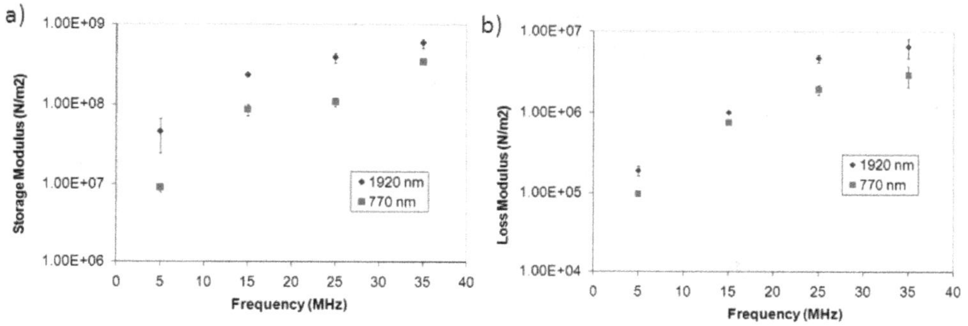

Fig. 8. a) Storage and b) loss modulus as a function of harmonic frequency for 770 nm and 1920 nm thick SU8 layer at 5, 15, 25 and 35 MHz (error bars are smaller than symbols when not visible)

c. Determination of mechanical and geometrical properties of an antibody layer

Third set of experiments were done by adsorbing an antibody layer on MTSM sensor surface under static conditions at 5, 15, 25 and 35 MHz. Antibodies play crucial importance in many applications such as biosensing (Hanbury et al. 1996) and drug delivery (Morrison et al., 1995). The sensor surface was saturated with antibody to form a uniform protein layer on the surface. Change in the frequency and magnitude responses at 15, 25 and 35 MHz are presented in figure 9. At the fundamental frequency (5 MHz), high fluctuations observed in sensor response are likely due to insufficient energy trapping as described by others (Li et al. 2004).

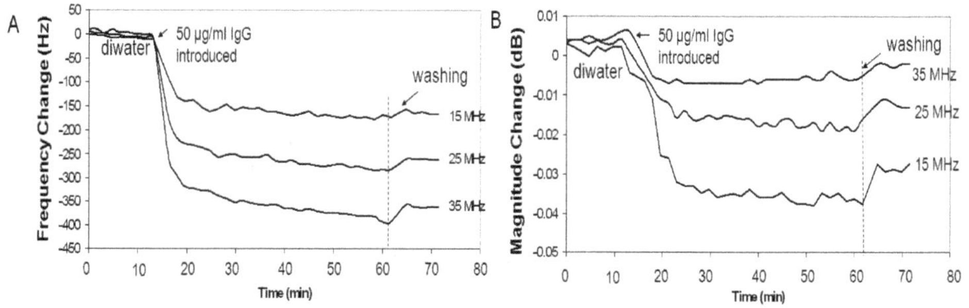

Fig. 9. Time response of A. resonant frequency and B. maximum magnitude responses of MTSM sensor to antibody binding at 15, 25 and 35 MHz

The properties of the medium were determined at t_1 = 10 and t_2 = 70 minutes. At t_1 = 10, the system is modeled as MTSM sensor loaded with semi-infinite Newtonian medium (DIwater) (fig 10A). The height of the column (2 mm) was much higher than the penetration depth of the acoustic wave at 5 MHz (~250 nm in DI water).
At t_2 = 10 min., the MTSM/GA determined properties of the layer at 15, 25 and 35 MHz are presented in table 5. The variations in the determined thickness values were very high (ranging from 300 nm to 5 μm due to the fact that column height was much larger than the penetration depth. Solution range for thickness values was set to be between 1 nm to 10 μm in genetic algorithm. Thus any thickness value larger than the penetration depth will satisfy the solution because the MTSM sensor is not sensitive to the changes beyond the penetration depth. However, the solutions were always higher than penetration depth as expected. Due to the high fluctuations in thickness values, it was not presented here. In contrast the solutions for ρ_1, η_1 and C_1 match with the literature values very well. (Literature values are ρ_1 = 1000 kg/m^3, η_1 = 0.001 kg/m.s and C_1 = 0 N/m^2 at room temperature (Greczylo and Deboswka 2005)).

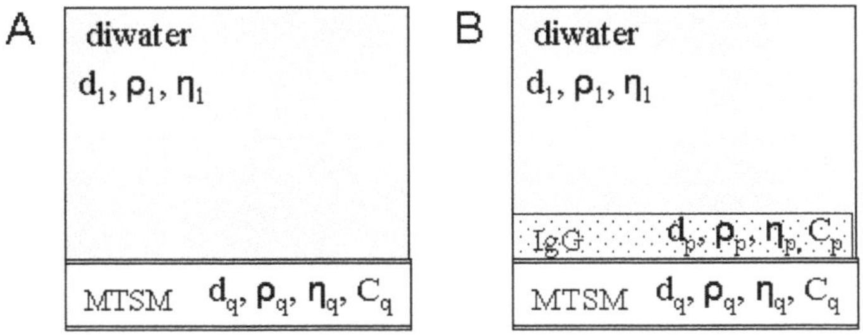

Fig. 10. Physical model for MTSM sensor system at A) t=10 and B) t=70

MTSM Frequency	Density (kg/m^3)	C^I (N/m^2)	η (kg/m.s)
15	1006±5	(2.00±1.00) x10^2	(1.05±.004) x10^{-3}
25	1003±2	(5.00±3.00) x10^2	(1.08±0.03) x10^{-3}
35	1004±4	(1.00±1.00) x10^2	(1.06±0.04) x10^{-3}

Table 5. Determined properties for semi-infinite Newtonian medium layer by MTSM/GA at 15, 25 and 35Hz

At t = 70 min., the physical model is presented in fig 10b. A viscoelastic layer (protein layer) with finite thickness and semi-infinite Newtonian medium were loaded on MTSM sensor. The properties for diwater layer were entered into the algorithm as known variables and the unknown properties (ρ_p, η_p, C_1 and d_p) of viscoelastic layer were determined using the MTSM/GA method. The results are presented in table 6. The thickness of the layer was determined to range from 10.3 to 11 nm for the harmonics. This number correlates well with the values presented by the other researches. Westphal et al (Westphal and Bornmann 2002) calculated the height of antibody layer as 9.2 nm. Furthermore Liao et al (Liao et al 2004) measured the average height of the antibody layer as 10.1±3.3 nm.

The density of the antibody layer was also determined by the MTSM/GA to be 1030±14 kg/m³. This density value is close to the water density in which the antibodies were suspended. Hook et al. (2002) considered the density of antibody layer as 1050 kg/m³ when the antibodies were not attached to gold surface. After the cross-linking, the density value was1300 kg/m³, this is closer to the density value of dry protein. Voros (2004) also showed that the wet density of antibody layer is significantly different than the dry protein density value due to the solvent present in the adsorbed proteins. Therefore we believe that the determined value of the density is in a reasonable range.

MTSM Frequency	Thickness (nm)	Density (kg/m³)	G^{I} (N/m²)	G^{II} (N/m²)
15	11±0.3	1050±10	(5.20±0.5) x10⁴	(4.80±0.58) x10⁵
25	10.4±0.6	1080±12	(5.00±0.13) x10⁴	(9.50±1.40) x10⁵
35	10.3±0.4	1040±14	(5.60±0.12) x10⁴	(1.52±0.31) x10⁶

Table 6. Determined properties for antibody layer by MTSM/GA at 15, 25 and 35 MHz

As seen from the table 6, the adsorbed antibody layer has low storage modulus (<1e5 N/m²), and relatively higher loss modulus. While storage modulus was same for each harmonic, loss modulus changed with frequency. It has been experimentally shown that the adsorbed protein layers on TSM sensor, such as antibody, vesicles and cells do not behave like "rigid and thin" films (Voinova et al, 2002). Therefore the linear relationship between resonant frequency shift and mass deposition is not observed. Saluja et al. (2005) indicated low concentrations (less than 60 mg/ml) of antibody suspension behave like Newtonian medium. But it should not be expected that the properties of adsorbed layer will not be the same as the properties of antibody suspension. The effect of the binding between protein layer and gold layer should be considered. No literature value was found for direct comparison. Therefore we believe that MTSM/GA technique will lead to development of a quantitative tool for study of biological interfacial processes.

5. Conclusions

It was shown that MTSM sensor combined with genetic algorithm can be used to extract mechanical and geometrical properties of biological layers. The developed technique was first experimentally tested with SU8-2002 polymer layers with known properties having two different thicknesses. It was shown that the developed technique was successfully determined the mechanical and geometrical layers of thin polymer layers. MTSM/GA technique was then applied to extract the properties of antibody layer coated on MTSM sensor. The obtained data support our hypothesis about use of MTSM/GA technique can be a powerful tool for quantitative characterization of interfacial biological interfacial processes.

6. Acknowledgments

We are thankful to Dr. Moses Noh for providing supplies and micro-fabrication facilities for polymer coating.

7. Appendix I

A. The depth of penetration of a shear wave (δ)

The depth of penetration of a shear wave (δ) in a Newtonian medium is given by the equation shown below:

$$\delta_n = \frac{1}{\left(-\omega_n\left(\frac{\rho_m^2}{C_m^2+(\omega_n\eta_m)^2}\right)^{1/4}\right)Sin\left(-\frac{1}{2}\left(\arctan\left(\omega_n\frac{\eta_m}{C_m}\right)\right)\right)} \qquad (2)$$

ρ_m = density of medium (kg/m³),
η_m = viscosity of medium (kg/m.s),
C_m = stiffness of medium (N/m²),
ω = angular frequency (rad/s) and
n = harmonic number

B. Basic terminologies of a genetic algorithm

Individual: A solution to the problem is called an individual.
Population: The total number of solutions is called population.
Chromosome: Each individual has a number of chromosomes that represent each parameter (i.e. variables to be determined) of the problem.
Genes: Each chromosome contains a fixed number of genes, the number of genes per chromosome determine the resolution of the total solution. The number of genes per chromosome is mostly determined by the broadness of the range in which each chromosome lies.
Fitness: Every individual has to be weighed according to its fitness. The individual fitness value determines its survival and breeding probability. A higher fitness individual has higher probability of survival.

C. Mason's transmission line model

As seen in fig. 11, the biological process consist of a piezoelectric layer (MTSM sensor) and a non-piezoelectric biological layer. In this model, each layer of load can be represented as a T-network of impedances.

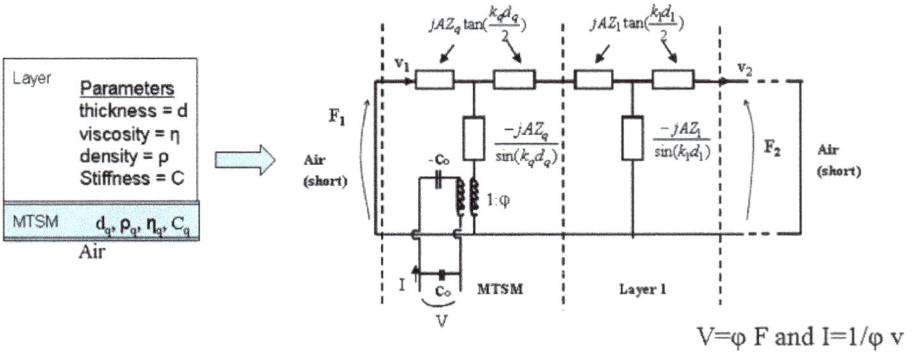

Fig. 11. Mason model representation of non-piezoelectric layers loaded on piezoelectric plate

8. Appendix II (Symbols)

F_1 = input force (N)
F_2 = output force (N)
v_1 = input particle velocity (m/s)
v_2 = output particle velocity (m/s)
A = area of active electrode of MTSM (m^2)
k = propagation constant (m^{-1})
d = thickness (m)
Z = acoustic impedance (acoustic ohm)
I = current (C)
C_0 = static capacitance of MTSM (F)
φ: transformer ratio

9. References

Alig et al, 1996. Ultrasonic shear wave reflection method for measurements of the viscoelastic properties of polymer films. *Rev. Sci. Instrumen*. 68. 1536-1542

Ayad M. M., Nagy L. Torad. 2009. Alcohol vapours sensor based on thin polyaniline salt film and quartz crystal microbalance. *Talanta*. 78. 1280-1285

Behling C, Ralf Lucklum and Peter Hauptmann. 1997. Possibilities and limitations in quantitative determination of polymer shear parameters by TSM resonators. *Sensors and Actuators A: Physical* V. 61, pp. 260-266

Cho N. J., J. Nelson D'Amour, Johan Stalgren, Wolfgang Knoll, Kay Kanazawa, Curtis W. Frank. 2007. Quartz resonator signatures under Newtonian liquid loading for initial instrument check. *Journal of colloid and Interface Science*. 315. 248-254

Cote G. L., R. M. Lec and M. Pishko. 2003. Emerging biomedical sensing technologies and their applications. *IEEE Sensors Journal*. 3. 251-266.

Desa J., Zhang Q., Ergezen E., and Lec. R., 2010 Microparticle manipulation on the surface of a piezoceramic actuator. Measurement Science and Technology. 21 (10) 105803

Dylkov M.S., Sanzharovskii A.T., Zubov P.I. 1966. The effect of thickness on the strength of polymer films. *Mekhanika Polimerov*. 2. 940-942

Edvardsson M., Michael Rodahl, Bengt Kasemo, and Fredrik Holo1k. 2005. A Dual-Frequency QCM-D Setup Operating at Elevated Oscillation Amplitudes. *Anal. Chem*. 77. 4918-4926.

Ergezen E, Appel M, Shah P, Kresh JY, Lec RM, Wootton DM, Real-Time Monitoring of Adhesion and Aggregation of Platelets using Thickness Shear Mode (TSM) Sensor, Biosensors and Bioelectronics, V. 23, #4, pp. 575-82, 2007.

Ergezen E. 2010 Multi-Resonant Thickness Shear Mode (MTSM) Measurement Technique for Quantitative Characterization of Biological Interfacial Processes Drexel University, PhD Thesis

Fredriksson, C., S. Kihlman, M. Rodahl, and B. Kasemo. 1998. The piezoelectric quartz Crystal mass and dissipation sensor: a means of studying cell adhesion. *Langmuir*, 14, 248-251.

Galipeau D.W., Vetelino J.V., Lec R. M. and Freger C.1991. The Study of Polyimide Film properties and Adhesion Using a Surface Acoustic Wave Sensors, ANTEC '91,

Conference Proceedings, Society of Plastic Engineers and Plastic Engineering, Montreal, pp. 1679-1984.

Gautam Garaia and B.B. Chaudhurib.2007. Adistributed hierarchical genetic algorithm for efficient optimization and pattern matching. *Pattern Recognition*. 40. 212-228

Greczylo T. and Deboswka E. 2005. Finding Viscosity of liquids from Brownian motion at students' laboratory. *European Journal of Physics*. 26. 827-833

Hanbury C. M., Miller G. W. and Harris B. R. 1996. Antibody characteristics for a continuous response fiber optic immunosensor for theophylline. *Biosensors and Bioelectronics*. 11. 1129-1138.

Hong S., Ertan Ergezen, Ryszard Lec, Kenneth A. Barbee. 2006. Real-time analysis of cell-surface adhesive interactions using thickness shear mode resonator. *Biomaterials*, 27, 5813-5820

Hook F., Larsson C., Fant C., 2002. Biofunctional Surfaces Studied by Quartz Crystal Microbalance with Dissipation Monitoring. *Encyclopedia of Surface and Colloid Science*. 774-790.

Jiang L., Hossenlopp J., Cernosek R., Josse F. 2003. Characterization of Epoxy Resin SU-8 Film Using Thickness-Shear Mode (TSM) Resonator. *Proceedings of IEEE International Frequency Control Symposium*. 996-982

Johannsmann D., Ilya Reviakine, Elena Rojas, and Marta Gallego. 2008. Effect of Sample Heterogeneity on the Interpretation of QCM(-D) Data: Comparison of Combined Quartz Crystal Microbalance/Atomic Force Microscopy Measurements with Finite Element Method Modeling. *Anal. Chem*. 80. 8891-8899.

K. Kanazawa and Nam-Joon Cho. 2009. Quartz Crystal Microbalance as a Sensor to Characterize Macromolecular Assembly Dynamics. *Journal of Sensors*. doi:10.1155/2009/824947

Katz A. and Ward M. D. 1996. Probing solvent dynamics in concentrated polymer films with a high frequency shear mode quartz resonator. *J. Applied Physics*. 80. 4153

D. Kroger, A. Katerkamp, R. Renneberg and K. Cammann. 1998. Surface investigations onthe development of a direct optical immunosensor. *Biosens. Bioelectron*. 13, 1141–1147.

M. Kunze, Kenneth R. Shull, and Diethelm Johannsmann. 2006. Quartz Crystal Microbalance Studies of the Contact between Soft, Viscoelastic Solids. *Langmuir*. 22. 169-173

Kwoun S., 2006. A Multi-Resonant Thickness Shear Mode (MTSM) Sensor for Monitoring The Formation of Biological Thin Films. PhD Thesis, Drexel University, pp. 125.

Li J, Thielemann C, Reuning U, Johannsmann D. 2004. Monitoring of integrin-mediated adhesion of human ovarian cancer cells to model protein surfaces by quartz crystal resonators: evauation in the impedance analysis mode. *Biosens Bioelectron*. 20. 1333–7.

Liao W., Wei F., Qian X. M., Zhao S. X. 2004. Characterization of protein immobilization on alkyl monolayer modified silicaon (111) surface. *Sensors and Actuators B*. 101. 361-367

Lin Z., Yip M. C., Joseph S. I., Ward D. M. 1993. Operation of an ultrasensitive 30 MHz quartz crystal microbalance in liquids. *Anal. Chem*. 65. 1546-1551.

Liu M., Sun J., Bock C., Chen Q. 2009. Thickness dependent mechanical properties of Polydimethylsiloxane. *J. Micromech. Microeng*. 19. 1-4

Lucklum R. and Hauptman P. 1997. Determination of polymer shear modulus with quartz crystal resonators. *Faraday Discussions*. 107. 123-140.

Lucklum R., Peter Hauptmann. 2000. The quartz crystal microbalance: mass sensitivity, viscoelasticity and acoustic amplification. Sensors and Actuators B 70 _2000. 30–36

Luo C., Schneider T., White R., Currie J., Paranjape M. 2003 A simple deflection testing method to determine Poission's ratio for MEMs applications. *J. Micromech. Microeng*. 13. 129-133

Mads Bruun Hovgaard, Mingdong Dong, Daniel Erik Otzen, and Flemming Besenbacher. 2007. Quartz Crystal Microbalance Studies of Multilayer Glucagon Fibrillation at the Solid-Liquid Interface. *Biophysical Journal*. 93. 2162-2169

Martin S.J and G.C. Frye. 1991. Polymer film characterization using quartz resonators. *Ultrasonic. Symp*. 393-398.

Morrison L.S. and Shin S. U. 1995. Genetically engineered antibodies and their applications to drug delivery. *Advance Drug Delivery Reviews*. 15.147-175.

Nam-Joon Cho, Kay K. Kanazawa, Jeffrey S. Glenn, and Curtis W. Frank. 2007. Employing Two Different Quartz Crystal Microbalance Models To Study Changes in Viscoelastic Behavior upon Transformation of Lipid Vesicles to a Bilayer on a Gold Surface. *Anal. Chem*. 79. 7027-7035.

Ndiritu J.G. and Daniell T. M. 2001. An improved genetic algorithm for rainfall-runoff model calibration and function optimization. *Mathematical and Computer Modeling*. 33. 695-706.

Newton M. I., Evans, C.R Simons, Hughes DC. 2007. Semen quality detection using time of flight and acoustic wave sensors. *Applied Physics Letters*. 90 (15) 154103

F. N. Nunalee, Kenneth R. Shull, Bruce P. Lee, Phillip B. Messersmith. 2006. Quartz Crystal Microbalance Studies of Polymer Gels and Solutions in Liquid Environments. *Anal. Chem*. 78. 1158-1166.

Roach, P; McHale, G; Evans, CR, Shirtcliffe NJ, Newton MI. 2007. Decoupling of the liquid response of a superhydrophobic quartz crystal microbalance. *Langmuir*, 23, 9823-9830

Rosenbaum J.F. 1998. Bulk acoustic Wave Theory and Devices. Boston, MA: Artech House Publishers

Saluja A., Kalonia S. D. 2005. Application of Ultrasonic Shear Rheometer to Characterize Rheological Properties of High Protein Concentration Solutions at Microliter Volume. *Journal of Pharmoceutical Sciences*. 94. 1161-1168

Sapper A., Wegener J., Janshoff A., 2006. Cell Motility Probed by Noise Analysis of Thickness Shear Mode Resonators. *Anal. Chem*. 78, 5184-5191.

Sivanandam S. N. and Deepa S. N. 2008. Introduction to Genetic Algorithms. Springer Berlin Heidelberg. New York. 34-35

Song Wang and Atam P. Dhawan. 2008. Shape-based multi-spectral optical image reconstruction through genetic algorithm based optimization. *Computerized Medical Imaging and Graphics*. 32. 429–441

Su H and Thompson M. 1996. Rheological and interfacial properties of nucleic acid films studied by thickness - shear mode sensor and network analysis. *Can. J. Chem*., 74, 344-358

Sun K., R. M. Lec, Cairncross R. A., Shah P., Brinker C. J. 2006. Characterization of Superhydrophobic Materials Using Multiresonance Acoustic Shear Wave Sensors. *IEEE Transactions on Ultrasonics Ferroelectrics and Frequency Control*. 53. 1400- 1403

Szabad, Z. Sangolola, B. and McAvoy, B. 2000. Genetic optimisation of manipulation forces for co-operating robots. *IEEE International Conference on Systems, Man, and Cybernetics*. 5. 3336-3341

Urbakh M. and Leonid Daikhin. 2007. Surface morphology and the quartz crystal microbalance response in liquids. *Colloids and Surfaces A. Physicochemical and Engineering Aspects*. 134. 75-84

Voinova M., Jonson M., Kasemo B., 2002. Missing Mass effect in biosensor's QCM Application. *Biosensors and Bioelectronics*. 17, 835-841

Voros J. 2004. The Density and Refractive Index of Adsorbing Protein Layers. *Biophysical Journal*. 87. 553-561

Yang D., Huang C., Lin Y., Tsaid D., Kao L., Chi C. Lin C., 2003 Tracking of secretory vesicles of PC12 cells by total internal reflection fluorescence microscopy. *Journal of Microscopy*, 209, 223-227

Young-Doo Kwon, Soon-Bum Kwon, Seung-Bo Jin and Jae-Yong Kim. 2003. Convergence enhanced genetic algorithm with successive zooming method for solving continuous optimization problems. *Computers and Structures*. 81. 1715-1725

Wegener, J, et al., 2000. Analysis of the composite response of shear wave resonators to the attachment of mammalian cells. *Biophys. J*. 78, 2821–2833.

Werner C. 2008. Interfacial Phenomena of Biomaterials. *Polymer Surfaces and Interfaces*. Dresden, Germany. Springer-Verlag Berlin Heidelberg. pp. 299

Westphal S. and Bornmann A. 2002. Bimolecular detection by surface plasmon enhanced Ellipsometry. *Sensors and Actuators B*. 84. 278-282

Zhang Q, Desa J., Lec R., Yag G., and Pourrezaei K. 2005, Combination of TSM and AFM for Investigating an Interfacial Interaction of Particles with Surfaces. *Joint IEEE International Frequency control Symposium (FCS) and Precise Tie and Time Interval (PTTI) Systems and Applications Meeting*. 4490454

Zhuang H., Pin Lu, Siak Piang Lim, and Heow Pueh Lee. 2008. Effects of Interface Slip and Viscoelasticity on the Dynamic Response of Droplet Quartz Crystal Microbalances. *Anal. Chem*. 80. 7347-7353

12

Analysis of Biological Acoustic Waves by Means of the Phase–Sensitivity Technique

Wojciech Michalski[1], Wojciech Dziewiszek[2] and Marek Bochnia[2]
[1]*Technical University of Wrocław,*
[2]*Medical University of Wrocław*
Poland

1. Introduction

The analysis of hearing mechanisms and research on the influence of various internal (pathologies, ageing) and external (trauma, vibration, noise) factors on sound perception are usually done using acoustic waves induced in the external ear canal. Stimuli which have been used for this purpose are: clicks, tone bursts, half-sine-waves, single tones or pairs of tones. The Corti organ's responses to the external stimuli have either an electric or acoustic character. In the former case, these are cochlear microphonics (CMs) picked up from the surface or from the inside of the cochlea, which are usually used as an indicator of damage to the organ of Corti in animals. In the latter case, these are acoustic waves that appear in the external ear canal as a result of stimulation. The acoustic waves have an important clinical value. Taking into account the presence of nonlinear distortions in the cochlea, the waves that appear after stimulation with a pair of tones are called distortion product otoacoustic emissions (DPOAE).

In studies on CMs, the origin of stimulating waves is often a single earphone (controlled by a generator of defined, often periodical, electrical signals) placed in the external auditory canal. In studies on DPOAE, a probe with two miniature earphones and one microphone is placed in the external auditory canal. The earphones are controlled by two generators of frequencies f_1 and f_2 and the microphone converts the returned DPOAE wave with a combination frequency, e.g. $f_3 = 2f_1 - f_2$, into an electrical signal.

The acoustic wave which induces CM signals is usually a periodic wave, while the waves inducing DPOAE signals consist of two pure tones. Thanks to the easy access to the output(s) of the generator(s) the phase-sensitive detection (PSD) technique can be used to measure both CM and DPOAE signals. Very weak (even below single microvolts) CM and DPOAE signals originating from the unimpaired cochlea can be measured in this way. Thanks to this technique signals obscured by other disturbing sources (even thousand times larger) can be measured accurately. This is possible because the phase-sensitive detector singles out the input signal with a specific reference frequency while signals with frequencies different from the reference are rejected. The fundamentals of this technique and its measuring potential are described in section 2.

In section 3, the authors' own experiments aimed at determining the effect various factors on the electrical function of the Corti organ are described. The factors include: vibration

(3.3), ototoxic medicines (3.4) and laser beams used in ear microsurgery (3.6). The role of the signal phase in the measurements is given special attention.
In section 4, experiments involving acoustic waves being nonlinear products of the Corti organ are presented. The individual subsections describe the way in which the PSD technique is applied (4.1), compare the latter with the previously used methods (4.2) and discuss the authors' own experiments in which the phase-sensitive technique is employed to measure DPOAE signals (4.3) and to measure simultaneously DPOAE signals and CMDP (cochlear microphonic distortion product) signals (4.4). It is shown that the phase of DPOAE signals plays an essential role in otoacoustic emission studies.
All the experiments described in sections 3 and 4 were carried out on coloured guinea pigs, each weighing 500-650 g, being under general ketamine/xylazine (15 mg/kg and 10 mg/kg body weight, respectively) anaesthesia. A Homoth measuring probe was placed in the external auditory meatus of the animals. The probe contained two mini earphones and a standard microphone. Prior to the measurements the probe had been graduated in a Brüel&Kraej artificial ear 4144, using a measuring amplifier 2607 made by the same company. Permission to carry out the experiments had been given by the Bioethical Committee in Wrocław.
Section 5 presents the final conclusions and discusses the future of the phase sensitive detection technique in investigations into the function of the cochlea exposed to various hazards.

2. Phase-sensitive detection technique

2.1 Fundamentals

The measurement apparatus based on the phase-sensitive detection technique is called a *lock-in amplifier* or a *lock-in nanovoltmeter*. Lock-in measurements require a frequency reference which should be strictly connected with a fixed frequency of the function generator used in the experiment. The reference signal can be either a square wave or a sinusoid. A block diagram of a typical lock-in amplifier is shown in fig. 1.

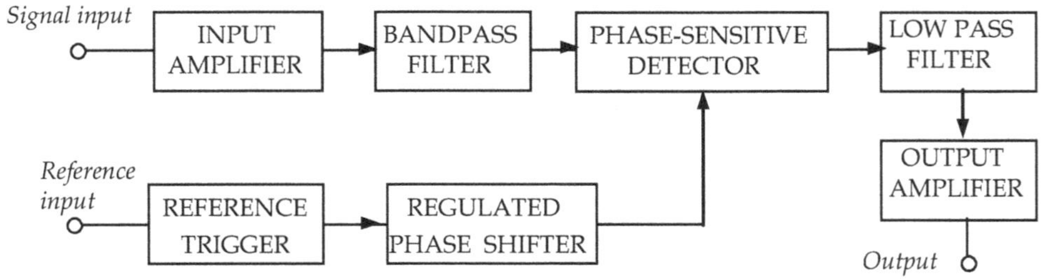

Fig. 1. Block diagram of lock-in amplifier with single phase-sensitive detector

Let us assume that the input signal can be described as:

$$V_{sig} = A_{sig} \cos(\omega_1 t + \alpha_0), \tag{1}$$

and the reference signal as:

$$V_{ref} = A_{ref} \cos(\omega_2 t + \beta_0) \tag{2}$$

The two signals have different amplitudes, frequencies and initial phases. At the phase-sensitive detector inputs there are signals with unchanged frequencies, but with different amplitudes and phases: $V_{sig} = A_{sig1} \cos(\omega_1 t + \alpha_{01})$ and $V_{ref} = A_{ref1} \cos(\omega_2 t + \beta_{01})$. The output of the PSD is simply the product of the two sine waves

$$V_{PSD} = A_{sig1} \cos(\omega_1 t + \alpha_{01}) \cdot A_{ref1} \cos(\omega_2 t + \beta_{01})$$

$$= 0{,}5 A_{sig1}\, A_{ref1} \left[\cos\left[(\omega_1 - \omega_2)t + (\alpha_{01} - \beta_{01})\right] + \cos\left[(\omega_1 + \omega_2)t + (\alpha_{01} + \beta_{01})\right]\right] \tag{3}$$

At the output of the PSD there are two signals: a slow-changing signal with differential frequency $(\omega_1 - \omega_2)$ and a signal with overall frequency $(\omega_1 + \omega_2)$. If the PSD output signals are passed through a low pass filter, the fast AC signal will be removed. When the frequencies of the two signals are approximately equal ($\omega_1 \approx \omega_2$), the filtered PSD output is a slowly changing DC signal proportional to the signal amplitude and $\cos(\alpha_{01}-\beta_{01})$. When $\omega_1 = \omega_2$, the filtered signal is exactly a DC signal. By adjusting the phase of the reference signal one can make $(\alpha_{01}-\beta_{01})$ equal to zero, in which case only B_{sig} can be measured. This is true if both initial phases α_0 and β_0 do not change over time, otherwise $\cos(\alpha_{01}-\beta_{01})$ will change over time and V_{out} of the lock-in amplifier will not be a DC signal.

The phase dependency of the output voltage of the lock-in amplifier with one PSD unit can be eliminated by adding a second PSD multiplying the same measured signal by a reference signal shifted by 90^0. A block diagram of the lock-in amplifier with a double PSD is shown in fig. 2.

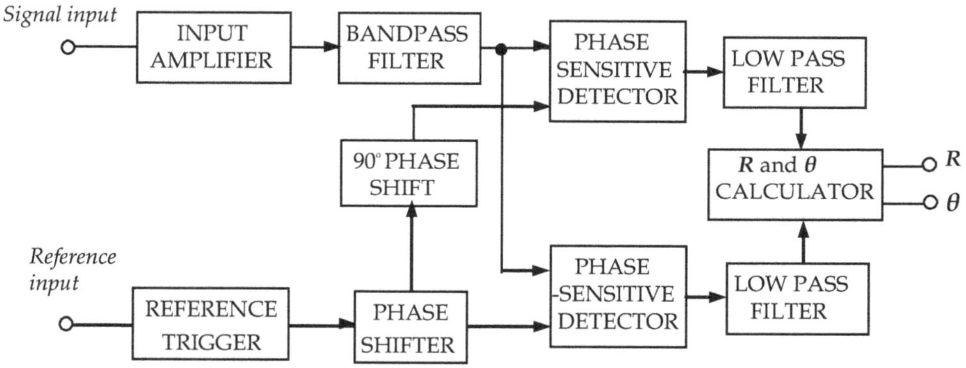

Fig. 2. Block diagram of lock-in amplifier with double PSD

The output of the second PSD, filtered by the low pass filter, is proportional to the signal amplitude and $\sin(\alpha_{01} - \beta_{01})$. Now there are two outputs: $X = A_{sig1} \cos(\alpha_{01} - \beta_{01})$ and $Y = A_{sig1} \sin(\alpha_{01} - \beta_{01})$. When

$$R = \sqrt{\left(A_{sig1} \cos(\alpha_{01} - \beta_{01})\right)^2 + \left(A_{sig1} \sin(\alpha_{01} - \beta_{01})\right)^2} = A_{sig1} \tag{4}$$

the phase dependency is removed. Phase difference (α_{01}-β_{01}) can be measured according to

$$\alpha_{01} - \beta_{01} = \tan^{-1}(Y/X) \tag{5}$$

The first lock-in amplifiers were based on analogue technology. The measured signal and the reference were analogue voltage signals and they were multiplied in an analogue PSD. The results of multiplication were filtered through a multistage RC filter. In such lock-ins the reference signal phase at the PSD input had to be manually adjusted to the phase of the measured signal so that $\cos(\alpha_{01} - \beta_{01}) = 1$. It was technically difficult to perform measurements by means of such lock-ins and it was practically impossible to register the amplitude and phase changes of the measured signals. Digital technology made it possible to build lock-in amplifiers in which both signal and reference inputs were multiplied and filtered digitally. Dual phase-sensitive detection eliminated the need for manual phase adjustments and enabled the simultaneous measurement of signal amplitude and phase. Such simultaneous measurements can be performed in *real time*, practically without any delay to the inducing signal. It also became possible to register short (below 0.1s) and slow changes in amplitude and phase over time.

2.2 Application of double-phase detection

The PSD technique offers greater measuring possibilities owing to the fact that:

1. the signal fed to the examined object may have various periodical waveforms,
2. the examined object can be linear or nonlinear,
3. the reference signal frequency can be equal to the frequency of the signal being delivered to the examined object, but it also can be an integral multiplicity (or submultiplicity) of this frequency.
4. two coherent signals can be introduced to the examined (usually nonlinear) object; the reference signal can be used at a frequency that is a linear combination of the frequencies of the inducing signals.

The basic experimental setup is shown in fig.3.

The generator used in the setup has two synchronous outputs. One of them (sync. output) supplies a TTL signal. Depending on the frequency of the reference signal one can measure the first harmonic, higher harmonics and subharmonics. In switch position 1 (fig.3), the reference signal is taken directly from the generator's synchronous output whereby the first harmonic can be measured. In switch position 2, the synchronous signal is multiplied by integral number n whereby the n-harmonic can be measured. In order to measure the n-subharmonic the generator's synchronic output must be divided by integral number n (the switch in position 3).

In the simplest case, the waveform of the signal directed to the examined object is sinusoidal. When the examined object is linear, using the PSD technique one can very precisely (with an accuracy of 1 nanovolt) measure the electrical response of the object. If the signal is a simple square wave or another periodical wave with frequency f, the examined linear object does not change the signal spectrum and the filtered PSD output is a DC signal proportional to the root mean square (rms) of the first component of the signal.

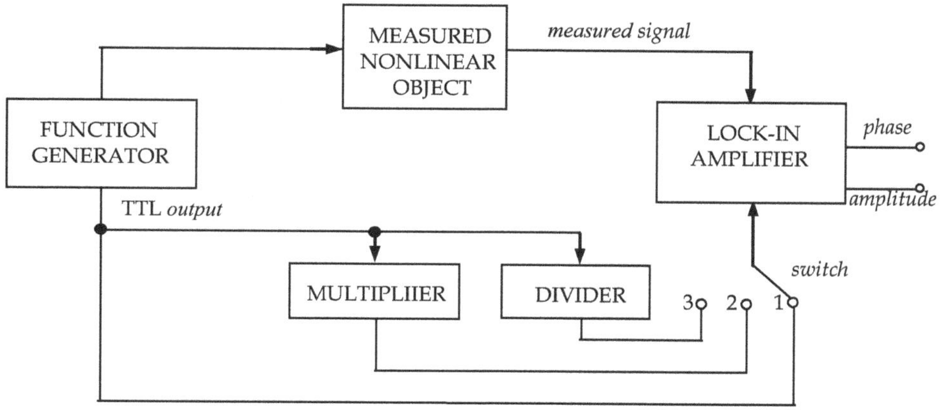

Fig. 3. Basic experimental set-up with lock-in amplifier for measuring first harmonic (1), higher harmonic (2) and subharmonic (3)

The situation becomes more complicated when the examined object is nonlinear. The signal spectrum at the PSD input differs from the one at the generator output (the nonlinear object changes the input signal spectrum). This is true for all the signal waveforms, including the sinusoidal one. Then the lock-in amplifier measures the rms of both the first harmonic and the n-harmonic (n-subharmonic). When harmonic or subharmonic distortion is measured, the function generator supplies a pure sinusoidal signal without any harmonics. The basic experimental setup shown in fig. 3 was used to carry out experiments described in section 3.

2.3 Nonlinear object testing with two synchronous signals

The double phase-sensitive detection technique and modern digital technologies offered new possibilities of examining nonlinear objects. An example of the measuring systems which have been developed is shown below (fig. 4). The main component of the setup is a digital sinus generator of three signals with synchronous frequencies. Two of the signals are delivered to the examined object while the third one serves as a reference signal. The frequency of the third signal is a linear combination of the frequencies of the other two signals.

Let us assume that the output-input function for the examined object can be described by the formula:

$$V_{out} = C \cdot V_{in}^{3}, \tag{6}$$

where C is a constant value. The input signal is the sum of signals with different amplitudes, phases and frequencies and so:

$$V_{out} = C\left[A_1 \cos(\omega_1 t + \alpha_{01}) + A_2 \cos(\omega_2 t + \alpha_{02})\right]^3 \tag{7}$$

After trigonometric conversions it is possible to receive a signal frequency spectrum at the object's output. The frequencies, amplitudes and phases of the particular spectral components (assuming that the examined object does not change its phase relations, i.e. it is characterized by pure resistances) are shown in table 1.

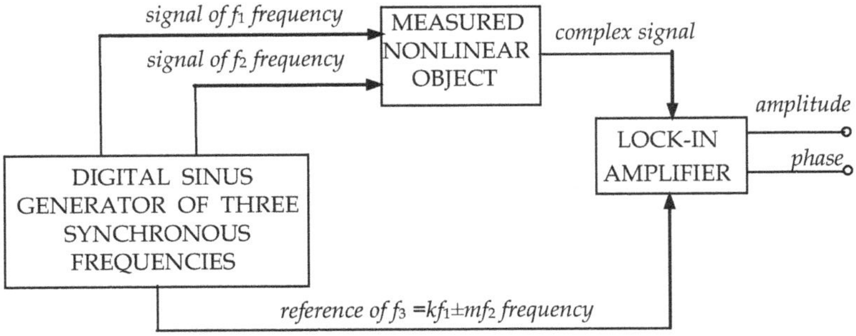

Fig. 4. Block diagram of experimental setup for more complicated studies of nonlinear objects

Frequencies of spectral components	Amplitudes of spectral components	Phase of spectral components
ω_1	$1{,}5A_1C(0{,}5A_1^2+A_2^2)$	a_{01}
ω_2	$1{,}5A_2C(0{,}5A_2^2+A_1^2)$	a_{02}
$3\,\omega_1$	$0{,}75CA_1^3$	$3\,a_{01}$
$3\,\omega_2$	$0{,}75C\,A_2^3$	$3\,a_{02}$
$2\,\omega_1+\omega_2$	$0{,}75CA_1^2\,A_2$	$2a_{01}+a_{02}$
$\omega_1+2\,\omega_2$	$0{,}75CA_1\,A_2^2$	$a_{01}+2\,a_{02}$
$2\,\omega_1-\omega_2$	$0{,}75CA_1^2\,A_2$	$2a_{01}-a_{02}$
$\omega_1-2\,\omega_2$	$0{,}75\,CA_1\,A_2^2$	$a_{01}-2\,a_{02}$

Table 1. Exemplary spectrum at output of object with 3rd-order nonlinearity, tested by pair of pure tones

The amplitude of each of the spectral components can be measured using this technique if a proper reference signal frequency is selected. The interpretation of phase shifts between the particular spectral components is much more complicated and requires taking into account the phase shifts introduced by the examined object. Moreover, the phase shifts introduced by the examined object may be a function of frequency and so they may be different for each spectral component. The technique was used to examine nonlinear distortions during the stimulation of the cochlea by a pair of pure tones. The results of this research are presented in section 4.

3. Using PSD technique to study cochlear potentials

3.1 Measuring techniques

Cochlear potentials are biopotentials of the inner ear. They are described as electrical signals arising in response to the acoustic stimulation (usually by a click or a tone) of the organ of

Corti. For the first time they were registered by Wever and Bray in 1930 (Wever &Bray, 1930). The discovery of the signals made it much easier to examine the function of the inner ear and made it possible to assess the impact of various external and internal factors on this function. It is widely believed that cochlear microphonics (CMs) are generated mainly by outer hair cells (OHCs). Therefore it seems reasonable to use CMs as an indication of the OHC function. On the basis of measurements performed over a long period (e.g. a few weeks or months) one can assess if given hearing damage is temporary or permanent. The CM signal originating from different places in the human ear (or the animal ear) can be recorded. In humans CMs are usually picked up from the round window during surgical procedures performed on patients with various hearing pathologies. There are much fewer reports describing the reception of CM signals from the promontory or the ear canal near the eardrum. The past and present studies of the mechano-electrical cochlear function (based on the reception of CMs) are conducted mainly on animals, using: *in vivo* preparations of anaesthetized animals with positive Preyer's reflex, *in vitro* preparations of the cochlea or *in vitro* preparations of the hair cells. As regards the research into the impact of various external and internal factors on the hearing organs, the *in vivo* studies seem to be most clinically valuable.

In the 1930s and 1940s CMs were measured at the cochlea's round window (Wever & Bray, 1930). In most animals the round window is relatively easily accessible and so measuring electrodes were usually placed on it or in its direct proximity. In the first years after the discovery of inner ear potentials, CM signals were measured by a single active probe. Several years later the first mapping of CMs on the cochlear surface was described (Thurlow, 1943). It was probably the first attempt ever to place the probe so close to the source of cochlear microphonics. CM potentials are continued to be measured at the cochlea's round window today (Brown, 2009). This measuring technique was not abandoned after the introduction of very sensitive (but invasive) procedures (Tasaki et al., 1952). Tasaki monitored CMs using a pair of active intracochlear electrodes in the basal turn (one electrode in scala tympani, the other in scala vestibuli). The electrodes were connected to a balanced differential amplifier. The reference electrode was placed on the neck muscles. This enables the measurement of the potentials very close to the organ of Corti and eliminates the auditory nerve potentials. The largest drawback is the mixing of perilymph and endolymph when the probe is introduced.

A new recording technique has been described by Carricondo (Carricondo at al., 2001). In this technique, CM potentials are recorded by subcutaneous electrodes in animals or by surface electrodes in humans. Two active electrodes are placed on the mandibular muscles while the reference electrode is located on the head's vertex. All the three electrodes are connected to a differential amplifier. The signal coming from the amplifier's output is filtered and subsequently averaged through in-phase synchronization with the sound stimuli.

3.2 History of CM studies in Wroclaw

In 1960 at the Wroclaw University of Technology an oscillograph was built and used as part of an experimental setup for registering cochlear microphonics (CMs). In the following years Jankowski and Giełdanowski started a series of experiments on animals - first on cats, later on guinea pigs (Jankowski et al., 1962). A Biopotentials Research Workshop was founded, where biopotentials were measured after damage to the inner ear or the skull, in acoustic trauma, in hypothermia or hypoxia, after the administration of certain drugs, and so on. Figure 5 shows a schematic of the experimental setup.

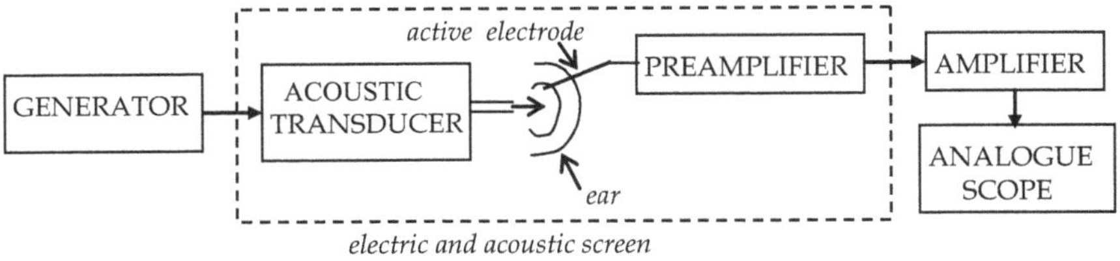

Fig. 5. Schematic of experimental setup used by Jankowski & Giełdanowski

The experimental animals (under urethane anaesthesia, which does not diminish CM voltages) underwent ear surgery: the bones were exposed and drilled until the round window was revealed. Platinum electrodes (platinum wires 0.1 mm in diameter, coated with PMMA) were used for CM measurements. The bare end (not coated with PMMA) of the platinum wire was brought into contact with the round window membrane (without damaging it). An injection needle was used as the other electrode. It was inserted into the muscles around the surgical wound. An example of an CM oscillogram from Ziemski's work (Ziemski, 1970) is shown in fig. 6.

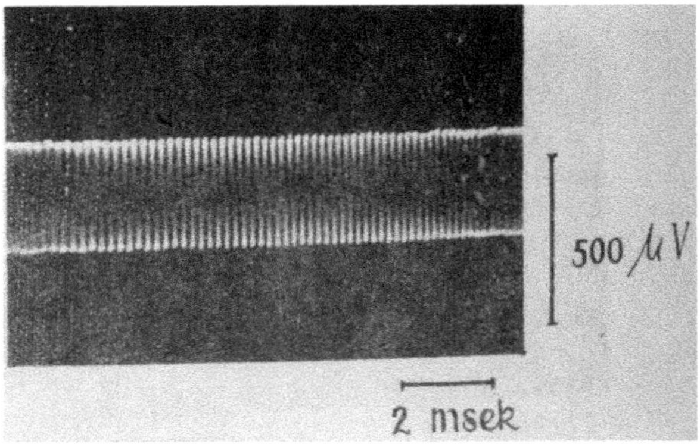

Fig. 6. Example of CM oscillograph record (stimulus tone parameters: f=4096 Hz, acoustic pressure level - 60dB)

In the middle of the 1990s a new surgical approach, making it possible to expose the whole cochlea in guinea pigs, was proposed. Skin was cut from the occipital part of the skull, around the angle of the mandible up to the place of about 0.5 cm above the angle of the animal's snout. The temporal muscles were removed, the mandible partially resected and the stylo-hyoid muscle cut. Bones were drilled, leaving the tympanic membrane intact. In the 1990s the present authors used the phase-sensitive detection technique to measure CMs. A patent was applied for in 1996 and the technique was patented in December 2000 (Patent PL no. 180060). A report on the use of a similar method was published by Kobayashi (Kobayashi, 1997). A schematic of the first setup used to measure CMs by the PSD technique is shown in fig. 7.

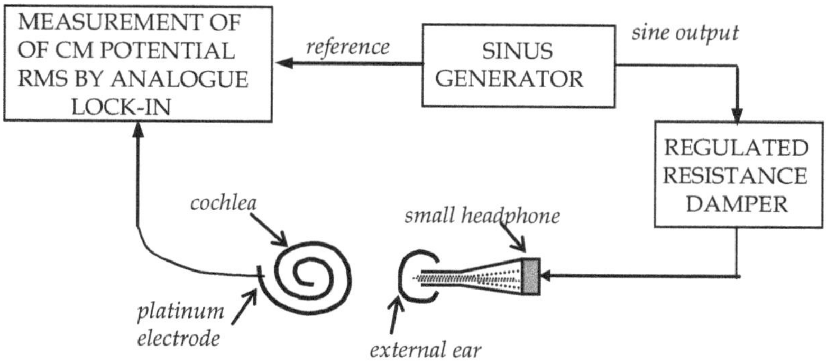

Fig. 7. Schematic of first setup used to measure CMs by phase-sensitive detection technique

Thanks to the new surgical approach combined with the phase-sensitive detection technique it became possible to create a map of CM potential amplitude and phase distributions on the cochlear surface. For this purpose, the active electrode would be fixed in six different points on the cochlea's surface (on the apex, one point at the third turn, two points at the second turn and two points at the basal turn). When the electrode was fixed in any of the six points, the frequencies of the stimulating acoustic wave would be successively selected from the measurable range. Three different acoustic wave pressures (60 dB, 80 dB and 95 dB) would be set for each of the measuring frequencies. This means that 18 different amplitude-phase (A-P) values of the CM signal were measured in each of the six points on the cochlea's surface. In this way 18 different A-P distribution patterns would be obtained for the guinea pig cochlea. Six of them (three for f = 260 Hz and three for f = 8 kHz) are shown in fig. 8. Phase in each measuring point is related to phase on apex at 60 dB.

It was found that CMs had a different phase and amplitude in the different parts of the cochlea, which was due to the fact that each of the six measuring points was located at a different distance from the CM sources inside the organ of Corti.

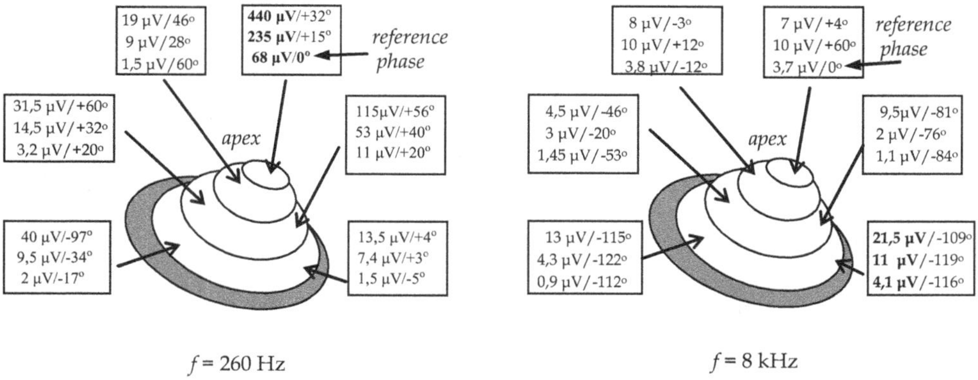

Fig. 8. CM potential amplitude and phase in six different points on cochlea surface, measured at 260 Hz and 8 kHz. At each of six points there are three amplitude/phase values: upper for 90 dB, central for 80 dB and lower for 60 dB

In order to unify CM measurements, two distinctive and universal measurement points were established: the cochlea's apex for the frequencies of 260, 500, 1000 and 2000 Hz and the cochlea's base for 4000 and 8000 Hz. Phase in each measuring point is related to phase on apex at 60 dB.

3.3 Influence of whole-body vibration on inner ear

Vibration is one of the most widespread injurious factors in the environment of civilized man (Palmer et al., 2000a, 2000b). The energy absorbed can have a pathological effect on all the tissues and organs of the body, although the consequences of exposure to vibration do not present a uniform clinical picture (Jones, 1996; Seidel & Heide, 1986). Because all machines and vibration devices also produce noise, usually the combined effect of the two factors is examined (Castelo Branco, 1999). There is a prevalent view that mechanical vibrations exert only a weak, additionally traumatic influence on the hearing organ (Seidel, 1993). Several experimental investigations into the harmfulness of vibration were carried out on animals (Hamernik et al., 1980, 1981). Changes in the hearing organ most often would be found in the hair cells (Rogowski, 1987). This made us undertake our own research in the 1990s. In order to determine the impact of long-term general vibration on the inner ear it was necessary to: 1) design and built noiseless vibration apparatus, 2) subject several groups of animals to general vibration (defined by controlled parameters over different periods of time) and 3) evaluate selected parts of the organ of hearing, using norms based on values derived from a control group.

In order to ensure proper experimental conditions, i.e. sinusoidal (10 Hz) vertical (5 mm) shaking, a device consisting of an electric impulse generator, a power amplifier and an impulse exciter was built (fig. 9). Experiments were carried out on young, coloured guinea pigs of both sexes weighing 240-360g. Fifty six animals with the normal Preyer reflex and without otoscopically detectable changes were used. The control group (group m0) consisted of 20 of the animals and served to establish functional and morphological norms. In order to avoid changes due to aging being interpreted as the effects of vibration, the control group was examined after a seven-month stay (6+1 months = duration of the longest experiment + a rest) in an animal house. The study group consisted of 36 guinea pigs divided into two subgroups of 18 animals each. Each subgroup was subjected to vibration over different periods, i.e. 30 (group m1) and 180 (m6) days. These were in fact respectively 22 days (5 days/week, 6 hours/day = 132 hours) and 132 days (792 hours). After the experiment and a one-month (30 day) rest, the animals which were in good general condition and without otoscopically detectable changes were qualified for functional and morphological investigations.

Cochlear microphonics were measured under urethane anaesthesia, using the PSD technique and the setup schematically shown in fig. 3 (the switch in position 1). CMs were picked up from the apex of the cochlea for the frequencies of 250 Hz, 500 Hz, 1 kHz and 2 kHz and from the region of the round window for 4 kHz and 8 kHz, using a platinum needle electrode. For the two study groups and the control group, a total of 6048 data values were taken for the bilaterally examined pulse wave frequencies (260 Hz-8 kHz) and intensities (55 dB-95 dB).

The results of the CM measurements were subjected to statistical analysis. The aim was to find out whether the experiment had any influence on CMs and, if so, what that influence was. The questions asked were: 1) are there statistically significant differences between the

CM voltages obtained from the control groups and the study groups, and 2) are there statistically significant differences in the CM voltages obtained within the study groups? The CM values obtained from the healthy animals showed considerable individual differences, and their distribution showed neither normalcy nor log-normalcy. Therefore all the experimental samples were examined using non-parametric tests. The K-S Lilliefors test showed: 1) for control group m0 compared with study groups m1 and m6, a significant decrease in CMs for the frequencies of 260 Hz, 1 kHz and 2 kHz, and 2) for m1 compared with m6, a decrease in CM for the frequencies of 260 Hz and 2 kHz. The Kruskall-Wallis test confirmed the results of the K-S Lilliefors test as regards the location and nature of the changes.

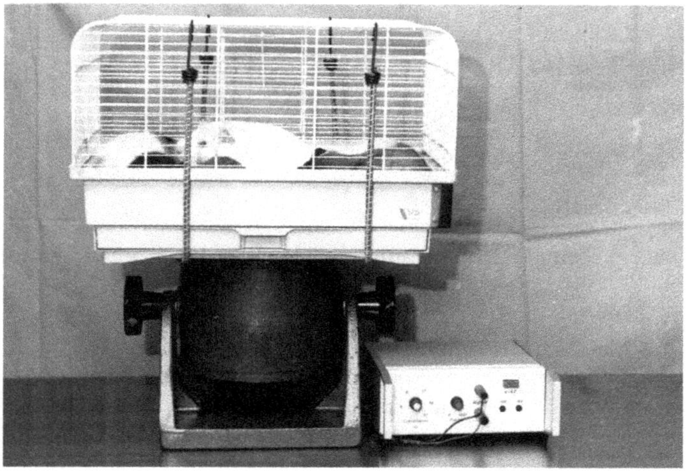

Fig. 9. Cage with animals exposed to vibrations

The results of the investigations indicated possible greater damage to the hair cells in the forth and third turnings of the cochlea. Further morphological examinations were needed to verify this observation. After the bilateral CM measurements the animals were decapitated and samples were prepared for SEM examinations of the sensorial epithelium. The samples were examined and photographed using a scanning DSM 950 microscope. The influence of general vibration on the organ of Corti was assessed on the basis of the condition of the hair cells, taking into consideration their disorganization, deformation, mutual adhesion and any reduction in the number of cilia.

SEM examinations were carried out on 20 cochleae from the control group animals and on all the animals in the two study groups. In the healthy animals, the sensorial epithelium was found to be normal in every case, but in each of the study groups the above mentioned damage was observed. It usually occurred in the OHC region of the apex, and its extent gradually increased in the direction of the cochlea's base (up to the second turning). OHC3 was found to be most susceptible to vibratory trauma. Cell damage decreased from the circumference to the modiolus, and the OHCs showed considerably greater resistance to vibration (fig.10). Undoubtedly, the observed damage to the sensorial epithelium resulted from mechanical vibration, and its severity clearly increased with the duration of the experiment. Consequently, the mechanism of deterioration in hearing in all the frequency ranges (especially at low and average frequencies) in persons subjected to whole-body vibration could be discovered by analyzing the observed changes.

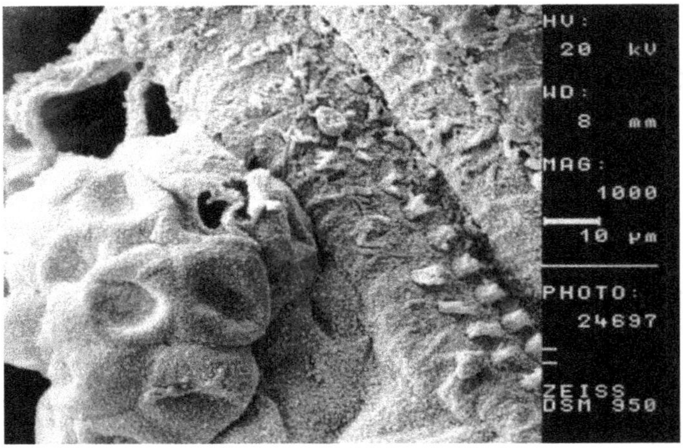

Fig. 10. Group M6, 4th cochlear turning: numerous lesions of hair cells and damage to Hensen's cells

3.4 Studies of gramicidin ototoxicity

Polypeptide antibiotics are used in a variety of clinical situations. Their molecules contain a specific chain of aminoacids and a non-aminoacidic part (e.g. fatty acids in polymyxins or glycopeptide in vancomycin). They are generally effective against Gram-positive bacteria, except for polymyxins which are effective against Gram-negative bacteria. They act by disrupting the selective permeability of bacterial cellular membranes. Despite their long history, polymyxins have had a limited clinical use due to the large number of side effects. Currently they are used primarily for topical treatment (Wadsten at all, 1985).

Since no descriptions of the effects of the systemic administration of gramicidin on the inner ear could be found in the literature, the authors decided to examine CMs and to compare the ototoxic effects after the systemic and topical administration of gramicidin. Also the inner ear of animals which received i.m. injections of gramicidin were examined using a DSM 950 scanning electron microscope (Bredberg at al., 1970; Davis, 1983) .

The research was conducted on 70 young, coloured guinea pigs. All the animals showed the positive Preyer reflex and no pathologies under otoscopic examinations. The experimental animals (G) were divided into 5 subgroups, depending on the drug administration mode and the administered dose. Each experimental subgroup (G1-G5) consisted of 8 randomly chosen animals. Subgroups G1-G3 received respectively 2, 5 and 10 mg of gramicidin/kg i.m., once per day, for 14 consecutive days. The animals from subgroups G4 and G5 were administered a 0.25% and 10% solution of gramicidin suspended on a haemostatic sponge placed on the round window.

The control group (K) consisted of 30 animals randomly divided into 2 subgroups (K1 and K2). The animals in control subgroup K1 were injected with normal saline solution once per day for 14 consecutive days. The animals in subgroup K2 were administered normal saline solution placed on the round window. One day after the last injection (the 15th day of the study) electrophysiological measurements were carried out on the animals in subgroups G1-G3 and K1. Then their cochleae were removed for SEM examinations. In the case of the animals belonging to subgroups G4, G5 and K2, CM measurements were performed after removing the haemostatic sponge from both ears and allowing the round windows with their surroundings to dry (Gale & Ashmore, 1977).

Cochlear microphonics (CMs) were investigated under urethane anaesthesia, using the PSD technique and the setup schematically shown in fig. 3 (the switch in position 1). CMs were picked up from the apex of the cochlea for the frequencies of 260 Hz, 500 Hz, 1 kHz and 2 kHz and from the region of the round window for 4 kHz and 8 kHz by means of a platinum needle electrode. As regards study subgroups G1-G5 and control subgroups K1 and K2, a total of 7560 data values were taken for the examined frequencies (260 Hz-8 kHz) and intensities (55 dB-95 dB). The results of the CM measurements were subjected to statistical analysis (the t-Student test).

Gramicidin administered systemically in a dose of 2 mg/kg led to a significant (38%) decline in CM voltage in K1 subgroup animals for the frequencies of 260 Hz and 2 kHz. For the other frequencies the drop in CMs amounted to about 15%, except for the 4 kHz at which a slight improvement was observed for sound levels between 55 and 70 dB. A significant drop in CMs was observed in subgroup G2 at 2 kHz and sound levels above 70 dB. At 95 dB the decline in CMs was 30% larger than in the G1 animals. The changes in the G2 animals relative to G1 were even more significant at 500 Hz, 1 kHz and 8 kHz. The animals receiving 10 mg/kg of gramicidin showed lower CMs than the ones registered in all the examined frequency ranges for control subgroup K1. The largest drop was registered at 2 kHz (31% lower than in the K1 control subgroup). The smallest changes were observed at 8 kHz. In subgroups G1-G3, the largest differences in CMs were observed at 4 kHz for all the sound levels.

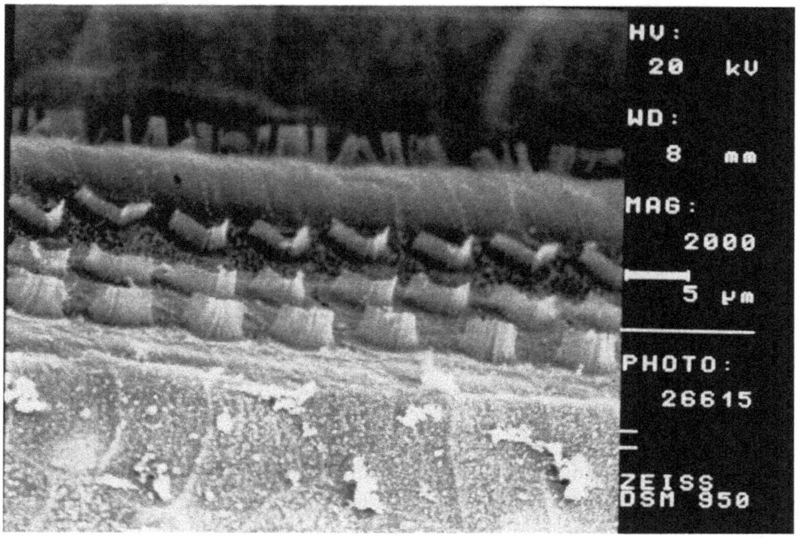

Fig. 11. Group K1, 2nd cochlear turn: unchanged sensory epithelium

In the animals receiving topical 0.25% gramicidin solution (G4), a significant drop in CMs (in comparison with control K2) was observed at 1 kHz and 2 kHz. In group G5 (where the animals were administered 10% gramicid in solution on the round window) a drop in CMs was observed also at 4 kHz and 8 kHz. At low sound levels the largest falls in CMs were observed in subgroup G4.

In the G1 and G2 animals no damage to the sensory epithelium was found under SEM. The destruction of cochlear hair cells occurred in the G3 animals. The changes were most visible in OHC3 cells in the cochlea's third turning.

To sum up, the systemic administration of gramicidin leads to greater disruptions of the bioelectric functions of the inner ear than local, topical administration (Linder at al., 1995).

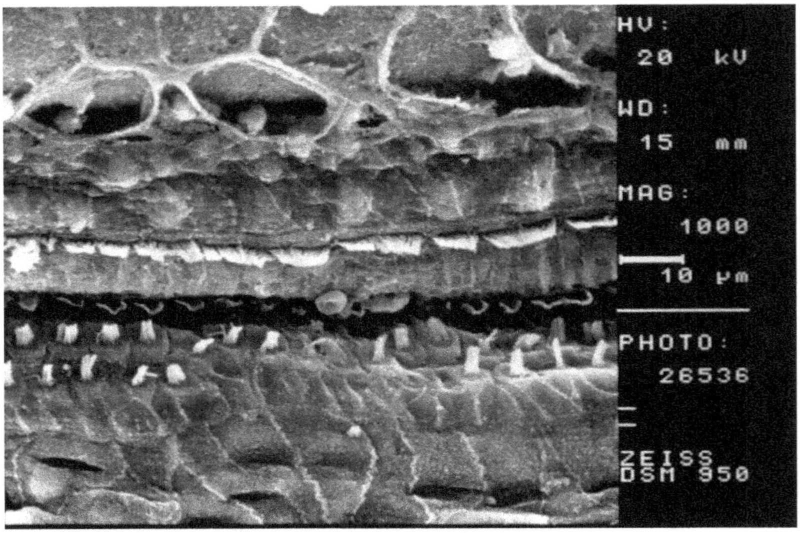

Fig. 12. Group G3, 3rd cochlear turn: numerous lesions in OHC3 cells and structural changes in cilia

3.5 CM amplitude and phase changes caused by changes in intensity of stimulating acoustic wave

Another important improvement in CM measurement came with the introduction of a lock-in amplifier with double phase-sensitive detection. In December 2003 a device for the phase-sensitive measurement of inner cochlea microphonic potentials was registered at the Patent Office. It was patented in November 2010. The device can measure harmonic, subharmonic and linear distortion products of the cochlea after dual-tone stimulation. Figure 13 shows a schematic of the measuring device.

The amplitude and phase of CMs in a given point on the surface of the cochlea depend on the intensity (L) and frequency (f) of the sound. When the frequency is fixed, the two CM potential parameters (amplitude and phase) depend on only parameter L. Typical changes in amplitude and phase over time registered at two different acoustic wave frequencies (260 and 8000 Hz) for the same guinea pig are shown in fig. 14. For this data, graphs of CM potential rms and phase depending on the level of sound intensity are shown in Fig. 15.

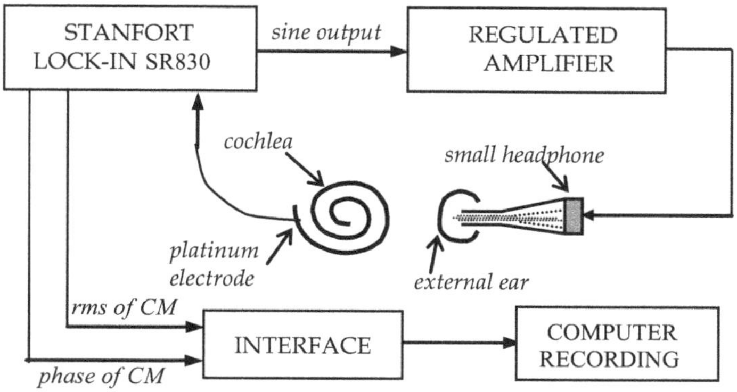

Fig. 13. Experimental setup for measuring CM potentials

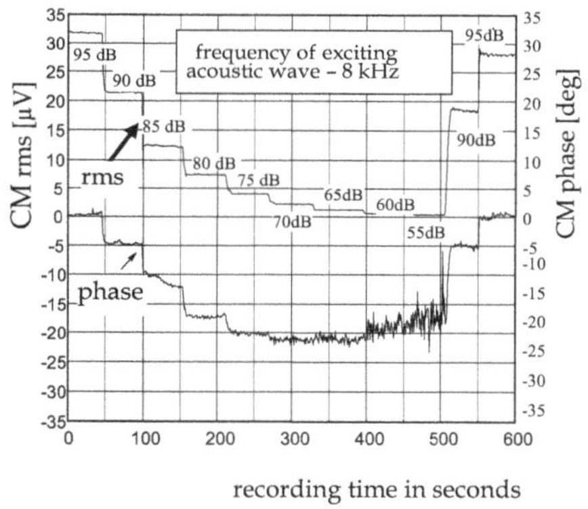

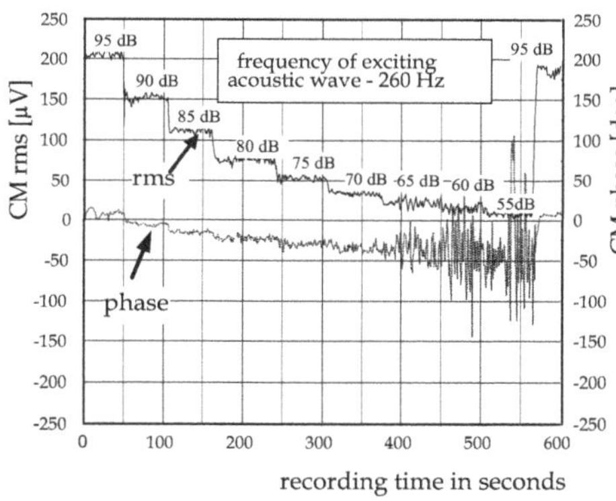

Fig. 14. Exemplary changes in CM rms and phase depending on sound intensity (sound levels were changed by 5 dB every 50 seconds)

Cochlear microphonic potentials are believed to be generated by the outer hair cells (OHCs). The latter are situated in three rows on the basilar membrane. All the OHCs have tiny strands (numbering about a hundred) called stereocillia. The apex of each single stereocillium lies in the tectorial membrane. In the resting state the stereocillia of each single cell form a conical bundle. During the acoustic excitation of the cochlea the stereocillia may dance about wildly. This alternating motion causes the channels in the stereocillia to open and close, providing a route for the influx of K^+ ions. The upper part of the OHCs acts as a resistor whose resistance changes according to the mechanical movements of the stereocillia. Changes in this resistance cause changes in extra-cellular currents. The measured CM potential is the result of the flow of extra-cellular currents through the input resistance of the lock-in amplifier.

The place theory suggests that a tone of a defined frequency excites mainly the OHCs located on the basilar membrane in a place specific for the given frequency (CF). The OHC electrical activity picked up from a given place on the cochlea surface is the vector sum of the extra-cellular currents generated by the particular OHC cells belonging to the given CF area (probably oval in shape). As the excitation wave intensity increases, extra-cellular currents are generated by an increasing number of OHC cells within the same CF area, which results in an increase in CM amplitudes. The phase changes registered then probably correspond to the shifts of the centre of the extra-cellular currents within the CF area.

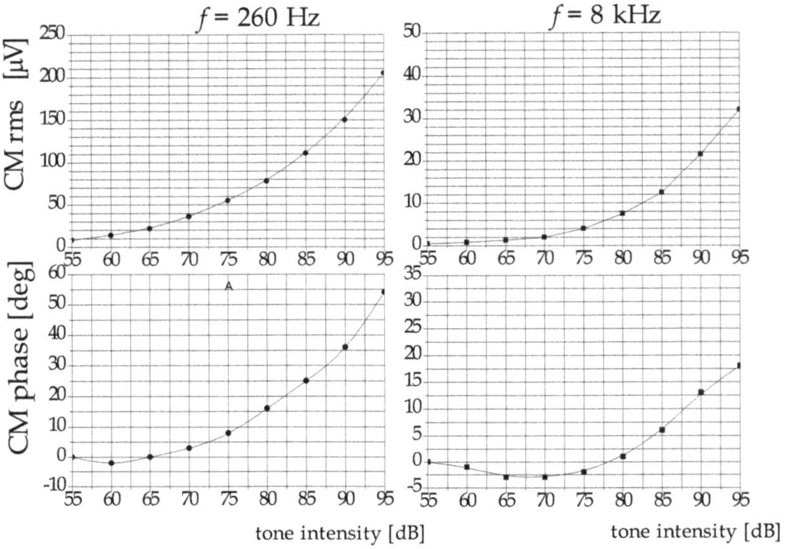

Fig. 15. Output-input characteristic obtained from traces shown in Fig. 14

3.6 Changes in amplitude and phase of CM potentials as result of laser irradiation

A focused laser beam can be a precise surgical scalpel. Perkins was the first to describe the use of a laser (an argon laser to be precise) in the surgical treatment of otosclerosis (Perkins, 1980). Since that time several kinds of laser (Ar, KTP, CO_2, Er) have been used in ear microsurgery. Vollrath and Schreiner were the first to use the rms of cochlear microphonics to estimate the effect of the argon laser beam on the electrical response of the cochlea in guinea pigs (Vollrath & Schreiner, 1982). The PSD technique enables the recording of the simultaneous changes in amplitude and phase of the CM potential during laser irradiation. The information about cochlear activity acquired in this way is more detailed.

Studies of the effect of Ar laser irradiation on the electrical activity of the cochlea have been described by us in several papers. We used the double PSD technique to record CM potentials prior to, during and after argon laser irradiation of the cochlea in guinea pigs. The goal of the studies was to determine safe laser parameters for argon laser stapedotomy, taking into account changes in not only the rms of CM potentials but also in their phase. In our experiments we used a CW argon laser with adjusted output power (0.1 - 3.0 W). An electronically controlled mechanical chopper was used to obtain laser light pulses differing in their parameters (the duration of a single laser pulse, the time interval between the successive pulses, the number of pulses in a series). Via a 200 μm optical lightguide the laser pulses would be delivered to the cochlear bone (near the round window) of an anaesthetized guinea pig with the surgically opened bulla. Exemplary traces selected from many different recordings are shown in fig. 16.

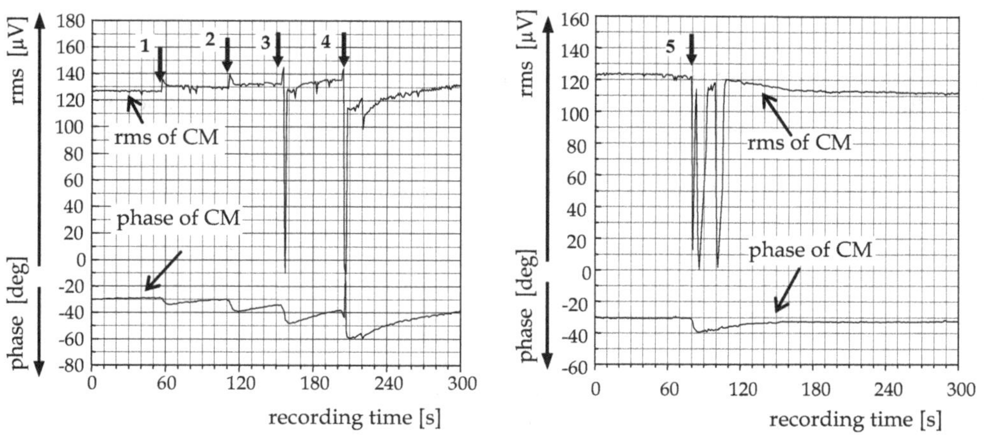

Fig. 16. Changes in rms and phase of CM potentials evoked by 80 dB acoustic wave of 1 kHz frequency during Ar laser pulse irradiation of 0.27 W (left) and 0.48 W (right) peak power. Irradiation parameters: 1 - single pulse of 0.5 s duration, 2 - single pulse of 0.5 s duration, 3 - single pulse of 1 s duration, 4 - two pulses of 1s duration with 1s interval between them, 5 - single pulse of 0.5s duration

It was found that in each registration the phase and amplitude of CM potentials changed during a laser pulse. The characteristic of the phase changes is always the same and diminishes relative to the initial (prior-to-irradiation) phase (in fig. 16 the initial phase was assumed to be equal to -30^0). The character of changes in CM rms depends on the peak power of the pulses used. Two characteristic peak power levels: P_1 and P_2 can be distinguished. When the peak power of the pulses is lower than P_1, laser irradiation results in a small increase in CM rms. This may be due to the slight increase in the temperature of the cochlea and to a biostimulating effect. After the first peak, but still below the second one (P_2), a sharp drop (even down to zero) in CM rms occurs. The drop is temporary and the cochlea quickly recovers its initial activity. Beyond P_2, changes in the electrophysiological activity of the cochlea are irreversible. As for today, the observed changes in the phase of CM potentials are hard to explain. It remains unknown why low-level laser radiation activates other groups of OHC cells in the CF area.

4. Double PSD technique in studies of DPOAE

4.1 Evoked otoacoustic emission

Evoked otoacoustic emissions (EOAE) are acoustic waves present in the external auditory canal after the cochlea is stimulated with an acoustic excitation wave. Depending on the excitation, different kinds of emission can be distinguished. If the stimulating signal is constant, then the emission is called *simultaneous evoked otoacoustic emission* (SEOAE). When pulse stimulating (clicks) sounds are used and the emission is registered between the clicks, the emission is called *transiently evoked otoacoustic emission* (TEOAE). If dual-tone stimulation (by two sinusoidal waves with respectively frequencies f_1 and f_2 and levels L_1 and L_2) is used, then the emission is called *distortion product otoacoustic emission* (DPOAE).

Otoacoustic emission was predicted by Gold as early as in 1948 (Gold, 1948). Thirty years later Kemp published a paper in which he described experiments proving the existence of this phenomenon (Kemp, 1978). He used clicks of 0.2 ms duration at a repetition rate of 16/s. In-between the successive pulses he recorded (with an electret microphone) acoustic wave pressure fluctuations at the outlet of the external acoustic canal. By applying an averaging procedure to the two-minute recordings he was able reduce the noise level to 0 dB SPL and reveal the backward signal which originated from the cochlea stimulated by the click. A few hundreds of works on this subject have been published since the first paper by Kemp. New experimental data are reported but their interpretations are not always explicit and mutually consistent. Despite the fact that the DPOAE mechanism is not yet fully understood, DPOAE signal estimation is a method of testing the human peripheral auditory function. The method is widely used in newborn hearing screening tests.

The presence of components which are absent in the stimulating acoustic wave is distinctive of DPOAE. The components result from the mechanical activity of the organ of Corti and are transmitted in the reverse direction through the middle ear and the tympanic membrane. Among the few possible products of cochlear nonlinearity, the acoustic wave $f_3 = 2f_1 - f_2$ is most widely examined because of its highest acoustic pressure level.

All the DPOAE acoustic waves are studied after their transduction into electric signals by a microphone. The microphone must be of high sensitivity and with a linear dynamic reserve (about 80 dB). The same requirements apply to the input preamplifier and the lock-in voltmeter amplifier since the measured DPOAE electrical signals cannot result from measuring system nonlinearity. The microphone placed in the external auditory canal transduces acoustic waves into electrical signals: both primary tones of 60-70 dB and reverse DPOAEs of 0 - 20 dB. Also floor noise occurs in the external ear canal. The apparatus used for measuring DPOAE must eliminate all undesirable signals with frequencies different than the frequency of the signal to be measured.

Otoemissions are examined after they have been converted in very accurate electric microphones. The biggest problem faced when examining DPOAEs is their extremely low level in comparison with the excitation waves. The difference may reach 30-60 dB. The phase-sensitive detection of DPOAE is therefore very useful. A basic experimental setup for measuring DPOAE signals is shown in fig. 17.

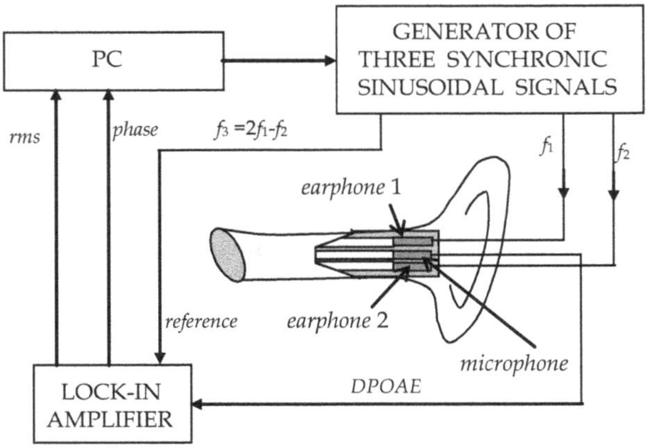

Fig. 17. Basic experimental setup for measuring DPOAE signals, using double PSD technique

The main unit of the experimental setup is a generator of three synchronous sinusoidal signals. The three frequencies are synchronized by a 18 MHz clock, whereby weak DPOAE signals can be measured using the PSD technique. The DPOAE response is measured by means of a probe which contains two miniature earphones and a low noise microphone. From the generator, pure tones with frequencies f_1 and f_2 are fed to the earphones. The two primary tones are digitally synthesized. The amplitudes, phases and frequencies of the tones are regulated by a PC with dedicated software. The software also enables the acquisition of the amplitude and phase of the DPOAE signals during measurements.

4.2 Previous techniques of measuring DPOAE

As mentioned earlier, DPOAE signals are of very low level, even if they are evoked in an unimpaired ear. Several signal processing techniques for the measurement of DPOAE signals under a large amount of noise and primaries of 60-70 dB have been developed. Initially, the Fast Fourier Transform (FFT) was used as the main signal processing tool for improving the signal-to-noise ratio in order to better estimate the level of DPOAE signals. In this method, the signals are first divided into data blocks and then averaged over time. For better reduction of the overall background noise long measurement time is required, which increases the amount of recorded data to be averaged. The FFT method requires about 10 seconds of block data. During long DPOAE recording, transient artefacts (e.g. talking, head movements) may occur, which when averaged together with the measuring signal may degrade the accuracy of the signal. Besides, the averaging method is incapable of measuring rapid changes of DPOAE signals.

In the first decade of the 21st century several novel methods of measuring DPOAE signals were developed (Ziarani & Konrad, 2004; Li et al., 2003). In comparison with the conventional methods, the new methods offer a shorter measurement time, which is of significance for clinical examinations. In addition, these methods are more immune to artefact and background noise. Thanks to the new methods it is possible to continuously record DPOAE signals. Besides offering the above advantages, the double PSD technique enables the simultaneous measurement of amplitude and phase of DPOAE signals. The two DPOAE parameters can be measured in a very short time, even below 10 ms.

In screening protocols typically a few pairs of primary tones with fixed acoustic levels are used and the responses are analyzed one after another (sequentially). In order to reduce the examination time the multiple-tone pairs method can be employed. In this method, DPOAE signals are evoked simultaneously by three or four pairs of two-tones. This method reduces measurement time but has a limited use .

4.3 DPOAE measurement using double PSD technique

Before the PSD technique was introduced to measure DPOAE signals it had been assumed that the amplitude and frequency of DPOAE signals depended on four acoustic parameters of the stimulating signals (primaries), i.e. the amplitude and frequency of each of the two signals. The DPOAE phenomenon itself is investigated according to the procedure described below. First the f_2/f_1 ratio (usually 1.22) and the stimulating signal intensity levels (e.g. L_2/L_1 = 60dB/65dB) are fixed. For the frequency of one of the stimulating waves (usually f_2) several discrete values are set while the frequency of the other wave is changed in small steps, e.g. 1/3 octave-bands centred around the fixed f_2 (Wagner at al., 2008). When examining the effect of different internal (e.g. age, gender) and external (e.g. industrial noise, medicines) factors, DPOAE is measured in the same stimulation conditions before and after the stimulus acts. Most experimental works in this field describe measurements of solely the amplitude of DPOAE signals. Some works also dealt with the phase of DPOAE signals, but it was measured in an indirect way, using signal processing methods. The phase-sensitive technique enables the simultaneous measurement of the amplitude and phase of DPOAE signals, with no need to use complex signal processing methods. The measurement takes place in real time.

Figure 18 shows an exemplary record of the simultaneous changes in the amplitude and phase of DPOAE signals caused by changes in the acoustic parameters of the stimulating waves. The recording was made in real time using the measuring setup shown in fig. 17. In the whole course of recording the combination frequency (f_3) remained constant at 3749 Hz while the other parameters were changed every 20 seconds in a specified sequence. The whole 980 second long recording time had been divided into seven 140 long time intervals in which parameter $k = f_2/f_1$ assumed the consecutive values: 1.10, 1.15, 1.20, 1.25, 1.39, 1.35, 1.40. The following seven combinations of stimulating wave levels: 1 - (55 dB, 55 dB), 2 - (55 dB, 60 dB), 3 - (60 dB, 55 dB), 4 - (60 dB, 60 dB), 5 - (65 dB, 60 dB), 6 - (60 dB, 65 dB) and 7- (65 dB, 65 dB) were fixed for each value of parameter k. In each of the combinations, the dB SPL of primary f_1 is in the first place. During the 980 second long recording the parameters of the primaries were changed 49 times in total.

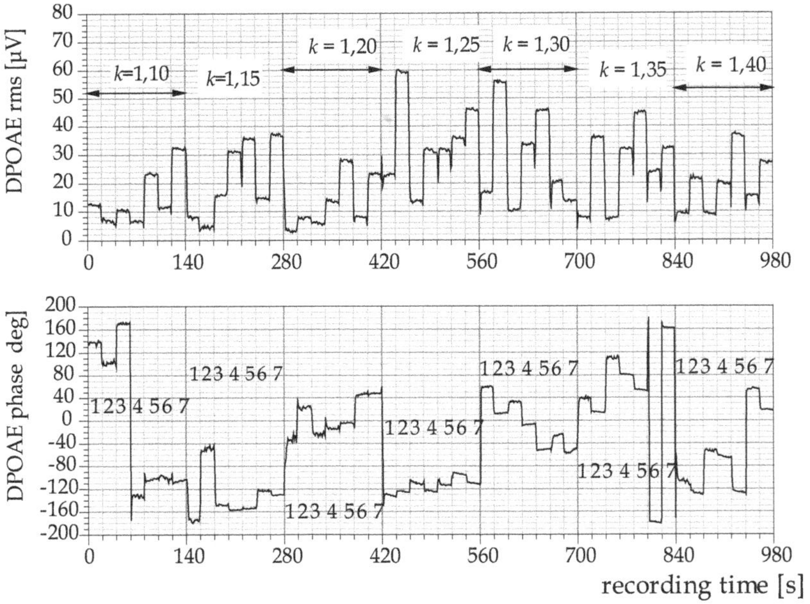

Fig. 18. Simultaneous changes in rms and phase of DPOAE signals, caused by fixed sequence of changes in parameters of primaries. Numbers 1 – 7 denote following combinations of primary frequencies L_{1dB} / L_{2dB} levels: 1 - (55 dB, 55 dB), 2 - (55 dB, 60 dB), 3 - (60 dB, 55 dB), 4 - (60 dB, 60 dB), 5 - (65 dB, 60 dB), 6 - (60 dB, 65 dB) and 7 - (65 dB, 65 dB). The same combinations of levels were used for each value of parameter $k=f_2/f_1$

The measurements showed that each change in the value of one of the parameters of the primaries results in a change of both the amplitude and phase of the DPOAE signal. Moreover, the character of the changes depends on the on the ontogenetic traits.
Thanks to the phase-sensitive technique one can determine the effect of the initial phase of each of the primaries on the amplitude and phase of DPOAE signals. For this purpose a generator of three synchronous sinusoidal signals was incorporated into the setup shown in fig. 17. The generator offers the possibility of fixing not only the amplitude and frequency of each of the primaries, but also the initial phase of each of the signals.
Five parameters of the primaries, i.e. the amplitude and frequency of each of the signals and the initial phase of one of the signals were fixed. The sixth parameter, i.e. the initial phase of the second primary was changed in a range of 0 - 360 degrees. The phase was changed in steps of 22.5 degrees. Exemplary measurements are shown in figs 19 and 21. Each of the figures comprises four panels. On the left side of each of the figures there are two panels showing experimentally determined changes in the amplitude and phase of DOPAE signals, caused by changes in the initial phase of primary f_1 (fig.19) or primary f_2 (fig(21). The data were obtained for combination frequency f_3 = 3749 Hz, parameter $k = f_2/f_1 = 1.25$, intensity levels L_1/L_2 = 65/55dB and the zero initial phase of primary f_1 (fig.19) or f_2 (fig.21).
The graphs in the panels on the right side of each of the figures were plotted on the basis of formulas (8) - (10), but the values of some of the constants in the formulas were matched to obtain agreement with the experimental traces.

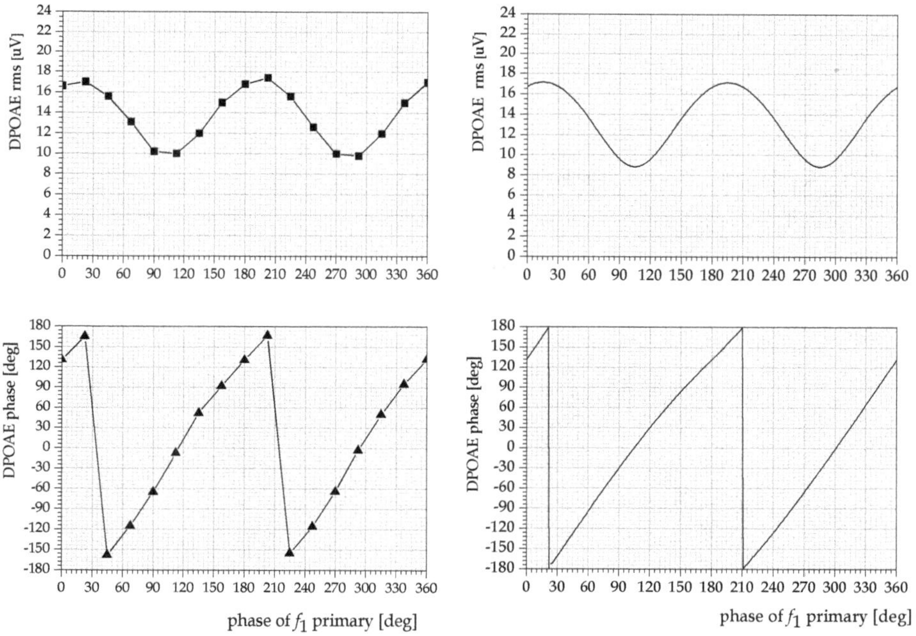

Fig. 19. Simultaneous changes in amplitude (upper panels) and phase (lower panels) of DPOAE signals, caused by changes in initial phase of primary f_1, obtained from experiment (left) and theoretically (right) (details in text)

Currently, it is generally believed that the DPOAE signal induced in the external acoustic canal by a double-tone is composed of two backward travelling waves (e.g. Knight &Kemp, 2000). The primary wave arises in the place where the two regions (CF_1 and CF_2) characteristic of frequency f_1 and f_2 overlap (but much more closer to CF_2). The wave propagates in the basilar membrane towards both the cochlea's base and its apex . The wave directed towards the apex bounces off in the region characteristic of frequency f_3 (CF_3) and propagates towards the base. Thus two waves with the same frequency f_3, but shifted in phase relative to each other, propagate towards the cochlea's base. Depending on the difference between the two waves, destructive or constructive amplitude interference occurs.

There are two different theories in the literature, concerning how the waves propagate backward from their generation places (He at all, 2007). According to one theory, the two waves propagate as compression waves to the cochlear base via the cochlear fluids. According to another theory, the two waves are transverse waves slowly propagating along the basilar membrane. Currently the prevailing view is that two backward waves, being transverse waves in the basilar membrane, arise in the cochlea excited by two tones. Taking into consideration this view and the previously determined sites where the backward waves arise, the schematic shown below (fig. 20) was drawn.

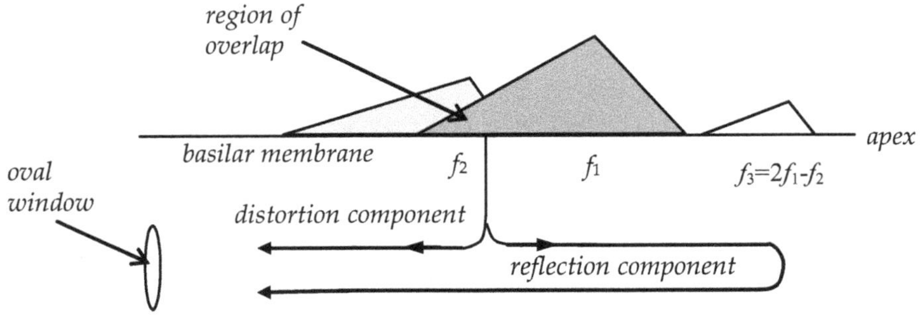

Fig. 20. Schematic diagram of source of two backward travelling waves whose interference produces DPOAE wave in auditory canal

The resultant wave near the oval window can be written as

$$A_w(t) = A_m \cos(\omega_3 t + \alpha_{10} + \kappa + 2\beta_1 - \beta_2), \tag{8}$$

where

$$A_m = A_{1m}\sqrt{1 + K^2 + 2K\cos\alpha_r}, \tag{9}$$

$$\kappa = \mathrm{a}\tan\left(\frac{\sin\alpha_r}{K^{-1} + \cos\alpha_r}\right) \tag{10}$$

A_{1m} - the amplitude of the primary wave,

$K = A_{2m}/A_{1m}$,

α_{10} - the initial phase of the primary wave,

β_1, β_2 - the phases induced by the initial phases of the primaries,

α_r - a phase difference between the primary and secondary wave, due to the path length distance.

It follows from formula (9) that the amplitude of the resultant wave does not depend on the the initial phases of the primaries, and the phase of the resultant wave:

$$\Omega = \alpha_{10} + \kappa + 2\beta_1 - \beta_2 \tag{11}$$

(directly measured by the lock-in amplifier) is a linear function of the stimulating waves phase. However, experimental results do not corroborate the above dependence. The amplitude of the DPOAE signal turns out to be a function of the initial phases of the stimulating signals, and the measured phase is only approximately a linear function of the initial phases (panels on the left side of figs 19 and 21). If it is assumed that angle α_r changes in the same way as the initial phase of the primary, full agreement between the experimental traces and the ones determined from formulas (8) and (10) is obtained. This is shown in the panels on the right side of figs 19 and 21. The graphs on the right side of fig. 19 were plotted on the basis of formulas (8) and (10), assuming $\alpha_r = 2\beta_1$, while the graphs on the right side of fig. 21 were plotted assuming $\alpha_r = -\beta_2$.

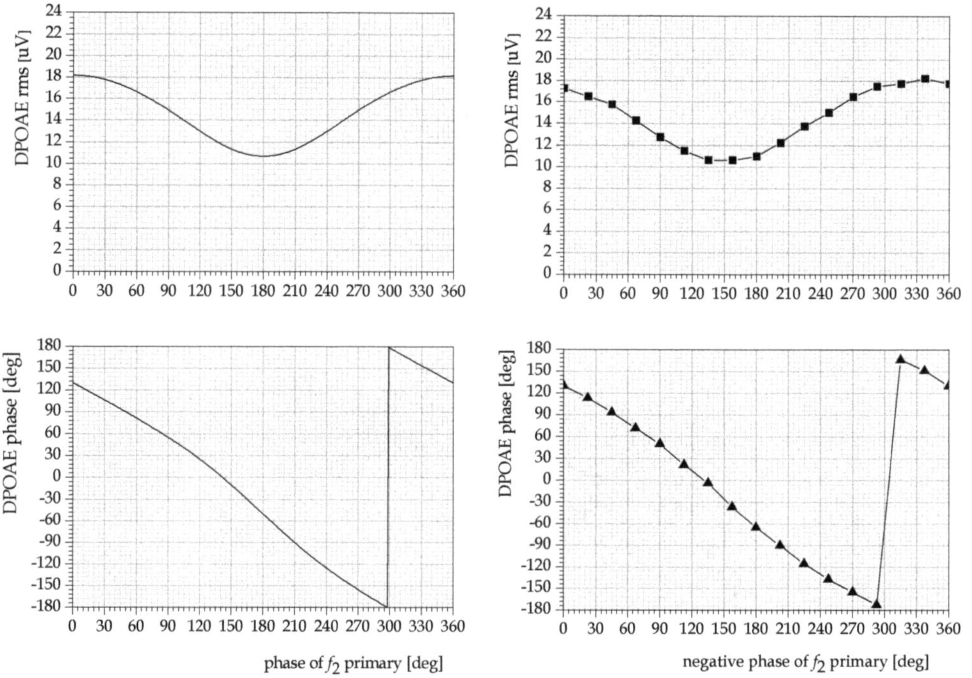

Fig. 21. Simultaneous changes in amplitude (upper panels) and phase (lower panels) of DPOAE signals, caused by changes in initial phase of primary f_2, obtained in experiment (left) and theoretically (right) (details in text)

It follows from formula (9) that the ratio of the maximum amplitude of the DPOAE wave (A_{wmax}) to the minimum value (A_{wmin}) amounts to

$$\frac{A_{w\max}}{A_{w\min}} = \frac{1+K}{1-K} \quad \Rightarrow \quad K = \frac{A_{2m}}{A_{1m}} = \frac{A_{w\max} - A_{w\min}}{A_{w\max} + A_{w\min}} \tag{12}$$

For the measurement conditions for which the traces shown in figs 19 and 21 were determined, it was calculated from formula (12) that K = 0.34 and K = 0.27 when respectively initial signal phase f_1 and f_2 is changed. This means that about 11.5% and 7.3% of the primary wave energy is reflected from region CF_3 in respectively the former and latter case.

The preliminary measurements shows that there is a certain mechanism in the Corti organ, which is responsible for the fact that a change in the phase of one of the stimulating signals (i.e. phase modulation) causes the amplitude modulation of the DPOAE signal. Further research is needed to explore this mechanism, but already at this stage one can say that the PSD technique proposed by the authors will play a major role in the exploration of this mechanism.

4.4 Simultaneous measurements of DPOAE and CMDP, using double PSD technique

Much of the experimental research reported in the world literature indicates that the main source of CM signals and DPOAE waves are OHCs. Thanks to the use of phase-sensitive detection in the measurement of each of the signals one can observe the simultaneous changes in the amplitude and phase of the two signals, resulting from changes in the parameters of the primaries. The measuring setup used for this purpose is shown in fig. 22. The setup incorporates two patents developed by the authors.

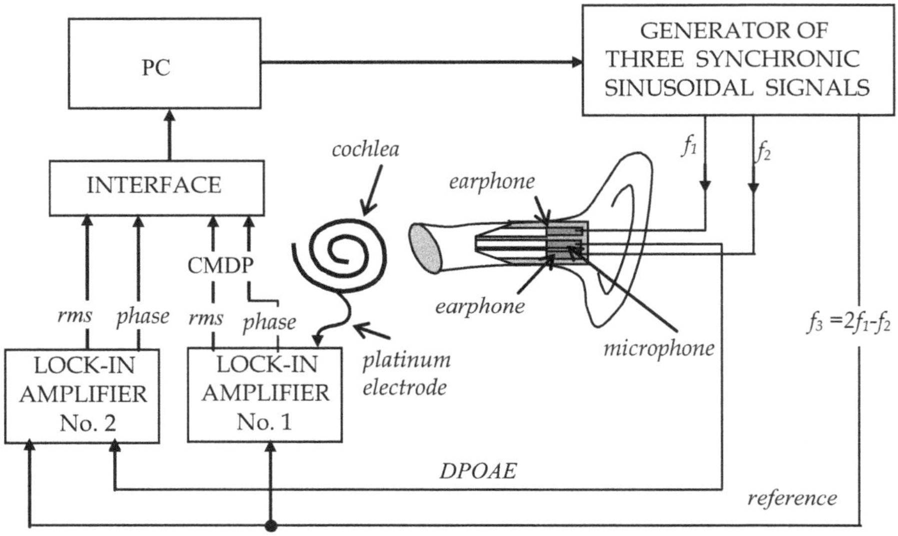

Fig. 22. Experimental setup for simultaneous measurement of amplitude and phase of DPOAE and CMDP signals

Two lock-in amplifiers, one for measuring the rms and phase of the DPOAE signal (amplifier No. 2) and the other for measuring the rms and phase of the CMDP signal (amplifier No. 1), have been incorporated into the setup. The CMDP signal is the distortion product in cochlear microphonics. The same signal (with combination frequency f_3) from the generator is fed to the reference inputs of each of the amplifiers. The reference input of lock-in No.2 can also be successively fed signals with stimulation frequencies f_1 and f_2 and combination frequency f_3 and the rms and phase of three CM signals with different frequencies can be measured. Measurements made in this way may provide a fuller picture of the cochlea functions.

The above setup was used to measure changes in the rms and phase of both DPOAE and CMDP signals, caused by changes in the excitation parameters. Four anaesthetized guinea pigs with the positive Preyer reflex were subjected to the experiments. Recordings were made at the following combination frequencies: 1312, 1875, 2671, 3749 and 5342 Hz. The parameters of the primaries were changed as in sect. 4.3, i.e. one of the parameters of the primaries was changed every 20 seconds in a specified way. Exemplary traces recorded for frequency f_3 = 1875 Hz are shown in fig. 23.

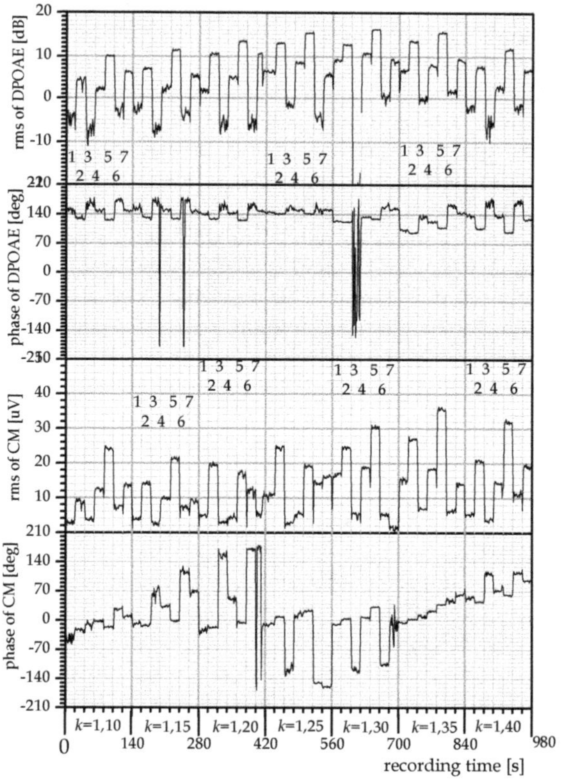

Fig. 23. Simultaneous changes in rms and phase of both DPOAE and CMDP, induced by changes in parameters of primaries

Nearly 100% correlation between the DPOAE rms and the CMDP rms was found, i.e. when after a change in one of the parameters of the primaries the DPOAE rms increased, then CMDP rms would also increase. Unfortunately, there was no such correlation between the phases of the two signals. This situations is well illustrated by the records of the changes, shown in fig. 23.

5. Conclusion

Practically all the ways of measuring biological acoustic waves in the cochlea, in which the phase-sensitive detection technique can be applied, have been described. Exemplary experimental results coming from many different measurement cycles carried out by the authors on guinea pigs in the last nearly 20 years were presented to demonstrate the measuring possibilities offered by the PSD technique. The latter's main advantage is that very weak (even below the ambient noise level) electrical signals can be measured in a very short time (in the order of milliseconds). A minor limitation of this technique is that it is applicable to objects to whose input periodical signals are fed from the outside.

In many investigations into the electrophysiological function of the cochlea it is essential not only to simultaneously measure the amplitude response, but also the phase response to the stimulation. This is undoubtedly another advantage of the PSD technique.

Much more difficult than the measurement of the phase is the interpretation of its changes. As for now, it is not always possible to interpret the observed changes, which particularly applies to DPOAE. This phenomenon has been known for over 30 years, but it still has not been fully explored. The great worldwide interest in this subject is reflected in the large number of publications devoted to it. The interest stems from the fact that for many years DPOAE measurements have been part of hearing screening tests during which DP-grams are recorded. This especially applies to newborns and people with mental disabilities, in which cases it is impossible to record audiograms. Besides gaining an insight into the nature of the DPOAE phenomenon, it is essential to determine the correlation between the DP-gram and the audiogram. In the authors' opinion, the phase-sensitive detection technique represents a new tool for investigating electrophysiological phenomena in the cochlea and it will contribute to the better understanding of the phenomena taking place in this organ.

6. References

Bredberg, G., Lindeman, H. H., Ades H. W., West R. & Engstrom H. (1970). Scanning electron microscopy of the organ of Corti. *Science*, Vol. 170, No.960, pp. 861-863, ISSN Print 0036-8075

Brown, D.J., Hartsock, J. J., Gill, R.M., Fitzgerald, H.E. & Salt, A.N. (2009). Estimating the operating point of the cochlear transducer using low-frequency biased distortion products. *J. Acoust. Soc. Am.*, Vol. 126, No. 4, pp.2129 - 2145, ISSN Print 0001-4966

Carricondo F., Sanjuan-Juaristi J., Gil-Loyzaga P. & Poch-Brotto Joaquin. (2001). Cochlear microphonic potentials: a new recording technique. *The Annals of otology, Rhinology & Laryngology*, Vol.110, No. 6, pp. 565 - 573, ISSN Print 0003-4894

Castelo Branco (1999). The clinical stages of vibroacoustic disease. *Aviation, Space, and Environmental Medicine* , Vol. 70, Suppl 3, A32-39, ISSN 0095-6562

Davis, H. (1983). An active process in cochlear mechanics. *Hearing Research*, Vol. 9, No,1, pp. 79-90, ISSN Print 0378-5955

Gale, J. A. & Ashmore J. F. (1997). An intrinsic frequency limit to the cochlear amplifier. *Nature*, Vol. 389, No.6646, pp.63-66, ISSN 0028-0836

Gold, T. (1948). The physical basis of the action of the cochlea. *Proceedings the Royal of Society*, Vol.135, No.881, pp. 492-498, ISSN 1471-2954

Hamernik, R.P., Henderson, D., Coling D. & Slepecky N. (1980). The interaction of whole body vibration and impulse noise. *J. Acoust. Soc. Am.*, Vol. 67, No.3, pp. 928-934, Hamernik, R.P., Henderson, D., Coling, D. & Salvi R. (1981). Influence of vibration on asymptotic threshold shift produced by impulse noise. *Audiology*, Vol.20, No.3. pp.259-269, ISSN Print 0020-6091

Hamernik, R.P., Henderson, D., Coling, D. & Salvi R. (1981). Influence of vibration on asymptotic threshold shift produced by impulse noise. *Audiology*, Vol. 20, No.3. pp.259-269, ISSN 1499-2027

He, W., Nuttall, A.L. & Ren, T. (2007). Two-tone distortion at different longitudinal locations on the basilar membrane. *Hear. Res.*, Vol. 228, No.1-2, pp.112-122, ISSN Print 0378-5955

Jankowski, W. Giełdanowski, J. & Birecki W. (1962). Effect of some vasoconstrictor drugs on the microphonic potential of the cichlea. *The Polish OtoLaryngology*, Vol.16, pp. 321 – 329, PL ISSN 0030-6657

Jones, C.M.(1996). ABC of work related disorders. Occupational hearing loss and vibration induced disorders. Occupational. *Br Med J* , Vol. 313, No. 7051, pp. 223-226, ISSN 0959-8138

Kemp, D.T. (1978). Stimulated acoustic emissions from within the human auditory system. *J. Acoust. Soc. Am.*, Vol. 64, No.5, pp. 1386-1391, ISSN Print 0001-4966

Knight, R.D. & Kemp D.T. (2000). Indications of different distortion product otoacoustic emission mechanisms from a detailed f_1, f_2 area study, *J. Acoust. Soc. Am.*, Vol. 107, No.1, pp. 457-473, ISSN Print 0001-4966

Kobayashi, T., Rong Y., Chiba T., Marcus C.D., Ohyama K. & Takasaki T. (1997). Ototoxic effect of erythromycin on cochlear potentials in the guinea pig. *Ann. Otol. Rhinol. Laryngol*, Vol.106, No.7, pp. 599-603, ISSN Print 0003-4894

Li, X., Wodlinder, H. & Sokolov Y. (February 2003). A new method for measuring DPOAEs and ASSRs, Available from
http://www.hearing review.com/issues/articles/2003-02 05.asp

Linder, T.E., Zwicky, S. & Brändle P. (1995). Ototoxicity of ear drops: a clinical perspective. *Am. J. Otol.*, Vol. 16, No. 5, pp. 653-657, ISSN Print 0192-9763

Palmer, K.T., Griffin M.J., Bendall H., Pannett B. & Coggon D.(2000a). Prevalence and pattern of occupational exposure to hand transmitted vibration in Great Britain: findings from national survey. *Occup. Environ. Med.* , Vol. 57, No.4, pp. 218-228, ISSN Print 1351-0711

Palmer, K.T., Griffin M.J., Bendall H., Pannett B. & Coggon D. (2000b). Prevalence and pattern of occupational exposure to whole body vibration in Great Britain: findings from national survey. *Occup Environ Med*, Vol. 57, pp. 229-236, ISSN Print 1351-0711

Perkins, R. (1980). Laser stapedotomy for otosclerosis. *Laryngoscope*, Vol.90, No.2, pp.228-241, ISSN 1531-4995

Rogowski, M. & Chodynicki, S. (1987). Einfluβ von Vibrationen und Gentamycin auf das Gehörorgan des Meerschweinchens. *HNO-Prax* , Vol. 12, pp.219-223

Seidel, H., Heide R. (1986). Long-term effects of whole-body vibration: a critical survey of the literature. *Int Arch Occup Environ Health*, Vol.58, No.1, pp. 1-26, ISSN Print 0340-0131

Tasaki, I., Davis, H. & Legouix J.P. (1952). The space-time pattern of the cochlear microphonics (guinea pig) as recorded by differential electrodes. *J. Acoust. Soc. Am.*, Vol. 24, No.5, pp. 502 - 519, ISSN Print 0001-4966

Thurlow, W.R. (1943). Studies in auditory theory:II The distortion of distortion in the inner ear. *Journal of Experimental Physology: General*, Vol.32, No.4, pp. 344 - 350, ISSN 0096- 3445

Wadsten C.J., Bertilsson C.A., Sieradzki H. & Edström S. (1985). A randomized clinical trial of two topical preparations (framycitin/gramicidin and oxytetracycline /hydro-cortisone with polymyxin B) in the treatment of external otitis. *Arch Otorhinolaryngol* Vol. 242, No. 2, pp. 135-139. ISSN Print 0302-9530

Wever, E.G. & Bray, C.W. (1930). Auditory nerve impulses, *Science*, Vol.71, No.1834, p.215, ISSN Print 0036 8075

Vollrath, M., Schreiner, Chr.(1982). Influence of argon laser stapedotomy on cochlear potentials I: Alteration of cochlear microphonics (CM). *Acta Otolaryngol*, suppl **385**, pp.1-31, ISSN Print 365-5237

Ziarani, A.K. & Konrad A. (2004). A novel metod of estimation of DPOAE signals. *IEEE Transactions on biomedical engineering*, Vol.51, No.5, pp.864-868, ISSN 0018 9294

Ziemski, Z. (1970). Ototoxity of selected organic solvents of industrial plastics in experimental animals. *Papers of Medical University in Wroclaw*, Vol.15, No.1, pp.59-128

13

Photoacoustic Technique Applied to Skin Research: Characterization of Tissue, Topically Applied Products and Transdermal Drug Delivery

Jociely P. Mota, Jorge L.C. Carvalho, Sérgio S. Carvalho and Paulo R. Barja
UNIVAP
Brazil

1. Introduction

The photoacoustic (PA) effect basically consists in the production of acoustic waves due to the absorption of modulated (or pulsed) radiation by a sample. Graham Bell discovered the PA effect in 1880, when he noticed that the incidence of modulated light on a diaphragm connected to a tube produced sound. Thereafter, Bell studied the PA effect in liquids and gases, showing that the intensity of the acoustic signal observed depended on the absorption of light by the material.

In the nineteenth century, it was known that the heating of a gas in a closed chamber produced pressure and volume changes in this gas. However, there were many different theories to explain the PA effect. Rayleigh said that the effect was due to the movement of the solid diaphragm. Bell believed that the incidence of light on a porous sample expanded its particles, producing a cycle of air expulsion and reabsorption in the sample pores. Both were contested by Preece, who pointed the expansion/contraction of the gas layer inside the photoacoustic cell as cause of the phenomenon. Mercadier explained the effect conceiving what we call today *thermal diffusion mechanism*: the periodic heating of the sample is transferred to the surrounding gas layer, generating pressure oscillations.

The lack of a suitable detector for the PA signal made the interest in this area decline until the invention of the microphone. Even then, research in this field was restricted to applications in gas analysis up to 1973, when Rosencwaig started to use the PA technique in spectroscopic studies of solids and, together with Gersho, developed a mathematical model for the generation of the PA signal in solid samples - the Rosencwaig-Gersho (RG) Model (Rosencwaig & Gersho, 1976).

In condensed matter samples, one of the most important mechanisms for PA signal generation is the thermal diffusion, classically described by the RG model. According to this model, the (modulated) radiation absorbed by condensed matter samples is converted into heat, causing temperature modulation in the surrounding atmosphere. This eventually produces the mechanical effect of periodic expansion and contraction originating sound waves that can be detected by a microphone.

Since the publishing of the RG model and, soon after that, of the generalized theory for the PA effect by McDonald and Wetsel (1978), the PA technique has already proved its relevance in a large number of very different fields, from the polymerization of dental resins (Balderas-Lopez et al., 1999) to photosynthesis studies (Malkin & Puchenkov, 1997; Herbert et al., 2000).

1.1 Objectives

The purpose of this chapter is to present applications of the PA technique in skin research, both in the characterization of skin itself and in transdermal drug delivery studies. The basic experimental setup for such studies will be briefly presented, aiming to help those who may be interested in developing similar studies. Emphasis will be done to *in vivo* measurements, because of its importance in this field. Our objective is to show the usefulness of the PA technique in the biomedical field, particularly in skin research; finally, perspectives for future work in this field will be presented.

2. Photoacoustic measurements

2.1 Basic experimental setup

Figure 1 presents one scheme for a basic photoacoustic experimental setup.

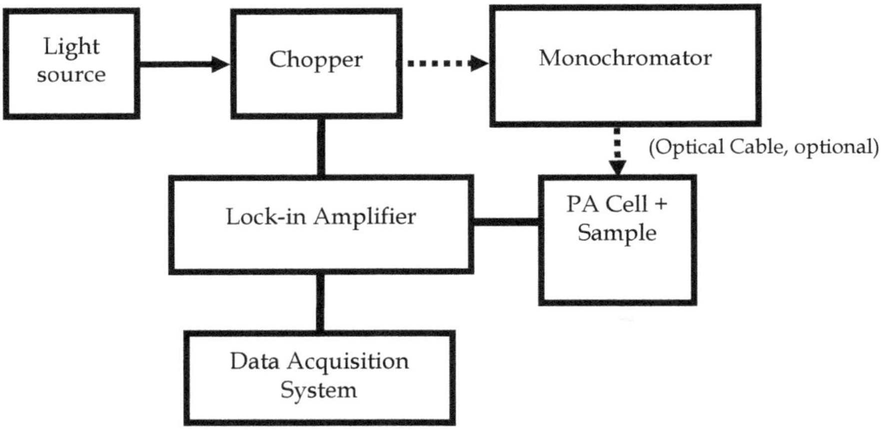

Fig. 1. Example of a basic photoacoustic experimental setup (scheme)

The experimental scheme in Figure 1 shows a (typically mechanical) chopper positioned in front of the light source, in order to modulate the radiation that comes into a monochromator (utilized in PA spectrocopy measurements). Light absorbed by the sample generates acoustic waves inside the PA cell; the PA signal is captured by a microphone (inside the PA cell) that sends it to the lock-in amplifier (also connected to the chopper, to receive information on the frequency modulation). The lock-in amplifier is connected to a microcomputer for data acquisition. In vivo, skin measurements are performed with an open-ended PA cell, in which it is the sample itself that closes the chamber.

2.2 Measurements as a function of time

The PA signal depends on the optical and thermal properties of the sample, which may vary with time due to different factors. When a sample undergoes changes in its composition or structure (as it occurs during the polimerization of a dental resin, for instance), the propagation of heat inside the sample is also modified, thereby altering the PA signal.

We must also mention the possibility of performing photosynthesis studies using PA measurements as a function of time (W.J. Silva et al., 1995). When PA measurements are performed in photosynthesizing samples as plant leaves, the PA signal presents, in addition to the photothermal component, a photobaric component, resulting from the gas exchanges associated to the photosynthesis process (Acosta-Avalos et al., 1996). This allows the study of the so-called photosynthetic induction, that is, the increase of the net photosynthetic rate that occurs when a plant is shifted from darkness to light (Sui et al., 2011).

As stated by Bodzenta et al. (2002) in their work on PA detection of drug diffusion into a membrane, PA measurements give the possibility for investigations in relatively long time periods. This makes the PA technique suitable for the monitoring of dehydration processes (Lopez et al., 2005) and of changes occurring in time in biological tissues such as skin. It is possible to study, for example, the kinetics of transdermal drug delivery through the analysis of PA measurements as a function of time. One example will be presented at the section 4 of the present chapter.

2.3 Studies on the modulation frequency: depth profile

In thermally thick samples (as skin tissue), only the light absorbed within the first thermal diffusion length (μ_T) of the sample/tissue contributes to the PA signal (Rosencwaig, 1980). As the thermal diffusion length depends on the modulation frequency (f) of the incident light by the relation

$$\mu_T = \sqrt{\frac{\alpha}{\pi f}} \tag{1}$$

where α is the thermal diffusivity of the sample, it is possible to perform depth-profile studies, with the evaluation of the penetration depth of a product (or even a microorganism) in tissue. The possibility of performing depth-profile studies is particularly interesting in the characterization of multilayer systems (as skin itself).

The frequency dependence analysis of the PA signal can also be employed in the determination of the thermal properties (thermal diffusivity, thermal effusivity) of a sample or material (Balderas-Lopez & Mandelis, 2001), including biological tissues as porcine skin (Gao et al., 2005; Qiu et al., 2008).

2.4 Measurements as a function of the wavelength: Photoacoustic spectroscopy

Photoacoustic spectroscopy (PAS) is already incorporated to the roll of useful photothermal techniques since the 1980s (Rosencwaig, 1980; Vargas & Miranda, 1988). Besides the possibility of rendering depth-profile analysis in multi-layered samples, PAS presents at least two additional advantages over other spectroscopy techniques: i) as transmitted and reflected light do not interfere in PAS measurements, it is a "more direct" technique, representing a direct measurement of the light absorption by the sample; ii) it allows the study of optically opaque and highly scattering samples (which could not be analyzed by conventional optical spectroscopy).

In PAS measurements, the emission spectra of the light source is typically obtained through measurements using black carbon powder (or other black material) as the sample, with all the remaining measurements being normalized with respect to the lamp spectrum.

PAS can also be employed in skin research. In 2004, Benamar and co-workers presented a PAS study on the effect of dihydroxyacetone, frequently employed for artificial tan. Measurements were carried out in the presence and absence of dimethylisosorbide (a solvent for dihydroxyacetone), on excised human skin. By monitoring the PAS signal intensity with time in the UV (300-400nm) range, these authors demonstrated that dihydroxyacetone in combination with dimethylisosorbide enhances the process of tanning (Benamar et al., 2004).

Recently, Melo et al. (2011) applied PAS to evaluate the penetration rate of *Helicteres gardneriana* extract, topically applied for anti-inflammatory purposes. Experiments were conducted *ex vivo* in mice. Croton oil was applied into both mouse's right and left auricles to induce inflammatory response, and the left auricle was treated with the extract. The strong anti-inflammatory effect observed for the *Helicteres gardneriana* extract was associated with the deep percutaneous penetration observed for the formulation, according to PA data (Melo et al., 2011).

2.5 Photoacoustic imaging and tomography

Photoacoustic imaging is based on the production of acoustic waves following irradiation by a short pulse of light whose absorption generates local heating and transient thermoelastic expansion (Balogun et al., 2009). According to Beard (2009), haemoglobin "represents the most important source of endogenous contrast" in PA imaging. This makes the technique particularly indicated to studying tissue abnormalities as tumors and other diseases related to changes in the structure and oxygenation status of the vasculature (Beard, 2009).

Recently, Hu and Wang (2010) presented "PA tomography" as a method combining high spatial resolution and optical absorption contrast, important in microvascular imaging and characterization. Reviewing the "major embodiments of PA tomography" (microscopy, computed tomography and endoscopy), they have analyzed the methods employed in different studies, including hemodynamic monitoring, determination of hemoglobin concentration, evaluation of oxygen saturation, studies of blood flow and tumor-vascular interaction.

Besides being applied to soft tissues, PA imaging can also be employed to hard tissues. Li and Dewhurst (2010) have applied a PA imaging system with a near-infrared (NIR) pulsed laser to obtain images from both soft tissue and post-mortem dental samples. They have also performed simulations (based on the thermoelastic effect) to predict initial temperature and pressure fields within a tooth sample, observing that values are maintained below the corresponding safety limits. In this way, the results presented by Li and Dewhurst show that the PA technique can be sucessfully applied to image both soft and hard tissues.

3. Photoacoustic measurements and the characterization of skin

Biological materials are sometimes difficult to study employing conventional techniques that require previous preparation of the samples, because these materials can have its properties significantly altered by preparation processes as solubilization, for example. The PA technique does not require previous preparation; it can be described as a non-invasive technique that allows even *in vivo* measurements.

In general, biological tissues can be characterized as highly scattering samples; however, this is not a problem for PA mesurements, in which the signal is based in the direct absorption of radiation. As pointed by Cahen and co-workers (1980), "the relative insensitivity to scattered light of the PA signal makes such measurement an attractive way to measure biological samples *in vivo*". These features explain the potential of the PA technique in the study of opaque materials and complex biological systems such as skin. PA measurements can be employed to determine the absorption characteristics of the skin itself or topically applied products, as well as kinetic changes related to transdermal drug delivery.
Skin diseases can also be studied through PA measurements. In 2010, Swearingen et al. developed a PA methodology to determine the nature of skin lesions (pigmented and vascular) *in vivo,* which is important because misdiagnosis may even lead to cancerous lesions not receiving proper medical care. These authors irradiated skin with two laser wavelengths (422 and 530nm), with the relative response at these two wavelengths (422nm/530nm) indicating whether the lesion is pigmented or vascular, due to the distinct absorption spectrum of melanin and hemoglobin (Swearingen et al., 2010).

3.1 Skin type classification

Skin type classification is important not only for medical or clinical purposes, but also for pharmaceutical and cosmetic industries, following the idea that an objective, precise characterization of skin could be useful in the design of new topically applied products and in defining more specific skin treatments according to each skin type.
However, in dermatology, there is still no universal agreement about the best method for classifying skin, as even the widely accepted method proposed by Fitzpatrick (1988) - defining the so-called "skin phototypes" - is based in clinical, subjective analysis.
More recently, Baumann (2006a, 2006b) proposed a new skin type classification, according to which 16 different skin types are defined from the combination of four parameters, as skin can be characterized as: i) pigmented or nonpigmented; ii) dry or oily; iii) sensitive or resistant; and iv) wrinkled or tight. Baumann´s skin typing is based on an extensive research, performed with 1400 volunteers. However, it relies essentially on the response of volunteers to a questionnaire; therefore, it does not fulfill "per se" the need of an objective classification, which would require experimental evaluation.
PA measurements have a potentially important role to play in an experimental approach to skin type classification. In 2000, Schmidt and co-workers conducted non-contacting, *in vivo* PAS measurements in skin (performed in 50 volunteers), in the VIS-NIR range, seeking an objective determination of pigmentation, blood microcirculation and water content of human skin (Schmidt et al., 2000). According to these authors, strong spectral variations observed within the same skin type are probably based on the natural variability of human skin and in the subjective clinical evaluation of the skin type; nevertheless, PAS results obtained show good correlation between PA data and (clinically evaluated) skin type, indicating that skin type determination could indeed be performed through the analysis of PA measurements.

3.2 Skin pigmentation analysis employing photoacoustic measurements

In 2004, Viator and co-workers proposed a method for the determination of the epidermal melanin content employing a PA probe using a Nd:YAG (neodymium, yttrium, aluminum, garnet) laser at 532nm (Viator et al., 2004). Ten human subjects with skin phototypes I-VI were tested using the PA probe and visible reflectance spectroscopy (VRS); melanin content was evaluated through each of these methods, and a good linear fit (r^2=0.85) was obtained for the plot of PA x VRS.

Pigmentation skin level can also be evaluated through simple, direct PA measurements employing non-laser light sources. Actually, PA measurements have been performed at the Laboratory of Photoacoustic Technique Applied to Biological Systems (FASBio), at UNIVAP (Brazil). The objective of such *in vivo* measurements was to classify different skin types according to the amplitude of the PA signal, which can be associated to the corresponding pigmentation level of the skin.

In the following subsections, we present this straightforward PA approach to skin characterization according to the level of pigmentation, employing PA measurements in volunteers. Experimental results are compared both to Fitzpatrick and Baumann clinical skin type evaluations.

3.2.1 Materials and methods

The PA setup employed in such *in vivo* skin measurements consisted of a 250W tungsten halogen lamp as light source (with wavelength range 400nm<λ<700nm and light intensity of about 20W/m^2), a mechanical chopper (SRS, model SR540), a lock-in amplifier (SRS, model SR530) and a microcomputer for data acquisition.

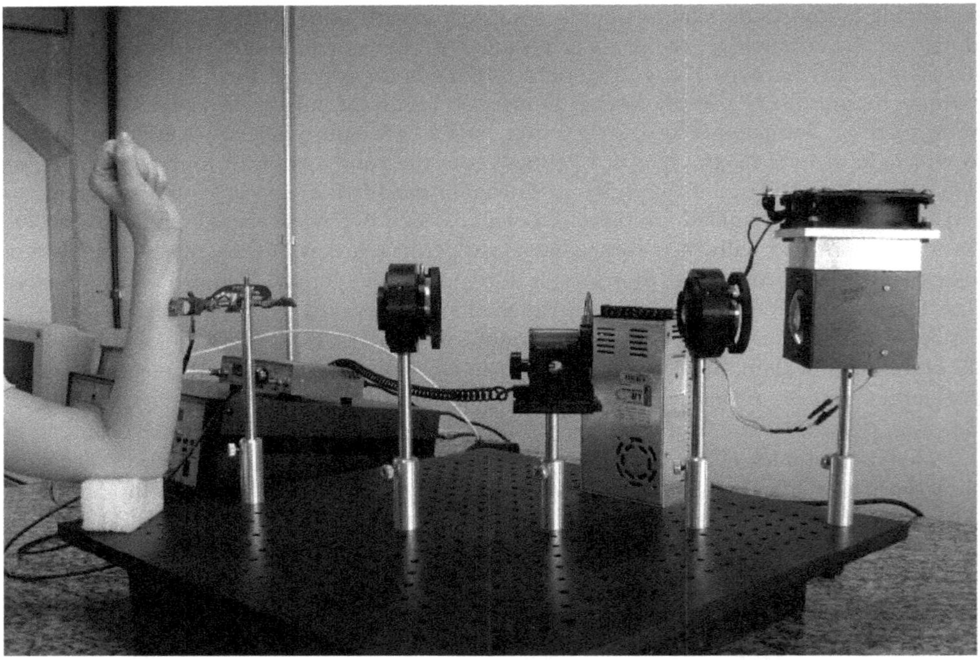

Fig. 2. Experimental setup with volunteer positioned for *in vivo* skin measurement

The double faced PA cell employed, with an electret microphone, was developed at UNIVAP; sensitivity was 15 mV/Pa for the frequency employed in skin measurements,

17Hz. The electret microphone structure was described by Marquezini et al. (1990). The PA cell has a cylindrical body and two opposite, parallel faces (one is closed by a thin glass layer and the other, by the sample itself). For the modulation frequency employed (17Hz), the thickness of the skin layer under study is about 30μm.

PA measurements were recorded as a function of time (200 readings for each measurement, in 0.5s intervals, up to a total of 100s per measurement). During measurements, one face of the PA cell was closed with a thin transparent window, while the forearm of the volunteer was gently pressed against the opposite face.

Figure 2 shows the PA experimental setup employed for in vivo skin measurements at the FASBio/UNIVAP, with a volunteer positioned for measurement.

Measurements were performed in 57 female volunteers, between 20 and 30 years-old. Initially, each volunteer answered a questionnaire according to their daily routine associated to skin care; volunteers were also clinically evaluated and, as a result, they were classified according to skin phototype, following Fitzpatrick classification (Fitzpatrick, 1988).

Before measurements, the skin area to be evaluated was cleaned with cotton embedded in alcohol 70%. The PA signal was then recorded for the inner and outer faces of both forearms. Volunteers were then classified according to the respective PA signal amplitude, and this classification was compared to the phototype classification.

3.2.2 Results and discussion

Initially, a comparison between the PA signal amplitude of the inner and outer faces was performed, showing a highly significant statistical difference (paired t-test, $p<0,005$), with higher PA amplitude being observed for the outer face of the forearm. This result can be attributed to the higher pigmentation level of the skin region continuously exposed to solar radiation, demonstrating that skin constitution and aspect are clearly influenced by the level of sun exposure.

After clinical evaluation of the volunteers for skin phototype (following Fitzpatrick classification), PA results were grouped according to the phototype of each volunteer. Results are presented in Table 1.

Skin phototype (Fitzpatrick)	PA signal amplitude (mV)
II	1.26 ± 0.05 [a]
III	1.59 ± 0.09 [ab]
IV	1.70 ± 0.10 [ab]
V	1.80 ± 0.10 [b]

Table 1. PA signal amplitude (mV) for the inner face of the forearm, for each skin phototype (average ± standard error). Different indexes (a, b) indicate significant statistical difference (comparison among groups: $p=0.009$, ANOVA)

Comparison among phototype groups was performed and significant statistical difference was verified (as we can see in Table 1), showing that the PA signal level (amplitude) for the inner forearm tends to scale with skin phototype, as defined by Fitzpatrick.

Afterwards, the PA signal amplitude for each volunteer (average values for the inner face of the forearm) allowed the division of the volunteers in two groups, "pigmented" (P) and "non-pigmented" (NP), following the Baumann proposal. As the average PA signal amplitude obtained for all measurements (inner face of the forearm) was 1.5mV, this was the cutoff value adopted for separating the volunteers into "P" (for PA signal amplitude above 1.5 mV) and "NP" (under 1.5 mV). Table 2 shows the division of each (clinically evaluated) phototype group into the (experimentally evaluated) P and NP groups. In this way, the PA technique allowed the comparison between two different skin classification forms.

Skin phototype	NP group (%)	P group (%)
II	93	7
III	46	54
IV	31	69
V	20	80

Table 2. Distribution of the volunteers of each phototype in the NP and P groups, according to the PA signal level

Table 2 shows that phototype II is highly related to the NP group, while phototypes IV and V concentrate in the P group. Phototype III appears in both groups, showing the variability of elements inside this classification.

The simple methodology presented here and the corresponding results obtained open perspectives for an objective, experimental classification of skin types, based upon PA measurements. Additional work in this field is currently being performed at FASBio/UNIVAP (Brazil).

4. Transdermal drug delivery

4.1 Topical application of drugs: advantages and requirements for evaluation

Topical application of drugs is known as an interesting alternative route to oral and intravenous administration, both aiming to systemic effects and local action, offering advantages such as ease of administration and lack of first-pass effect (Aqil et al., 2007). However, in studies of transdermal drug delivery (penetration of substances into skin), one must employ non-invasive techniques, in order to avoid second order effects that would at least bring difficulties to the interpretation of the results. *In vivo* measurements are particularly desirable in such studies, because the response of excised skin can be affected by dehydration, and the response of artificial skin differs significantly from that of *in vivo* skin at least in some cases - as when we talk about in-depth processes, in which even blood circulation may play a role.

As mentioned earlier in the present chapter, the PA technique can be applied without previous preparation of the samples and even for *in vivo* measurements; as such, transdermal drug delivery studies have been performed employing the PA technique in order to obtain the penetration rate of a wide range of different products topically applied to skin (Bernengo et al., 1998; Hahn et al., 2001; Savateeva et al., 2001; Pedrochi et al., 2005; Truite et al., 2007).

4.2 Substrates for transdermal drug delivery studies

Besides measurements in human skin, penetration rates of topically applied products are frequently evaluated through measurements performed in animal skin tissue. In this case, rabbit and pig skin are, by far, the most employed alternatives, because of the similarity to human skin.

As pointed by Simon & Malbach (2000), physiological and anatomical similarities between man and pig make this animal a good model for man in biomedical research. The correlation of quantitative data between pig skin and human skin can be frequently classified as very good (Benech-Kieffer et al., 2000); therefore, pharmacological (and even toxicological) skin research is often based on the knowledge of pig skin absorption and percutaneous permeation.

Recently, Nicoli et al. (2008) employed qualitative and quantitative analysis of stratum corneum lipids and permeation experiments to analyze the utilization of rabbit ear skin in transdermal permeation studies, using pig ear skin as a reference. Their results showed that the stratum corneum of both rabbit ear skin and pig ear skin present similar thickness. Probably due to its higher lipophilicity, rabbit ear skin was less permeable to hydrophilic compounds; however, the permeability to progesterone was comparable between isolated pig epidermis and rabbit ear skin. Nicoli and co-workers conclude that the rabbit ear skin can be sucessfully employed in skin permeatin studies.

4.3 Photoacoustic evaluation of topically applied products

Different pharmaceutical formulations for a topically applied drug may present very different transdermal delivery ratios, depending on the product composition (excipients usually play a major role in the penetration kinetics of topically applied products). These penetration rates may be evaluated through the analysis of the time-dependence of the PA signal after topical application of a given product in skin. This methodology can also be applied to the evaluation of sunscreens, that may be characterized in terms of their (photo)stability after topical application (in this case, the lower the rate, the better the product).

Gutierrez-Juarez et al. (2002) employed PA measurements in the analysis of substances topically applied to the human skin. To fulfill this purpose, these authors utilized a double-chamber PA cell; the absorption determination was obtained through the measurement of the thermal effusivity of the binary system substance-skin. The model employed by Gutierrez-Juarez and co-workers (that assumes that the effective thermal effusivity of the binary system corresponds to that of a two-phase system) was experimentally applied to study different topically applied substances, in different parts of the body. The corresponding relative concentrations of substances as a function of time were determined by fitting a sigmoidal function (for ketoconazol and sunscreen) or an exponential function (for nitrofurazona, vaseline and vaporub) to the experimental data.

Pedrochi and co-workers (2005) employed PAS measurements to evaluate the penetration rate of different sunscreens into human skin *in vivo*. Their results showed that the diminution rate of the sunscreen amount in the skin surface depends on the form of the product: sunscreens in cream form tend to present faster reduction after application in skin. This leads to the conclusion that sunscreens in gel form are more adequate (presenting longer protection against UV radiation).

Another transdermal drug delivery study is the work of Truite et al. (2007), which employed PAS measurements in the *ex vivo* determination of the penetration rate of different phytotherapic formulations (with and without salicylic acid) for treatment of vitiligo. Measurements were performed as a function of time in rabbits. PA depth monitoring showed that both formulations propagated through the skin up to the melanocytes region, leading the authors to suggest that the delivery of the active agents may occur even without the use of queratinolitic substances (that are not really recommended, since they are known to induce side effects in animals).

The PA technique can be employed to study the penetration kinetics of topically applied products not only as a function of product composition, but also according to the application method. Phonophoresis is the utilization of ultrasound (US) waves to enhance the delivery of topically applied substances (Byl, 1995). In physiotherapy practice, phonophoresis is one of the various strategies developed to overcome the skin's resistance in transdermal drug delivery, enhancing skin permeability (Duangit et al., 2011).

In the last years, comparative studies between massage and phonophoresis (in its different modes) as application methods for different anti-inflammatories have been in the front line of research at the FASBio/UNIVAP; in the experiments, transdermal drug delivery has been evaluated through PA measurements as a function of time after topical application of different drugs in the forearm region, using manual massage or phonophoresis.

Results indicate that different products present distinct absorption times (depending on the vehicle employed, for example); the application method also affects the typical time constant of drug penetration into skin, though not for all tested formulations.

In the following subsections, we present one experiment performed at FASBio/UNIVAP in which the penetration kinetics of the pharmaco *Cordia verbenacea* DC (Acheflan) in the human skin was evaluated through PA measurements as a function of time for each of the application methods: massage and phonophoresis.

4.3.1 Materials and methods

The pharmaco *Cordia verbenacea* DC (Acheflan) is a topic usage anti-inflammatory medication widely employed in medicine, having alpha-humulen and trans-caryophyllen as active agents. Our experiment aimed to evaluate the penetration kinetics of Acheflan in the human skin (massage *versus* phonophoresis) through *in vivo* PA measurements.

The survey was conducted in 10 volunteers (four men and six women) aged between 18 and 30 years. The following inclusion criteria were adopted: (i) absence of ulcers or any change in dermatology distal forearm and wrist; (ii) not being allergic to any component of the formula topically applied; (iii) absence of metal implants in the wrist or forearm; (iv) absence of stomach pain complaints; and (v) not being pregnant.

The protocol for cleaning prior to drug application consisted of cleaning the skin area to be evaluated (region near the distal forearm and right ulnar artery) with cotton soaked in 70% alcohol. The area of topical application was then demarcated and *Cordia verbenacea* DC (essential oil 5.0mg/g) was applied by rubbing the head of the ultrasound therapy equipment configured for continuous mode and intensity of $1.2W/cm^2$, but switched off, for five minutes. This procedure was repeated on the opposite forearm with the ultrasound therapy equipment turned on, also for five minutes.

PA measurements employed the same experimental system described in section 3.2.1. During measurements, the volunteers were positioned adjacent to the assembly with an aluminum foil (60μm thick) sealing the PA cell and the distal forearm positioned in direct contact with it, as proposed by Bernengo et al. (1998). For an aluminum foil with this thickness, the cutoff frequency is approximately 7kHz; in the present study, the modulation frequency employed was 17Hz, so that the foil can be considered as thermally thin (Rosencwaig, 1980).

Each measurement series consisted of 40 readings (one each two seconds, for a total time of 80 seconds), repeated 10 times for each application form, with rest intervals of 100 seconds between successive series (total time of 30 minutes for each volunteer and application method). The software "SISCOMF" (developed at UNIVAP) was employed for data acquisition.

In order to analyze the typical time constant for the penetration of the applied drug, the experimental curves were fitted by the Boltzmann equation:

$$PA(t) = \frac{A_1 - A_2}{1 + e^{(t-t_0)/dt}} + A_2 \qquad (2)$$

The Boltzmann curve is a S-shaped curve in which A_1 is the initial signal amplitude, A_2 the final signal amplitude, t_0 is the half-absorption time and dt, the time interval such as 67% of the penetration process occurs between t_0-dt e t_0+dt.

Analysis of the PA data was performed with the aid of the software Microcal Origin® 7.5 (employed for the generation of the fitting curves); statistical analysis was performed with GraphPad Instat® 3.0.

4.3.2 Results and discussion

Figure 3 shows an example of PA data measurements as a function of time fitted by a Boltzmann curve.

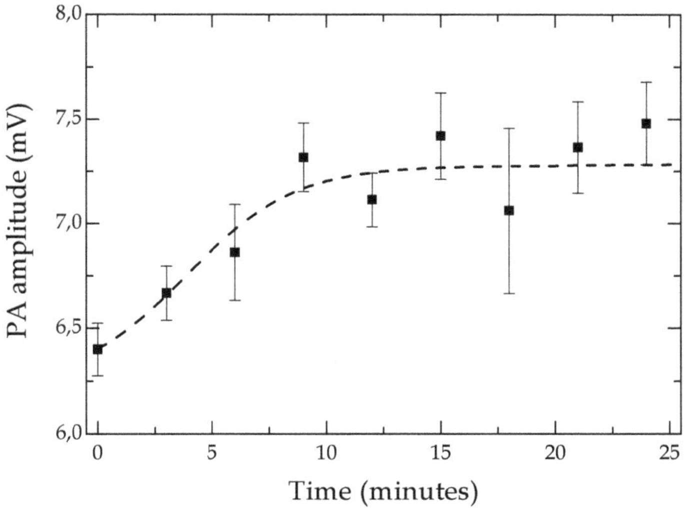

Fig. 3. Example of a Boltzmann curve (dashed line) fitting PA data obtained for one of the volunteers after phonophoresis application of the pharmaco *Cordia verbenacea* DC (Acheflan)

The experimental results obtained (average values for each parameter of the fitting curve) are summarized in Table 3; interpretation of such results must consider that the initial amplitude (A_1) of the PA signal corresponds to the system formed by the applied drug+skin, while the final amplitude (A_2) corresponds to the skin only, with the product having penetrated beyond the layer responsible for the generation of the PA signal (about 30μm, in the present case).

Application Method	A_1 (u.a)	A_2 (u.a)	t_0	Dt	t_s (t_0 + dt)
Massage	0,66 ± 0,04	0,70 ± 0,02	7 ± 2	4 ± 2	11 ± 2
Phonophoresis	0,65 ± 0,03	0,72 ± 0,04	5 ± 1	1,3 ± 0,3	7 ± 1

Table 3. Average values (± standard error) for A_1 and A_2 (in arbitrary units), t_0 and dt (in minutes), for each of the application methods (N=10)

In order to understand if penetration was effective, first of all, it is imperative to evaluate if the difference between A_1 and A_2 is statistically significant. Therefore, initially a paired t-test was carried on to verify if there was significant difference between A_1 and A_2 for each application method employed (indicating significant penetration of the applied product). This was verified for both application methods; however, this difference is more evident for phonophoresis application, in which the difference between the initial and the final signal presents $p=0{,}011$ ($p=0{,}066$ was found for massage application).

Statistical tests were also employed in the comparison between the two application methods (massage and phonophoresis); no statistical significance was found for t_0 and dt. Considering $t_S = (t_0 + dt)$ as the total effective penetration time for the epidermal layer under study, the results obtained (average ± error, N=10) are 11(±2) minutes (for massage application) and 7(±1) minutes (for phonophoresis application). The paired t-test for this parameter shows $p=0{,}073$.

Experiments performed at FASBio/UNIVAP show that the form of application can influence the kinetics of transdermal drug delivery, depending on the applied product. In the experiment presented here, significant penetration has been reported for both forms of administration (massage and phonophoresis); PA measurements showed that effective penetration is at least more evident after phonophoresis application, when compared to massage application.

5. Conclusion and perspectives

The possibility of performing *in vivo* studies brings great value to the use of the PA technique in experimental skin research. The potential and relevance of PA measurements in this field have been shown by a large range of experiments being performed by different research groups around the world - actually, the examples here presented must be seen as a sample of what has been done.

PA measurements *in vivo* are able to detect alterations in skin pigmentation. Even in skin regions normally protected from sun exposure, the PA signal level tends to follow clinical classification; actually, the use of the PA technique goes one step ahead, allowing comparative and quantitative research through simple, direct measurements.

Topical drug application has been employed in the treatment of many pathological processes; its efficiency is associated to the efficiency of transdermal drug delivery. PA measurements have been sucessfully employed in transdermal drug delivery studies, allowing a quantitative analysis of the kinetics and effectivity of drug delivery. Different PA experiments point to the fact that gel formulations tend to be more adequate for topical use. Depending on the topically applied product, the form of application can also be determinant in the kinetics of transdermal drug delivery.

Measurements already performed indicate various perspectives for future research, such as: i) the determination of the skin oiliness level; ii) analysis of skin lesions; iii) studies on the photostability of sunscreens and even determination of the sun protection factor (SPF) of sunscreens (through PAS measurements); and iv) further studies on formulation and application form of a wide range of topically applied products.

6. Acknowledgment

The authors acknowledge Fapesp and CNPq for financial support of biomedical research being developed at FASBio/UNIVAP, São José dos Campos (SP), involving the characterization of human skin and transdermal drug delivery of topically applied products.

7. References

Acosta-Avalos, D.; Alvarado-Gil, J.J.; Vargas, H.; Frías-Hernández, J.; Olalde-Portugal, V.; Miranda, L.C.M. (1996) Photoacoustic monitoring of the influence of arbuscular mycorrhizal infection on the photosynthesis of corn (Zea mays L.). *Plant Science*, v.119, n.1-2, pp.183-190; doi:10.1016/0168-9452(96)04454-8

Aqil, M.; Abdul, A.; Yasmin, S.; Asgar, A. (2007) Status of terpenes as skin penetration enhancers. *Drug Discovery Today (Oxford)*, v.12, n.23/24, pp.1061-1067.

Balderas-Lopez, J.A.; Moreno-Márquez, M.M.; Martínez, J.L.; Sánchez-Sinencio, F. (1999) Thermal characterization of some dental resins using the photoacoustic phase lag discontinuites. *Superfícies y Vacio*, v.8, pp.42-45.

Balderas-Lopez, J.A.; Mandelis, A. (2001) Thermal diffusivity measurements in the photoacoustic open-cell configuration using simple signal normalization techniques. *Joumal of Applied Physics*, v.90, n.5. pp.2273-2279.

Balogun, O.; Regez, B.; Zhang, H.F.; Krishnaswamy, S. (2009) Real time, full-field imaging of photoacoustic generated signals for biomedical applications, *ICPPP15 - Book of Abstracts, 15th International Conference on Photoacoustic and Photothermal Phenomena*, pp.77, Leuven, Belgium, July 19-23, 2009.

Baumann, L. (2006a) New Skin Typing System. *Skin & Aging* , v.14, n.2, pp.60-64.

Baumann, L. (2006b) *The Skin Type Solution*. Bantam Books, ISBN-10: 0-553-80422-7, New York, NY.

Beard, P. (2009) High resolution spectroscopic photoacoustic imaging for characteristic tissue structure and function, *ICPPP15 - Book of Abstracts, 15th International Conference on Photoacoustic and Photothermal Phenomena*, pp.79, Leuven, Belgium, July 19-23, 2009.

Benamar, N.; Laplante, A.F.; Lahjomri, F.; Leblanc, R.M. (2004) Modulated photoacoustic spectroscopy study of an artificial tanning on human skin induced by dihydroxyacetone. *Physiological Measurement*, v.25, n.5, pp.1199-1210.

Benech-Kieffer, F.; Wegrich, P.; Schwarzenbach, R.; Klecak, G.; Weber, T.; Leclaire, J.; Schaefer, H. (2000) Percutaneous Absorption of Sunscreens in vitro: Interspecies

Comparison, Skin Models and Reproducibility Aspects. *Skin Pharmacology and Physiology*, v.13, pp.324-335; doi: 10.1159/000029940

Bernengo, J.C.; Gasquez, C; Falson-Rieg, F. (1998) Photoacoustics as a tool for cutaneous permeation studies. *High Temperature-High Pressure*, v.30, n.5, pp.619-624.

Bodzenta, J.; Bukowski, R.J.; Christ, A.; Pogoda, T. (2002) Photoacoustic detection of drug diffusion into a membrane: theory and numerical analysis. *International Journal of Heat and Mass Transfer*, v.45, n.22, pp.4515-4523; doi:10.1016/S0017-9310(02)00155-2

Byl, N.N. (1995) The Use of Ultrasound as an Enhancer for Transcutaneous Drug Delivery: Phonophoresis. *Physical Therapy*, v.75, n.6, pp.539-553.

Cahen, D.; Bults, G.; Garty, H.; Malkin, S. (1980) Photoacoustic in life sciences. *J. Biochemical Biophysical Methods*, v.3 (5), 1980, pp.293-310.

Duangjit, S.; Opanasopit, P.; Rojanarata, T.; Ngawhirunpat, T. (2011) Characterization and *In Vitro* Skin Permeation of Meloxicam-Loaded Liposomes versus Transfersomes. *Journal of Drug Delivery*, v.2011, ID 418316; doi:10.1155/2011/418316

Fitzpatrick, T.B. (1988) The validity and practicality of sun-reactive skin types I through VI. *Archives of Dermatology*, v.124, pp.869–871.

Gao, C.M.; Zhang, S.Y.; Zhang, Z.; Shui, X.J. (2005) Thermal diffusivity of porcine tissues characterized by photoacoustic piezoeletric technique. *Journal of Physics IV (France)*, v.125, pp.777-779; doi:10.1051/jp4:2005125179

Gutierrez-Juarez, G.; Vargas-Luna, M.; Cordova, T.; Varela, J.B.; Bernal-Alvarado, J.J.; Sosa, M. (2002) *In vivo* measurement of human skin absorption of topically applied substances by a photoacoustic technique. *Physiological Measurement*, v.23, pp.521-532.

Hahn, B.D.; Neubert, R.H.H.; Wartewig, S.; Lasch, J. (2001) Penetration of compounds through human stratum corneum as studied by Fourier transform infrared photoacoustic spectroscopy. *Journal of Controlled Release*, v.70, n.3, pp.393-398.

Herbert, S.K.; Han, T.; Vogelmann, T.C. (2000) New applications of photoacoustics to the study of photosynthesis. *Photosynthesis Research*, v.66, n.1-2, pp.13-31; doi: 10.1023/A:1010788504886

Hu, S.; Wang, L.V. (2010) Photoacoustic imaging and characterization of the microvasculature, *Journal of Biomedical Optics*, v.15, pp.011-101; doi:10.1117/1.3281673

Li, T.; Dewhurst, R.J. (2010) Photoacoustic imaging in both soft and hard biological tissue. *Journal of Physics: Conference Series*, v.214, 012028; doi: 10.1088/1742-6596/214/1/012028

Lopez, T.; Picquart, M.; Aguilar, D.H.; Quintana, P.; Alvarado-Gil, J.J.; Pacheco, J. (2005) Photoacoustic monitoring of dehydration in sol-gel titania emulsions. *Journal of Physics IV (France)*, v.125, pp.583-585; doi:10.1051/jp4:2005125134

McDonald, F.A.; Wetsel G.C. (1978) Generalized theory of the photoacoustic effect, *Journal of Applied Physics*, v.49, pp.2313.

Malkin, S.; Puchenkov, O.V. (1997) The photoacoustic effect in photosynthesis. In *Progress in Photothermal and Photoacoustic Science and Technology: Life and Earth Sciences* (Mandelis, A., and Hess, P., editors), SPIE, ISBN 0-8194-2450-1, Washington, USA.

Marquezini, M.V.; Cella, N.; Mansanares, A.M.; Vargas, H.; Miranda, L.C.M. (1990) Open photoacoustic cell spectroscopy. *Measurement Science & Technology*, v.2, pp.396-401.

Melo, J.O.; Pedrochi, F.; Baesso, M.; Hernandes, L.; Truiti, M.; Baroni, S.; Bersani-Amado, C. (2011) Evidence of Deep Percutaneous Penetration Associated with Anti-Inflammatory Activity of Topically Applied Helicteres gardneriana Extract: A Photoacoustic Spectroscopy Study. *Pharmaceutical Research*, v.28, n.2, pp.331-336; doi: 10.1007/s11095-010-0279-3

Nicoli, S.; Padula, C.; Aversa, V.; Vietti, B.; Wertz, P.W.; Millet, A.; Falson, F.; Govoni, P.; Santi, P. (2008) Characterization of Rabbit Ear Skin as a Skin Model for in vitro Transdermal Permeation Experiments: Histology, Lipid Composition and Permeability. *Skin Pharmacology and Physiology*, v.21, pp.218-226; doi: 10.1159/000135638

Pedrochi, F.; Sehn, E.; Medina, A.N.; Bento, A.C.; Baesso, M.L.; Storck, A.; Gesztesi, J.L. (2005) Photoacoustic Spectroscopy to Evaluate the Penetration Rate of Three Different Sunscreens into Human Skin *in vivo*. *Journal of Physics IV (France)*, v.125, pp.757-759; doi:10.1051/jp4:2005125174

Qiu, P.F.; Zhang, S.Y.; Shui, X.J. (2008) Photoacoustic study of thermal properties of biological tissues detected by PVDF film transducer. *European Physics Journal - Special Topics*, v.153, pp.487-490.

Rezende, D.V.; Nunes, O.A.C.; Oliveira, A.C. (2009) Photoacoustic Study of Fungal Disease of Açai (*Euterpe oleracea*) Seeds. *International Journal of Thermophysics*, v.30, n.5, pp.1616-1625; doi: 10.1007/s10765-009-0655-6

Rosencwaig, A.; Gersho, A. (1976) Theory of the photoacoustic effect with solids. *Journal of Applied Physics*, v.47, pp.64-69.

Rosencwaig, A. (1980) *Photoacoustics and Photoacoustic Spectroscopy*. John Wiley & Sons, ISBN 0-471-04495-4, New York, USA.

Savateeva, E.V.; Karabutov, A.A.; Oraevsky, A.A. (2001) *Proceedings of SPIE* 4256, pp.61-69.

Schmidt, W.D.; Fassler, D.; Zimmermann, G.; Liebold, K.; Wollina, U. (2000) Non-contacting diffuse VIS-NIR spectroscopy of human skin for evaluation of skin type and time-dependent microcirculation. *Progress in biomedical optics and imaging*, v.1, n.31, pp.91-102.

Silva, W.J.; Prioli, L.M.; Magalhães, A.C.N.; Pereira, A.C.; Vargas, H.; Mansanares, A.M.; Cella, N.; Miranda, L.C.M.; Alvarado-Gil, J.J. (1995) Photosynthetic O_2 evolution in maize inbreds and their hybrids can be differentiated by open photoacoustic cell technique. *Plant Science*, v.104, n.2, pp.177-181; doi:10.1016/0168-9452(94)04026-D

Simon, G.A.; Maibach, H.I. (2000) The Pig as an Experimental Animal Model of Percutaneous Permeation in Man: Qualitative and Quantitative Observations – An Overview. *Skin Pharmacology and Applied Skin Physiology*, v.13, pp.229-234; doi: 10.1159/000029928

Sui, X.; Sun, J.; Wang, S.; Li, W.; Hu, L.; Meng, F.; Fan, Y.; Zhang, A. (2011) Photosynthetic induction in leaves of two cucumber genotypes differing in sensitivity to low-light stress. *African Journal of Biotechnology*, v.10, n.12, pp. 2238-2247.

Swearingen JA, Holan SH, Feldman MM, Viator JA. (2010) Photoacoustic discrimination of vascular and pigmented lesions using classical and Bayesian methods. *Journal of Biomedical Optics*, v.15, n.1, 016019; doi: 10.1117/1.3316297

Truite, C.V.R.; Philippsen, G.S.; Ueda-Nakamura, T.; Natali, M.R.M.; Dias Filho; B.P.; Bento, A.C.; Baesso, M.L.; Nakamura, C.V. (2007) *Photochemistry and Photobiology*, v.83, pp.1529-1536; doi: 10.1111/j.1751-1097.2007.00197.x

Vargas, H.; Miranda, L.C.M. (1988) Photoacoustic and related photothermal techniques. *Physics Reports*, v.161, n.2, pp.43-101.

Viator, J.A.; Komadin, J.; Svaasand, L.O.; Aguilar, G.; Choi, B.; Nelson, J.S. (2004) A Comparative Study of Photoacoustic and Reflectance Methods for Determination of Epidermal Melanin Content. *Journal of Investigative Dermatology*, v.122, pp.1432-1439.

14

Acoustic–Gravity Waves in the Ionosphere During Solar Eclipse Events

Petra Koucká Knížová and Zbyšek Mošna
Institute of Atmospheric Physic, Czech Academy of Sciences
Czech Republic

1. Introduction

Terrestrial atmosphere shows a high variability over a broad range of periodicities, which mostly consists of wave-like perturbations characterized by various spatial and temporal scales. The interest for short time variability in ionospheric attributes is related to the role that ionosphere plays in the Earth's environment and space weather. Acoustic-gravity waves (AGWs), waves in the period range from sub-seconds to several hours, are sources of most of the short-time ionospheric variability and play an important role in the dynamics and energetics of atmosphere and ionosphere systems. Many different mechanisms are likely to contribute to the acoustic-gravity wave generation: for instance, excitation at high latitudes induced by geomagnetic and consequent auroral activity, meteorological phenomena, excitation in situ by the solar terminator passages and by the occurrence of solar eclipses.

During solar eclipse, the lunar shadow creates a cool spot in the atmosphere that sweeps at supersonic speed across the Earth's atmosphere. The atmosphere strongly responds to the decrease in ionization flux and heating. The very sharp border between sunlit and eclipsed region, characterized by strong gradients in temperature and ionization flux, moves throughout the atmosphere and drives it into a non-equilibrium state. Acoustic-gravity waves contribute to the return to equilibrium. At thermospheric heights, the reduction in temperature causes a decrease in pressure over the totality footprint to which the neutral winds respond. Thermal cooling and downward transport of gases lead to neutral composition changes in the thermosphere that have significant influence on the resulting electron density distribution. Although the mechanisms are not well understood, several studies show direct evidence that solar eclipses induce wave-like oscillations in the acoustic-gravity wave domain.

Many different mechanisms are likely to contribute to wave generation and enhancement at ionospheric heights. Hence, it is difficult to clearly separate or differentiate each contributing agent and to decide which part of wave field belongs to the in situ generated and which part comes from distant regions. First experimental evidence of the existence of gravity waves in the ionosphere during solar eclipse was reported by Walker et al. (1991), where waves with periods of 30–33 min were observed on ionosonde sounding virtual heights.

1.1 Ionospheric sounding

As the solar radiation penetrates Earth's atmosphere it forms pairs of charged particles. Under a normal day-time conditions the ionization solar flux increases immediately after sunrise, reaches maximum around local noon and decreases again till sunset. Under such conditions concentration of charged particles significantly grows in the atmosphere and forms atmospheric plasma called ionosphere. Due to the composition of the neutral atmosphere together with the changing efficiency of the incoming solar radiation, ionosphere is stratified into the layers denoted D, E, F1 and F2. After sunset, electrons and ions recombine rapidly in the D, E and F1 layer. Due to slower recombination processes of atomic ions that dominate at heights approximately above 150km altitude, F2 layer remains present all the night. Special stratification Es, sporadic E layer, occurs sometimes at heights of E layer (Davies, 1990).

Ionosphere significantly affects propagation of the electromagnetic waves. According to a frequency of the wave with respect to a concentration of the ionospheric plasma, wave propagates through the medium or it is reflected. Electromagnetic waves with frequency lower than plasma frequency of the particular plasma parcel are reflected, which allows to estimate plasma frequency. Higher frequency waves propagate through plasma. An instrument called ionosonde (or digisonde) transmits electromagnetic wave of a defined frequency and detects it after reflection from the ionosphere. Typical ionosonde sounding range is 1 MHz - 20 MHz. For each sounding wave ionosonde records time of flight τ on the path transmitter - reflection point - receiver. Time of flight is simply converted into a virtual height $h_{virtual} = \frac{\tau.c}{2}$ that corresponds to wave propagation in the vacuum (c stands here for speed of light). Virtual height is equal or higher than the corresponding real height. The output of the measurement is height-frequency characteristics called an ionogram. Real height electron concentration profiles can further be inverted from ionograms using for instance programs POLAN (Titheridge, 1985) or NHPC (Huang and Reinish, 1996). Ionosphere represents inhomogeneous and anisotropic medium which leads to a wave splitting into an ordinary and extraordinary wave modes. Hence, two reflection traces are recorded by the ionosonde (as seen on ionograms in Figure 1). However, the extraordinary mode is not further used for electron concentration profile inversion.

Figure 1 shows typical day-time and night-time ionograms recorded by a digisonde in the observatory Pruhonice. Together with the ionograms there are plots of the real height electron concentration derived by NHPC routine. On the day-time electron concentration profile, three ionospheric layers E, F1 and F2 are present while on the night-time profile there is only F2 layer detected by the ionosonde.

Sequences of ionograms are widely used for analyses of variability of atmospheric plasma ranging from detection of rapid changes with periods of minutes to the study of long-term trends.

2. Basic theory of AGWs in the Earth's atmosphere

Most of the wave-like oscillations in the atmosphere can be described/parametrized using basic acoustic-gravity wave theory in the atmosphere. Details can be found, for instance, in works of Davies (1990), Bodo et al. (2001), Hargreaves (1982), Yeh & Liu (1974) among others. Here, we show brief derivation of the dispersion relation that any wave motion of the AGW type must satisfy. In a plane-stratified, isothermal atmosphere under gravity that is constant with height, two frequency domains exist in the atmosphere where atmospheric waves can propagate, acoustic and gravity wave. Atmosphere represents compressible gas that once compressed and then released would expand and oscillate about its equilibrium state. Its oscillation frequency is known as an acoustic cut-off frequency

$$\omega_a = \frac{c}{2H} \tag{1}$$

where c is speed of sound

$$c = \sqrt{\gamma g H} \tag{2}$$

γ is the ratio of specific heats at constant pressure and constant volume, g is the gravitational acceleration, and H is the scale height. For diatomic gas $\gamma \sim 1.4$.

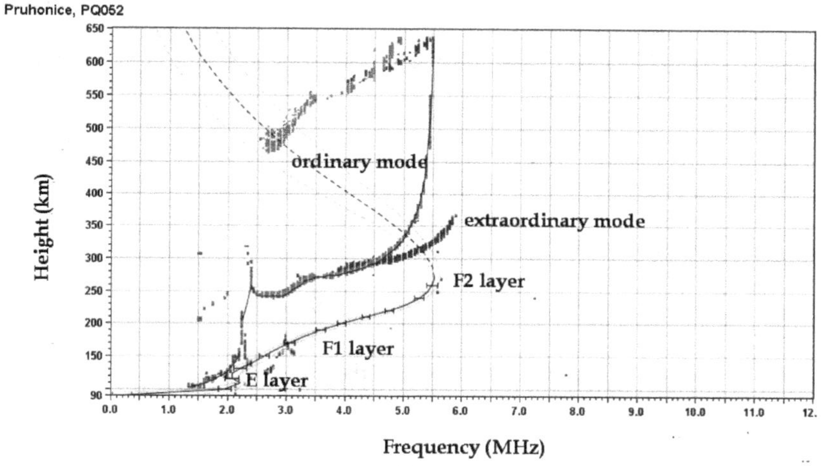

(a)

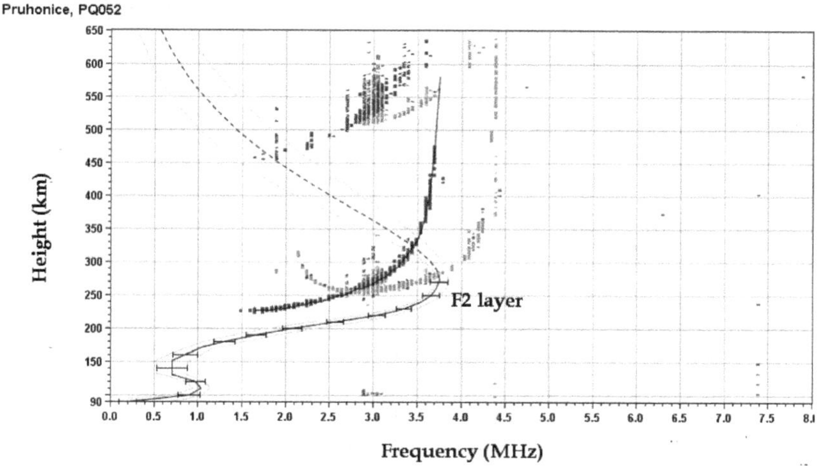

(b)

Fig. 1. Typical day-time (a) and night-time ionogram (b) measured by digisonde DPS 4 in the Observatory Pruhonice. On both plots, there is real hight electron concentration (solid line with error bars) provided as obtained by the NHPC routine.

Single element of fluid, parcel of the atmosphere, at height z with density ρ which is displaced in the vertical by Δz to a place where its density changes to ρ+Δρ, remains in pressure equilibrium with its surroundings. Displacement takes place adiabatically. This is valid when the motion is so slow that sound waves with speed

$$c = \sqrt{\frac{dp}{d\rho}} \tag{3}$$

where p stands for pressure can traverse the system faster than the time-scale of interest and the motion is so fast that the entropy is preserved. The parcel is no longer in equilibrium and starts to oscillate about its equilibrium height with buoyancy frequency.
The buoyancy force which acts on the parcel is balanced by inertial force (Newton's second law):

$$\rho \frac{d^2}{dt^2}(\Delta z) = -g\Delta\rho \tag{4}$$

where Δρ is the difference between internal and external densities.
Internal and external Δρ are derived as:

$$(\Delta\rho)_{internal} = \Delta p / c^2 = -\frac{g\rho}{c^2}\Delta z \tag{5}$$

which is due to compressibility of the fluid within the membrane and

$$(\Delta\rho)_{external} = -\frac{d\rho}{dz}\Delta z \tag{6}$$

is the change of background density at new position due to inhomogeneous nature of the atmosphere. Taking both the contributions of Δρ we get

$$\frac{d^2}{dt^2}(\Delta z) = \left(g\frac{d}{dz}(\ln\rho) + g^2/c^2 \right)\Delta z \tag{7}$$

which can be recast into

$$\frac{d^2}{dt^2}(\Delta z) + \omega_B^2 \Delta z = 0 \tag{8}$$

where

$$\omega_B^2 = -g\left(\frac{d}{dz}(\ln\rho) + g/c^2 \right) \tag{9}$$

If $\omega_B^2 > 0$, the solution is oscillatory and the fluid parcel will oscillate with characteristic buoyancy frequency ω_B called Brunt-Vaisala frequency.
More convenient form used for atmosphere is following:

$$\omega_B^2 = (\gamma - 1) g^2/c^2 + g / c^2 dc^2 / dz \tag{10}$$

This approximation is valid in the atmosphere-ionosphere system of our interest.

In isothermal atmosphere ω_B reduces to

$$\omega_B^2 = (\gamma - 1) g^2 / c^2 \tag{11}$$

In the terrestrial atmosphere the buoyancy period depends on the height. The height variance of the acoustic cut-off and buoyancy frequencies in the isothermal atmosphere is shown in Figure 2.

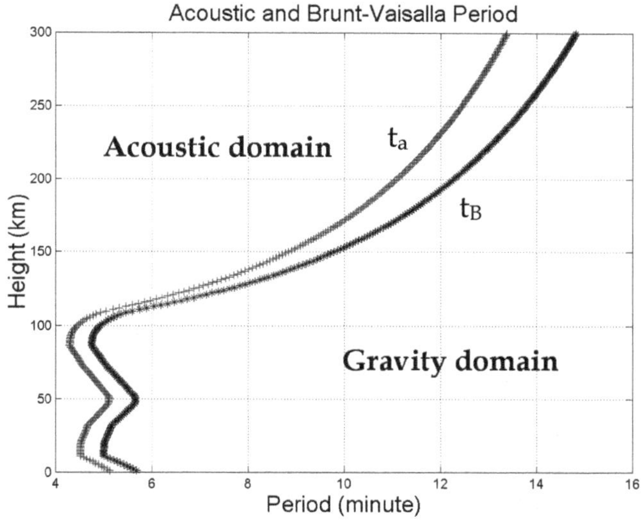

Fig. 2. Height dependence of acoustic cut-off period t_a and Brunt-Vaisala period t_B that represent limits dividing periods into acoustic and gravity wave domains. Period domain between acoustic cut-off and Brunt-Vaisala represents region where no AGW propagates.

Wave motion in the atmosphere can be described using mass conservation (continuity equation), and equation of motion:

$$\frac{\partial \rho}{\partial t} + \rho \nabla . \vec{u} + \left(\vec{u} . \nabla \right) \rho = 0 \tag{12}$$

$$\rho \left(\frac{\partial \vec{u}}{\partial t} + \left(\vec{u} . \nabla \right) \vec{u} \right) = -\nabla p + \rho g \tag{13}$$

where pressure gradients and gravity are the only forces causing the acceleration. Oscillation takes place adiabatically

$$\rho \left(\frac{\partial p}{\partial t} + \vec{u} . \nabla p \right) = \gamma p \left(\frac{\partial \rho}{\partial t} + \vec{u} . \nabla \right) \tag{14}$$

where ρ, p, γ and $\vec{u}$ are parameters of the atmosphere - density, pressure, ratio of specific heats, and velocity.

Applying the perturbation approach we are searching for wave-like solutions for the perturbation quantities. Further simplification comes from the assumption that the background state is of constant temperature T in which p_0/ρ_0 must be a constant.

$$p_0 / \rho_0 = c^2 / \gamma \tag{15}$$

Then the system (12), (13) and (14) reduces to:

$$\frac{\partial \rho'}{\partial t} + \rho_0 \vec{\nabla}.\vec{u}' - \rho_0 u_z / H = 0 \tag{16}$$

$$\rho_0 \frac{\partial \vec{u}}{\partial t} + \nabla p' - \rho' g = 0 \tag{17}$$

$$\rho_0 \left(\frac{\partial p'}{\partial t} - p_0 u_z' / H \right) = \gamma p_0 \left(\frac{\partial \rho'}{\partial t} - \rho_0 u_z' / H \right) \tag{18}$$

where index 0 denotes stationary (non fluctuating) component and the apostrophe denotes perturbation. These are the basic governing equations for the gravity waves. For a non-trivial solution the following prescription of the dispersion relation must be satisfied:

$$\omega^4 - \omega^2 \omega_a^2 - k_x^2 c^2 \left(\omega^2 - \omega_g^2 \right) - c^2 \omega^2 k_z^2 = 0 \tag{19}$$

From disperse relation, it is evident that between buoyancy frequency and acoustic cut-off frequencies one cannot have both k_x and k_z real. Figure 2 shows two period domains with border limits of acoustic cut-off period and buoyancy period.

An attenuation or growth in the wave amplitude must occur in either the vertical or the horizontal directions. We suppose that there is no variation in amplitude in horizontal directions so that k_x is purely real and k_z has an imaginary component. At frequencies exceeding acoustic cut-off ω_a, expression (19) becomes simple and the waves may be termed as ACOUSTIC WAVES. At frequencies smaller than Brunt-Vaisala frequency where gravity plays an important role, the waves are called GRAVITY or INTERNAL GRAVITY WAVES. Brunt-Vaisala frequency and acoustic cut-off frequency divide the frequency spectrum into two domains in which ω_g forms the high frequency limit for one class $\omega<\omega_g$ normally called internal gravity waves and ω_a is the low frequency limit for another class $\omega>\omega_a$ called the acoustic waves. A gap in the frequency spectrum exists between ω_g and ω_a where no internal waves can propagate.

Important approximations can be obtained under the assumption $|k_z| >> 1/2H$ and $\omega << \omega_g$ then:

$$k_z^2 = \left(\omega_g^2 / \omega^2 \right) k_x^2 \tag{20}$$

These approximations apply to much of the observed gravity waves. From (20) we see that the angle of ascent of the phase α is:

$$tg\alpha = k_z / k_x = \omega_g / \omega = \tau / \tau_g \tag{21}$$

The motions of the air parcels are, in general, ellipses in the plane of propagation and have components transverse to the direction of wave propagation. The ratio of the horizontal displacement ξ to its vertical displacement ζ is:

$$\frac{\xi}{\zeta}=\frac{\frac{ck_x}{\omega}}{\left(\frac{ck_x}{\omega}\right)^2-1}\left(\frac{ck_z}{\omega}-i\sqrt{\left(\frac{\omega_a}{\omega}\right)^2-\left(\frac{\omega_g}{\omega}\right)^2}\right) \tag{22}$$

On the frequencies just above the acoustic cutoff the air motion is essentially vertical. With acoustic waves on high frequencies the motion is radial as in sound waves. The motion is circular with horizontal propagation at a frequency $\omega_a\sqrt{2/\gamma}$. Gravity wave propagation is limited to angles between

$$\phi_{min}=\sin^{-1}\left(\frac{\omega}{\omega_g}\right)\ ,\ \phi_{max}=\pi-\sin^{-1}\left(\frac{\omega}{\omega_g}\right) \tag{23}$$

The sense of rotation of the air for gravity waves is opposite than for acoustic waves. As Φ approaches its asymptotic values the air motion becomes linear and transverse to the direction of propagation. Air parcel rotation is clockwise in case of acoustic waves while anticlockwise in case of gravity waves. Energy vector lies in the same quadrant as direction of propagation of acoustic waves. Energy flows up when phase travels down and vice versa in case of gravity waves propagation. This is important property since it accounts for the observed downward phase propagation when the source is below the level at which a disturbance is observed.
The horizontal u_x and vertical u_z components of the packet velocity are obtained from disperse relation:

$$\begin{aligned}u_x&=\frac{c^2k_x\left(\omega^2-\omega_g^2\right)}{\omega\left(2\omega^2-\omega_a^2-c^2k^2\right)}\\ u_z&=\frac{c^2k_z\omega^2}{\omega\left(2\omega^2-\omega_a^2-c^2k^2\right)}\end{aligned} \tag{24}$$

Due to coupling between neutral and charged components the initial wave-like oscillation in the neutral atmosphere induces wave-like perturbation in the ionosphere. Perturbation in the ion production is the most effective when solar ionizing rays are nearly in alignment with the initial wave front. Perturbations in the neutral atmosphere may cause perturbations in chemical processes. Presence of AGW influences the ionisation rate through changes in the local neutral density and temperature, and through changes in the ionisation radiation absorption (Hooke, 1970).

3. AGW in the ionospheric plasma

Acoustic-gravity waves are always present in the Earth's atmosphere. AGWs arise from many natural sources like convection, topography, wind shear, moving solar terminator, earthquakes, tsunami, etc. Increase in wave-like activity is associated also with human activity including coordinated experiments or unwilling accidents. AGWs influence on the upper atmosphere is not yet understood enough. They produce a great amount of variability and contribute to the background conditions in a specific parcel of the atmosphere. Gravity waves propagating from lower laying atmosphere have been long regarded as a very important source of the energy and momentum transfer in the upper atmosphere (Hines, 1960). The breaking of the upward propagating waves affects wind system, generates turbulence and heats the atmospheric gas.

Waves that reach upper atmosphere produce travelling atmospheric disturbances (TAD) or travelling ionospheric disturbances (TID) and even form the ionospheric inhomogenities which grow and finally break into the plasma instabilities observed by radar techniques that might cause scintillation of the communication signals propagating through the ionosphere. From the observation it is evident that the thermosphere is continuously swept by the acoustic-gravity waves. Statistically, the waves show a moderate preference for southward travel, with this preference being reduced or shifted to southeastward travel during disturbed times (Oliver et al., 1997). Experimental studies show that AGW activity in the ionosphere slightly increases during dawn and dusk periods of the day (Galushko et al., 1998; Somsikov & Ganguly, 1995; Sauli et al., 2005 among others). Influence of infrasonic waves generated by ground experimental sources on the ionosphere was reported for instance by Rapoport et al. (2004).

Solar eclipse represents well defined source of the AGW in the atmosphere and ionosphere systems. During solar eclipse event, solar ionization flux decreases producing well-defined cool spot in the atmosphere that moves through the Earth's atmosphere. Moving source in the atmosphere can emit both acoustic and gravity waves. Supersonic motion of the source forms wave field with bow wave. Both acoustic and gravity waves can be radiated in association with supersonic motion in the atmosphere. When the source is moving within atmosphere with subsonic velocity only gravity waves can be emitted (Kato et al., 1977).

4. Solar eclipse event – signatures in the ionospheric plasma

It has been proposed by Chimonas and Hines (1970) that solar eclipses can act as sources for AGWs. The lunar shadow creates a cool spot in the atmosphere that sweeps at supersonic speed across the Earth. The sharp border between sunlit and eclipsed regions, characterized by strong gradients in temperature and ionization flux, moves throughout atmosphere and drives it into a non-equilibrium state. Earth atmosphere shows variable sensitivity to the changes of ionization flux.

4.1 Experiments

Solar eclipse event represents phenomenon that can be precisely predicted, hence many observational campaigns are organised around the world. Effects of the solar eclipses on the ionospheric plasma are studied by mean of GPS techniques, radars, vertical ionospheric soundings etc. Study limitations lay mainly in the fact that there are no identical solar eclipse events. Moreover, solar eclipse induced effects are easily to be mixed with effects caused by geomagnetic field variations, diurnal changes of the ionosphere, seasonal variability of the atmosphere/ionosphere etc. In the upper atmosphere, AGWs can be observed either directly as neutral gas fluctuations or indirectly as induced ionospheric plasma variations. Despite intensive research many questions in the problem of the generation and propagation remain to be understood.

Studies by Fritts and Luo (1993) suggest that perturbations generated by the eclipse induced ozone heating interruption may propagate upwards into the thermosphere–ionosphere system where they have an important influence. Temperature fluctuations and electron density changes propagate as a wave, away from the totality path, cf. Muller-Wodarg et al. (1998). By means of vertical ionospheric sounding, Liu et al. (1998) detected waves excited during solar eclipse event at F1 layer heights and their generation and/or enhancement attributed to changes of temperatures and variations of the height of the transition level for the loss coefficient and the height of the peak of electron production. Studies reported by Farges et al. (2001) suggest a longitudinal diversity of the disturbances with respect to pre-noon and postnoon phases. Xinmiao et al. (2010) reported synchronous oscillations in the Es and F layer during the recovery phase of the solar eclipse. Ivanov et al. (1998) found that during solar eclipse with maximum obscuration of about 70% the F-region electron density decreased by 6-8% compared to a control day and detected travelling ionospheric disturbances. Additionally, they detected strong variations in the difference group delays with a period about 40 minutes associated with the start and end of the eclipse. Oscillations in the ionosphere, similar to gravity waves, were observed following some solar eclipse events (Chimonas and Hines, 1970; Cheng et al., 1992; Liu et al., 1998; Sauli et al., 2006). Investigation of the latitudinal dependence of NmF2 (the maximum electron density of the F2 layer) indicated that the strongest response was at middle latitudes (Le et al., 2009). The response of the sporadic-E (Es) layer also differed in each solar eclipse event. A remarkable decrease in Es layer ionization was observed during the eclipse of 20 July 1963 (Davis et al., 1964). Enhancement of Es layer ionization has also been reported and it has been suggested that it is related to internal gravity waves generated in the atmosphere during the solar eclipse (Datta, 1972).

4.2 Processes induced by solar eclipse

During the solar eclipse, on the time scale shorter than day-night change, the ionosphere reconfigures itself into a state similar to that of night situation. Photochemical ionization falls heavily almost to a night-time level. With the decreasing solar flux, atmospheric temperature falls in the moon shadow creating a cool spot with well defined border. Then the increasing solar flux starts ionization processes and warms the atmosphere again to daytime level.

Such changes in the ionization cause variation in the reflection heights, decrease/increase in electron concentration at all ionospheric heights, decrease/increase in the total electron content, rising/falling of the layer height. Such effects are characteristic for the processes during sunrise/sunset in the ionosphere. However, supersonic movement of the eclipsed region represents a key difference from the regular solar terminator motion at sunrise and sunset times. These changes in the neutral atmosphere and ionosphere induced by solar eclipse force the evolution of the ionospheric plasma toward a new equilibrium state. The return to equilibrium is likely accompanied by the eclipse induced wave motions excited in the atmosphere. Any moving discontinuity of gas parameters such as temperature, pressure etc. will generate transit-like waves. In the upper ionosphere, waves can be generated by a strong horizontal electron pressure gradient. Possible mechanisms contributing to the wave generation in the region of solar terminator are in detail discussed by Somsikov & Ganguly (1995).

Solar eclipse induces changes in all atmospheric regions extend from the upper atmosphere down to ground level. Despite the low magnitude of the eclipse induced effects at ground level, Jones et al. (1992) reported wave-like oscillation related to eclipse on the microbarometer pressure records. The cooling effect of the Moon's shadow may induce the powerful meridional airflow in the atmosphere, which accelerates the ionized clouds in the Es layer and forms the wind shear to raise the observed Doppler frequency shift and foEs values, respectively (Chen et al., 2010).

5. Solar eclipse observed by vertical ionospheric sounding in midlatitudes

Vertical sounding measurements provide local information on the electron density distribution of the bottomside ionosphere. Electron concentration in the plasma and its corresponding plasma frequency are related via following equation:

$$f_p^2 = \frac{Ne^2}{4\pi\varepsilon_0 m} \tag{25}$$

where f_p denotes plasma frequency and N, e, ε_0 and m stand for the electron concentration, the charge of electron, permittivity of free space, and the mass of the electron, respectively.

This section summarizes experimental results from the midlatitude ionospheric observatory Pruhonice (50N, 15E). At the observatory, the vertical sounding measurements were performed with ionosonde IPS 42 KEL Aerospace till the end of year 2003. Then this older equipment was replaced by digisonde DPS 4. Special campaigns of rapid sequence soundings were organized in order to study in detail ionospheric behavior during partial solar eclipses of 11 August 1999, 4 January 2011 and annular solar eclipse 3 October 2005. All three analyzed events were characterized by low geomagnetic activity; hence they represent a good occasion to observe mostly solar eclipse induced effects in the ionosphere. However, inconclusive results of the solar eclipse observations rise from the fact that different solar eclipses produce different plasma motions. Indeed, the travel cone geometry and its angular effects on the magnetized plasmas are different for each eclipse.

Solar eclipse of 11 August 1999 (as a total seen in place as close as 200 km from the measurement point) represents so far the event of the highest solar disc coverage observed in the Observatory Pruhonice. Figure 3 depicts sequence of raw ionograms measured during this event by IPS 42 KEL Aerospace equipment. The ionograms were recorded with the cadence of 1 minute. On the ionograms there is clearly seen that the eclipse event affects whole electron density profile. Critical frequencies in the E and F layer decrease before maximum disc occultation and then increase again. The electron density decrease in the E layer is much stronger than in the F layer due to different dominant type of the recombination. Electron density fall and increase occur simultaneously with occultation and de-occultation of the solar disc in the E and F1 layer while the F2 layer electron density reacts with slight delay. There are special structures of the spread F type developed on the profile after beginning of the solar disc occultation (clearly seen on ionograms at 9.14 UT and 9.16 UT). Shape of the F layer is affected as well. Unfortunately, effects in the F1 region cannot be discussed here in details because F1 layer is blanketed by strong sporadic E layer during part of the solar eclipse.

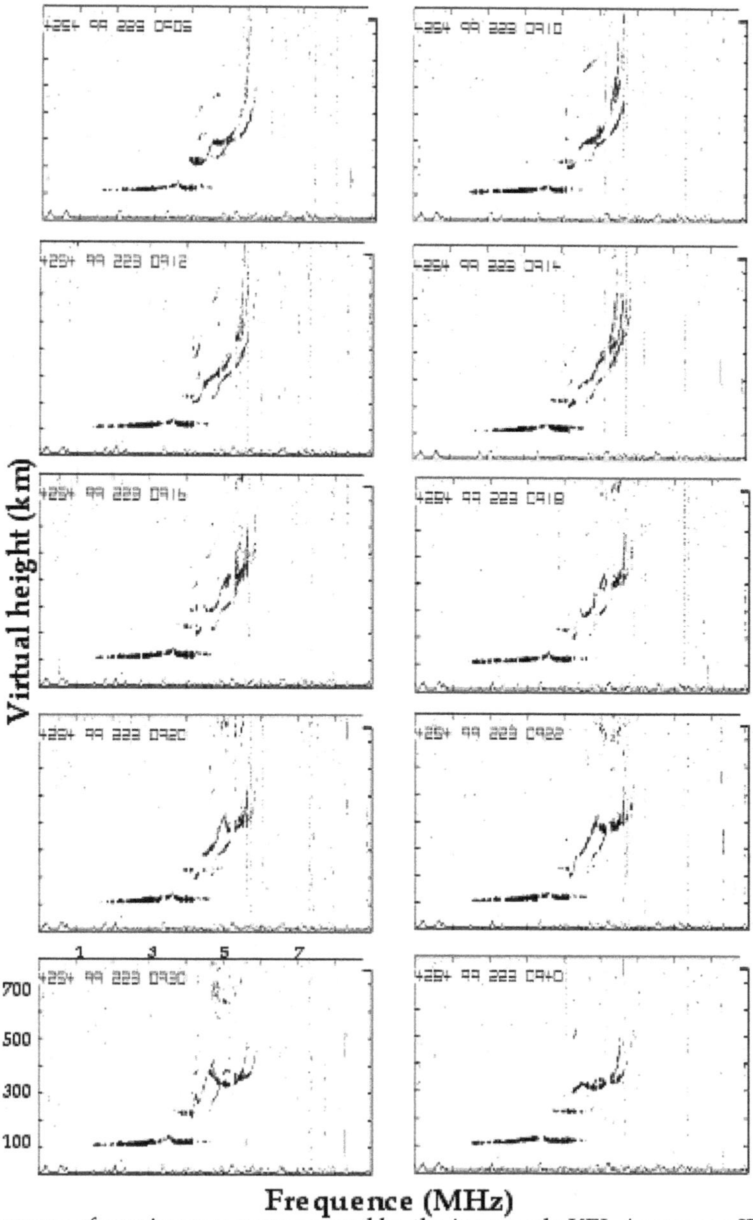

Fig. 3. Sequence of raw ionograms measured by the ionosonde KEL Aerospace IPS 42 at the observatory Pruhonice. During the special campaign ionograms were recorded with one-minute resolution in order to study rapid ionospheric changes during the solar eclipse.

Detail analysis of electron concentration by mean of spectral analysis reveals that within oscillation of electron concentration there occur several clear wave-like oscillations. It has been shown by Sauli et al. (2007) that wavelet spectral analysis is very convenient approach for such wave detection. The advantage of the wavelet based analysis is identification of the structure occurrence time which helps to associate particular wave-like structure to the agent. Figure 4 shows estimated wave parameters for selected structure that is coherent through all studied heights. Parametrization of the wave-like structure is based on AGW approximation described in Section 2. From Figure 4 it is evident that wave originates at height of about 200 km and propagates upward and downward from the source region.

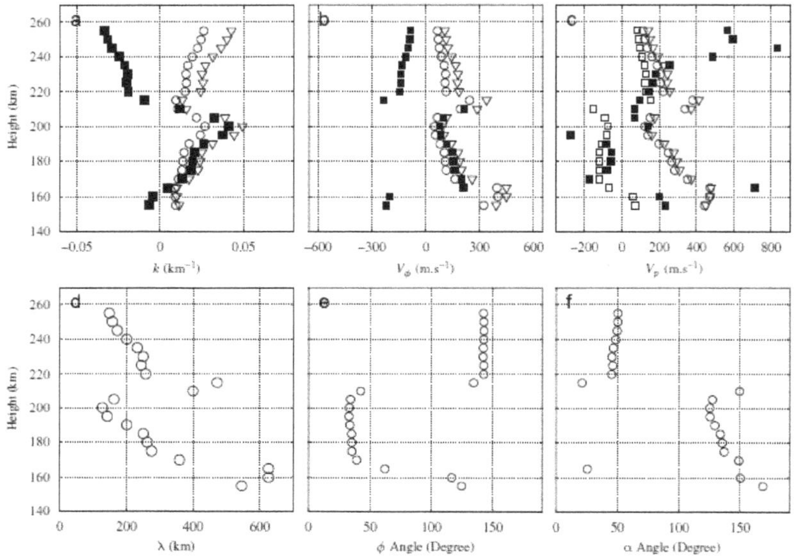

Fig. 4. Parameters of acoustic-gravity wave structure detected within ionospheric plasma during solar eclipse event 11 August 1999 (Sauli et al., 2007). Panels: wave vector (a), phase velocity (b), packet velocity (c), wave number (d), energy (e) and phase (f) angles. For the vectors of first row, the '□' correspond to the measured (full squares) and computed (empty symbols) z-components, the '○' correspond to the horizontal components while the '∇' are related to the modulus.

Another representation of the rapid changes in the ionospheric plasma is shown on the profilogram (Figure 5) measured during solar eclipse 3 October 2005 by DPS 4. Decrease in the plasma frequency at all heights is well developed. Within plasma frequency oscillation, several wave coherent structures were found that can be attributed to the eclipse event. These structures occur in the plasma at the maximum of the eclipse and after the event. In all cases we detected a component of upward energy progression. Due to the occurrence time and low geomagnetic activity the detected wave-like oscillations in the ionospheric plasma are likely signatures of bow shock and possibly waves excited by cooling of ozone in the lower laying atmosphere. Estimated velocities for one particular structure are shown in Figure 6.

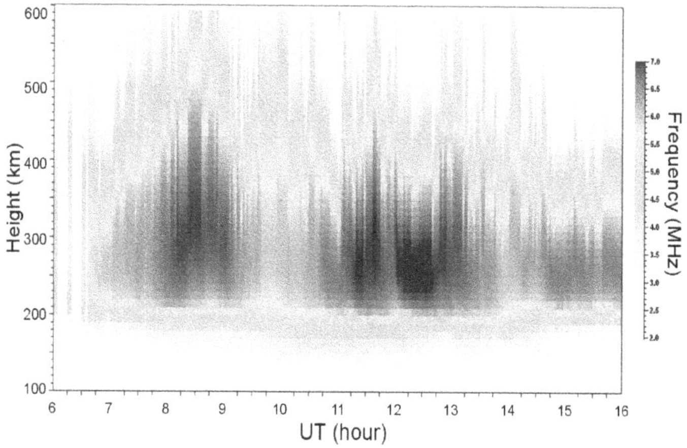

Fig. 5. Profilogram (height-time-plasma frequency development) during solar eclipse 3 October 2005 as measured by DPS 4. Ionograms were measured every 2 minutes. All ionograms were manually scaled and inverted into true-height profiles using True Height Profile Inversion Tool NHPC.

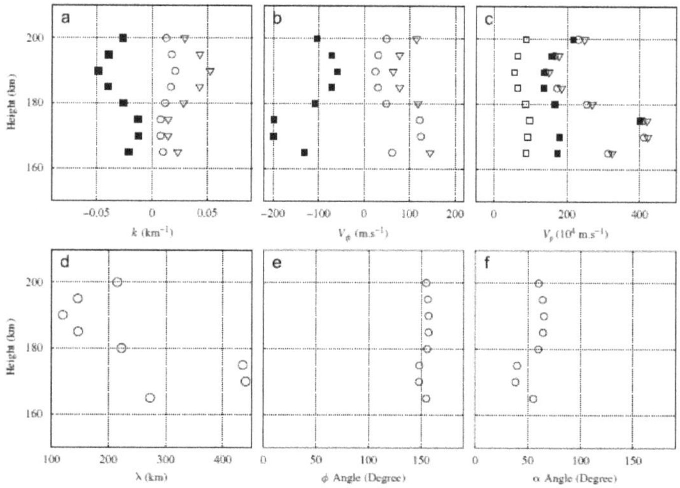

Fig. 6. Parameters of acoustic-gravity wave structure detected within ionospheric plasma during solar eclipse event 3 October 2005 (Sauli et al. 2007). Panels: wave vector (a), phase velocity (b), packet velocity (c), wave number (d), energy (e) and phase (f) angles. For the vectors of first row, the '□' correspond to the measured (black) and computed (empty) z-components, the '○' correspond to the horizontal components while the '▽' are related to the modulus.

Result of the annular eclipse is significantly different from the case of the total eclipse event of 1999 where the dominant AGW activity took place at the beginning of eclipse. The atmospheric cooling and decrease in radiation flux during an annular solar eclipse is not as strong as during a total eclipse and the ionospheric response occurs with time delay.

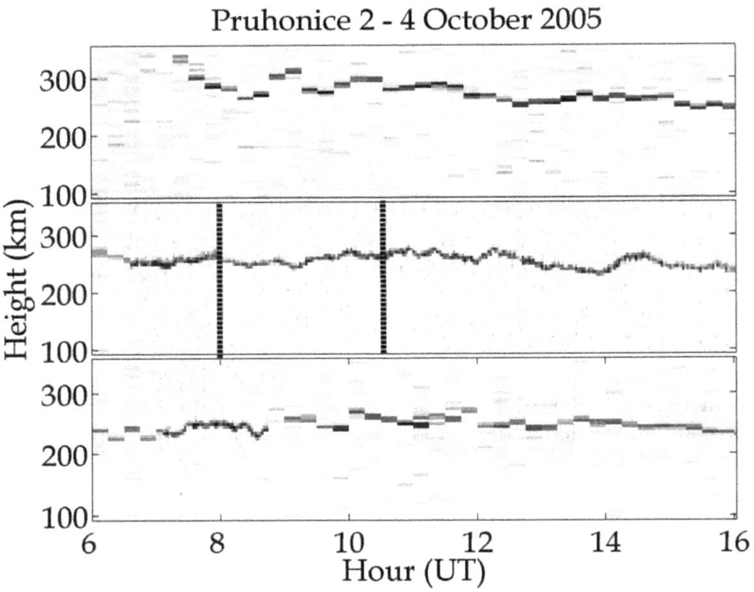

Fig. 7. Virtual reflection heights of plasma frequency in the range 4.2 - 4.3 MHz derived from raw ionograms. From up to bottom: day before, day of eclipse, day after eclipse. Vertical lines in middle panel depict beginning and end of the eclipse. Time resolution is different for day of solar eclipse (2 min) and days before/after (15 min).

In Figure 7 and Figure 8, there are plots of virtual reflection height variations at single frequency during three consecutive days, day of solar eclipse event and one day before and after the event. Variation of the reflection height during eclipse event of 3 October 2005 does not differ much from the corresponding variation during reference time span day before and day after. Wave-like oscillations excited by solar eclipse are of comparable magnitude as those induced by other sources preceding and consecutive day. On the contrary, clear difference in reflection height oscillation during reference days and solar eclipse event is perfectly seen in Figure 8. Records of virtual heights at fixed frequency from January 4, 2011 present strong ionospheric response which is exhibited as periodic changes in reflection height. Sharp changes in the reflection height develop immediately after the beginning of the solar disc occultation and last till the end of eclipse event. Higher wave-like activity remains remarkable whole day. In this partial solar eclipse event, wave-like oscillations can be very probably attributed to the solar eclipse.
Strong decrease in electron concentration in practically whole electron profile as well as the wave-like changes were observed during and after August 11, 1999 and January 4, 2011. Wave-like activity develops immediately after the start of the solar disc obscuration during

partial solar eclipse. During annular solar eclipse, significant acoustic-gravity wave type bursts develop around and after maximum phase of the eclipse.

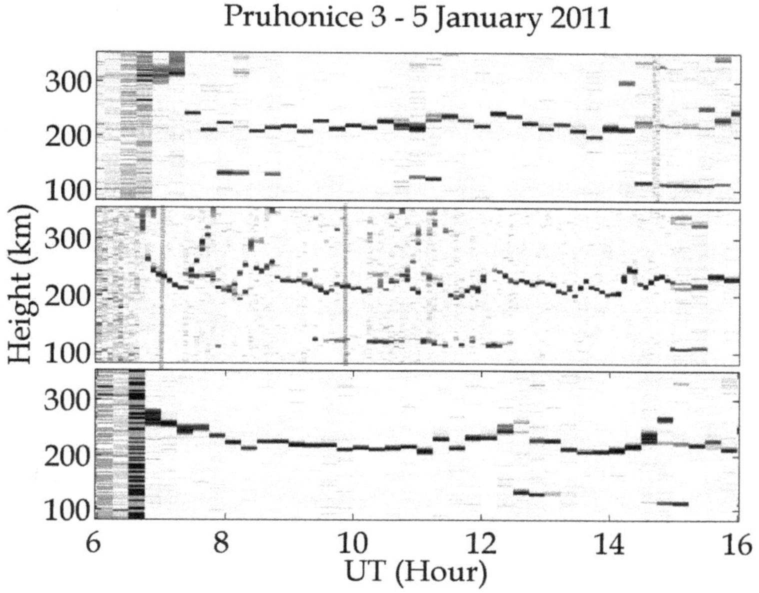

Fig. 8. Virtual reflection heights of plasma frequency range 3.4 – 3.5 MHz derived from raw ionograms. From up to bottom: day before, day of eclipse, day after eclipse. Vertical lines in middle panel depict beginning and end of the eclipse. Time resolution is different for day of solar eclipse (5 min) and days before/after (15 min).

6. Conclusion

Acoustic-Gravity waves play important role in the dynamic of the upper atmosphere. Vertical ionospheric sounding represents powerful tool that allows us to monitor acoustic-gravity wave activity in the ionosphere. Ionospheric observation of such a strong event as solar eclipse gives us an opportunity to better understand processes of creation and dissipation of the AGW in the area of the ionosphere. Although the acoustic-gravity waves are always present in the area of our interest, sharp temporally well-defined changes of solar flux during the solar eclipse give us a possibility to define sources of AGW.

It is rather uneasy to unambiguously assess causality between the solar eclipse events and the detected wave structures in the ionospheric plasma. Difficulties result from the fact that there are no two exactly identical solar eclipse events and from limitations of sounding techniques. Despite the fact that various AGW sources have been identified, many others remain to be found. Amongst irregular AGW bursts, regular increase in AGW activity were found to occur around sunrise and sunset hours, excited by Solar Terminator movement. Most of other sources (meteorological systems, geomagnetic and solar disturbances, etc.) and corresponding wave-like oscillations contribute to the irregular patterns of AGW activity observed in the ionospheric plasma.

As the solar eclipses, analyzed in the Section 5, occur sufficiently long time after the sunrise hours, one can assume that none of the reported waves are induced by solar terminator. During the analyzed sounding campaigns, no wave coming from auroral zone was expected, due to the quiet geomagnetic and solar activity. Additionally, meteorological analysis shows that meteorological systems very probably did not influence the ionosphere during studied events by means of AGW. The acoustic-gravity wave activity increases after a notably larger delay for the annular solar eclipse compared to the total solar eclipses: waves are found during the maximum phase of the eclipse only for the former while they occur during the initial phase for the latter. This discrepancy in gravity waves generation/occurrence can likely be explained by differences in the terrestrial atmosphere cooling: the border between sunlit and eclipsed region is much sharper in the case of total eclipse. Analyzing wave propagations, we observe predominantly upward propagating structures. The wave structure, that propagate upward and downward from the source region located around 200 km height, was created during an exceptional case related to the Solar eclipse of 11 August 1999.

7. References

Altadill, D., Sole, J.G. & Apostolov, E.M. (2001). Vertical structure of a gravity wave like oscillation in the ionosphere generated by the solar eclipse of August 11, 1999. *Journal of Geophysical Research*, 106 (A10), 21419–21428, ISSN 0148-0227.

Bodo, G., Kalkofen, W., Massaglia, S. & Rossi, P. (2001). Acoustic waves in a stratified atmosphere. III. Temperature inhomogenities. *Astronomy & Astrophysics*, 370, pp. 1088-1091, ISSN 0004-6361.

Chen, G., Zhao, Z., Zhou, C., Yang, G. & Zhang, Y. (2010). Solar eclipse effects of 22 July 2009 on Sporadic-E. *Annales Geophysicae*, 28, 353–357, ISSN 0992-7689.

Cheng, K., Huang, Y.N. & Chen, S.W. (1992). Ionospheric effects of the solar eclipse of September 23, 1987, around the equatorial anomaly crest region. *Journal of Geophysical Research*, 97, A1, 103–111. ISSN 0148-0227.

Chimonas, G. & Hines, C.O. (1970). Atmospheric gravity waves induced by a solar eclipse. *Journal of Geophysical Research*, 75, 4, pp. 875, ISSN 0148-0227.

Datta, R.N. (1972). Solar-eclipse effect on sporadic-E ionization. *Journal of Geophysical Research*, 77, 1, 260–262.

Davies, K. (1990). *Ionospheric radio.* Peter Peregrinus Ltd., ISBN 0 86341 186 X, London, United Kingdom.

Davis, J.R., Headrick, W.C. & Ahearn, J.L. (1964). A HF backscatter study of solar eclipse effects upon the ionosphere. *Journal of Geophysical Research*, 69 (1), 190–193. ISSN 0148-0227.

Farges, T., Jodogne, J.C., Bamford, R., Le Roux, Y., Gauthier, F., Vila, P.M., Altadill, D., Sole, J.G. & Miro, G. (2001). Disturbances of the western European ionosphere during the total solar eclipse of 11 August 1999 measured by a wide ionosonde and radar network. *Journal of Atmospheric and Solar-Terrestrial Physics*, 63, 9, pp. 915–924, ISSN 1364-6826.

Fritts, D.C. & Luo, Z. (1993). Gravity wave forcing in the middle atmosphere due to reduced ozone heating during a solar eclipse. *Journal of Geophysical Research*, 98, pp. 3011–3021, ISSN 0148-0227.

Galushko, V.G., Paznukhov, V.V., Yampolski, Y.M. & Foster, J.C. (1998). Incoherent scatter radar observations of EGW/TID events generated by the moving solar terminator. *Annales Geophysicae*, 16, pp. 821-827, ISSN 0992-7689.

Hargreaves, J.K. (1982). The upper atmosphere and solar-terrestrial relations. An introduction to the aerospace environment. Van Nostrand Reinhold, ISBN 0 521 32748 2, Cambridge, United Kingdom.

Hines, C.O. (1960). Internal atmospheric gravity waves at ionospheric heights. Canadian Journal of Physics, 38, pp. 1441–1481, ISSN 1208-6045.

Hooke, W.H. (1968). Ionospheric irregularities produced by internal atmospheric gravity waves. *Journal of Atmospheric and Solar-Terrestrial Physics*, 30, pp. 795-829, ISSN 1364-6826.

Huang X. & Reinish B.W (1996). Vertical electron density profiles from the digisonde network. *Advances in Space Research.*, 18, pp. 121□129, ISSN 0273-1177.

Ivanov, V.A., Ryabova, N.V., Shumaev, V.V., Uryadov, V.P., Nosov, V.E., Brinko, I.G. & Mozerov, N.S. (1998). Effects of the solar eclipse of 22 July 1990 at mid-latitude path of HF propagation. *of Atmospheric and Solar-Terrestrial Physics*, 60, 10, pp. 1013-1016, ISSN 1364-6826.

Jones, B.W., Miseldine, G.J. & Lambourne, R.J.A. (1992). A possible atmospheric-pressure wave from the total solar eclipse of 22 July 1990. *Journal of Atmospheric and Terrestrial Physics*, 54, 2, pp.113-115, ISSN 1364-6826.

Kato, S., Kawakami, T. & St. John, D. (1977). Theory of gravity wave emission from moving sources in the upper atmosphere. *Journal of Atmospheric and Terrestrial Physics*, 39, pp. 581–588, ISSN 1364-6826.

Le, H., Liu, L., Yue, X., Wan, W. & Ning, B. (2009). Latitudinal dependence of the ionospheric response to solar eclipses. *Journal of Geophysical Research*, 114, A07308, ISSN 0148-0227.

Liu, J.Y., Hsiao, C.C., Tsai, L.C., Liu, C.H., Kuo, F.S., Lue, H.Y. & Huang, C.M. (1998). Vertical phase and group velocities of internal gravity waves derived from ionograms during the solar eclipe of 24 October 1995. *Journal of Atmospheric and Solar-Terrestrial Physics*, 60, pp. 1679–1686, ISSN 1364-6826.

Muller-Wodarg, I.C.F., Aylward, A.D. & Lockwood, M. (1998). Effects of a mid-latitude solar eclipse on the thermosphere and ionosphere - a modeling study. *Geophysical Reseach Letters*, 25, 20, pp. 3787–3790, ISSN 0094-8276.

Oliver, W.L., Otsuka, Y., Sato, M., Takami, T. & Fukao, S. (1997). A climatology of F region gravity wave propagation over the middle and upper atmosphere radar. *Journal of Geophysical Research*, 102, A7, pp.14,499-14,512, ISSN 0148-0227.

Rapoport, V.O., Bespalov, P.A., Mityakov, N.A., Parrot, M. & Ryzhov, N.A. (2004). Feasibility study of ionospheric perturbations triggered by monochromatic infrasonic waves emitted with a ground-based experiment. *Journal of Atmospheric and Solar-Terrestrial Physics*, 66, pp. 1011-1017, ISSN 1364-6826.

Reinisch, B.W., Huang, X., Galkin, I.A., Paznukhov, V. & Kozlov, A. (2005). Recent advances in real-time analysis of ionograms and ionospheric drift measurements with digisondes, Journal of Atmospheric and Solar-Terrestrial Physics 67, pp. 1054-1062, ISSN 1364-6826.

Sauli, P., Abry, P., Boska, P. & Duchayne, L. (2006). Wavelet characterisation of ionospheric acoustic and gravity waves occuring during solar eclipse of August 11, 1999. *Journal of Atmospheric and Solar – Terrestrial Physics*, 68, pp. 586-598, ISSN 1364-6826.

Sauli, P., Roux, S.G., Abry, P. & Boska, J. (2007). Acoustic-gravity waves during solar eclipses: Detection and characterization using wavelet transforms. *Journal of Atmospheric and Solar-Terrestrial Physics*, 69, pp. 2465–2484, ISSN 1364-6826.

Sauli, P., Abry, P., Altadill, D. & Boska, P. (2005). Detection of the wave-like structures in the F-region electron density: two station measurement. *Studia Geophysica & Geodetica*, 50, pp. 131-146, ISSN 0039-3169.

Somsikov, V.M. & Ganguly, B. (1995). On the formation of atmospheric inhomogenities in the solar terminator region. *Journal of Atmospheric and Solar – Terrestrial Physics*, 57, 12, pp. 1513-1523, ISSN 1364-6826.

Titheridge J.E., 1985. Ionogram Analysis with the Generalised Program POLAN. *UAG Report-93*, 1985 (http://www.ips.gov.au/IPSHosted/INAG/uag_93/uag_93.html).

Walker, G.O., Li, T.Y.Y., Wong, Y.W., Kikuchi, T. & Huang, Y.N. (1991). Ionospheric and Geomagnetic effects of the solar eclipse of 18 march 1988 in East-Asia. *Journal of Atmospheric and Solar – Terrestrial Physics*, 53, 1-2, pp. 25–37, ISSN 1364-6826.

Xinmiao, Z., Zhengyu, Z., Yuannong, Z. & Chen, Z. (2010). Observations of the ionosphere in the equatorial anomaly region using WISS during the total solar eclipse of 22 July 2009. *Journal of Atmospheric and Solar – Terrestrial Physics*, 72, pp. 869–875, 1364-6826.

Yeh, K.C. & Liu, C.H. (1974). Acoustic-gravity waves in the upper atmosphere. *Reviews of Geophysics and Space Physics*, 12, 2, pp. 193-216, ISSN 8755-1209.

Permissions

The contributors of this book come from diverse backgrounds, making this book a truly international effort. This book will bring forth new frontiers with its revolutionizing research information and detailed analysis of the nascent developments around the world.

We would like to thank Marco G. Beghi, for lending his expertise to make the book truly unique. He has played a crucial role in the development of this book. Without his invaluable contribution this book wouldn't have been possible. He has made vital efforts to compile up to date information on the varied aspects of this subject to make this book a valuable addition to the collection of many professionals and students.

This book was conceptualized with the vision of imparting up-to-date information and advanced data in this field. To ensure the same, a matchless editorial board was set up. Every individual on the board went through rigorous rounds of assessment to prove their worth. After which they invested a large part of their time researching and compiling the most relevant data for our readers. Conferences and sessions were held from time to time between the editorial board and the contributing authors to present the data in the most comprehensible form. The editorial team has worked tirelessly to provide valuable and valid information to help people across the globe.

Every chapter published in this book has been scrutinized by our experts. Their significance has been extensively debated. The topics covered herein carry significant findings which will fuel the growth of the discipline. They may even be implemented as practical applications or may be referred to as a beginning point for another development. Chapters in this book were first published by InTech; hereby published with permission under the Creative Commons Attribution License or equivalent.

The editorial board has been involved in producing this book since its inception. They have spent rigorous hours researching and exploring the diverse topics which have resulted in the successful publishing of this book. They have passed on their knowledge of decades through this book. To expedite this challenging task, the publisher supported the team at every step. A small team of assistant editors was also appointed to further simplify the editing procedure and attain best results for the readers.

Our editorial team has been hand-picked from every corner of the world. Their multi-ethnicity adds dynamic inputs to the discussions which result in innovative outcomes. These outcomes are then further discussed with the researchers and contributors who give their valuable feedback and opinion regarding the same. The feedback is then collaborated with the researches and they are edited in a comprehensive manner to aid the understanding of the subject.

Apart from the editorial board, the designing team has also invested a significant amount of their time in understanding the subject and creating the most relevant covers. They scrutinized every image to scout for the most suitable representation of the subject and create an appropriate cover for the book.

The publishing team has been involved in this book since its early stages. They were actively engaged in every process, be it collecting the data, connecting with the contributors or procuring relevant information. The team has been an ardent support to the editorial, designing and production team. Their endless efforts to recruit the best for this project, has resulted in the accomplishment of this book. They are a veteran in the field of academics and their pool of knowledge is as vast as their experience in printing. Their expertise and guidance has proved useful at every step. Their uncompromising quality standards have made this book an exceptional effort. Their encouragement from time to time has been an inspiration for everyone.

The publisher and the editorial board hope that this book will prove to be a valuable piece of knowledge for researchers, students, practitioners and scholars across the globe.

List of Contributors

Zi-Gui Huang
Department of Mechanical Design Engineering, National Formosa University, Taiwan

V. I. Alshits
A.V. Shubnikov Institute of Crystallography, Russian Academy of Sciences, Moscow, Russia
Polish-Japanese Institute of Information Technology, Warsaw, Poland

A. Radowicz
Kielce University of Technology, Kielce, Poland

V. N. Lyubimov
A.V. Shubnikov Institute of Crystallography, Russian Academy of Sciences, Moscow, Russia

P. K. Karmakar
Department of Physics, Tezpur University, Napaam, Tezpur, Assam, India

Dorel Homentcovschi and Ronald Miles
Department of Mechanical Engineering, State University of New York at Binghamton, USA

A. H. Khater
Mathematics Department, Faculty of Science, Beni-Suef University, Beni-Suef, Egypt

M. M. Hassan
Mathematics Department, Faculty of Science, Minia University, El-Minia, Egypt

Marco G. Beghi
Politecnico di Milano, Energy Department and NEMAS Center, Milano, Italy

Dong Sik Gu and Byeong Keun Choi
Gyeongsang National University, Republic of Korea

P. Burgholzer, H. Roitner, J. Bauer-Marschallinger, T. Berer
Christian Doppler Laboratory for Photoacoustic Imaging and Laser Ultrasonics, Austria
Research Center for Non Destructive Testing (RECENDT), Austria

H. Grün
Research Center for Non Destructive Testing (RECENDT), Austria

G. Paltauf
Institute of Physics, Karl-Franzens-University Graz, Austria

Deli Gao and Qifeng Pan
China University of Petroleum at Beijing, Beijing, China

Georges Nassar
Université de Lille -Nord de France, France

Ertan Ergezen, Johann Desa, Matias Hochman, Robert Weisbein Hart, Qiliang Zhang, Sun Kwoun, Piyush Shah and Ryszard Lec
School of Biomedical Engineering, Health and Sciences, Drexel University, Philadelphia, USA

Wojciech Dziewiszek and Marek Bochnia
Medical University of Wrocław, Poland

Wojciech Michalski
Technical University of Wrocław, Poland

Jociely P. Mota, Jorge L.C. Carvalho, Sérgio S. Carvalho and Paulo R. Barja
UNIVAP, Brazil

Petra Koucká Knížová and Zbyšek Mošna
Institute of Atmospheric Physic, Czech Academy of Sciences, Czech Republic